THE MATTER FACTORY

THE MATTER FACTORY
A History of the Chemistry Laboratory

Peter J. T. Morris

REAKTION BOOKS

This book is dedicated to the memory of three departed friends who were true chemists and good historians: W. Alec Campbell (1918–1999), Colin A. Russell (1928–2013) and Frank Greenaway (1917–2013)

Published by
Reaktion Books Ltd
Unit 32, Waterside
44–48 Wharf Road
London N1 7UX, UK
www.reaktionbooks.co.uk

In association with
Science Museum
Exhibition Road
London SW7 2DD, UK
www.sciencemuseum.org.uk

First published 2015, reprinted 2016

Copyright © SCMG Enterprises Ltd 2015
Science Museum ® SCMG

All rights reserved
No part of this publication may be reproduced, stored in a retrieval system, or transmitted, in any form or by any means, electronic, mechanical, photocopying, recording or otherwise, without the prior permission of the publishers

Printed and bound in Great Britain
by TJ International, Padstow, Cornwall

A catalogue record for this book is available from the British Library

ISBN 978 1 78023 442 7

Contents

Introduction 9

1
Birth of the Laboratory: Wolfgang von Hohenlohe and Weikersheim, 1590s 18

2
Form and Function: Antoine Lavoisier and Paris, 1780s 39

3
Laboratory versus Lecture Hall: Michael Faraday and London, 1820s 63

4
Training Chemists: Justus Liebig and Giessen, 1840s 86

5
Modern Conveniences: Robert Bunsen and Heidelberg, 1850s 119

6
The Chemical Palace: Wilhelm Hofmann and Berlin, 1860s 146

7
Laboratory Transfer: Henry Roscoe and Manchester, 1870s
170

8
Chemical Museums: Charles Chandler and New York, 1890s
198

9
Cradles of Innovation: Carl Duisberg and Elberfeld, 1890s
232

10
Neither Fish nor Fowl: Thomas Thorpe and London, 1890s
269

11
Chemistry in Silicon Valley: Bill Johnson and Stanford, 1960s *291*

12
Innovation on the Isis: Graham Richards and Oxford, 2000s
311

Conclusion *337*

APPENDIX *348*
REFERENCES *353*
SELECT BIBLIOGRAPHY *393*
ACKNOWLEDGEMENTS *398*
PHOTO ACKNOWLEDGEMENTS *401*
INDEX *403*

1 Painting of an ICI analytical chemistry laboratory, 1957, by Ernest Wallcousins.

Introduction

The idea for this book was born in a seminar room in the Department of the History and Sociology of Science at the University of Pennsylvania, in Edgar Fahs Smith Hall. This building was named after a famous historian of chemistry and was formerly a pioneering hygiene laboratory – the Lea Institute of Hygiene – built in 1891. Sadly the building was demolished in 1995 to make way for a new wing of the neighbouring chemical laboratory, despite strenuous opposition from the historian of biochemistry Robert Kohler. The walls of the room were lined with PhD theses, and in 1985 I was reading about the chemistry department of the University of Illinois at Urbana in P. Thomas Carroll's thesis.[1] For the first time it struck me how the standings of chemistry departments are closely tied to their laboratories, and in particular the construction of new laboratories. I first developed this idea in a chapter on chemical practice in the nineteenth century, which I wrote for the *Enciclopedia Italiana* in the late 1990s with my good friend the late Alec Campbell.[2] The present book is a much-expanded version of my study of four laboratories for that chapter.

As a historian and curator I believe that the history of chemistry has to be more than just the history of chemical theories; it has to include the history of chemical practice and chemical culture. Laboratories are an important part of that practice and culture. They are the places where chemists are trained and where many of them spend their careers. They are where chemistry is carried out – from the freshman's first inorganic analysis to the most complex of organic syntheses. I have argued elsewhere that historians of chemistry have failed to meet the needs of chemists.[3] Chemists want to know how they 'got here', hence the need to write the history of recent chemistry, but they also want to learn more about the history of their professional home, the chemical laboratory. Furthermore, when the man in the street thinks of a laboratory, he usually thinks of a chemical laboratory with bubbling flasks, benches and bottle racks (reagent shelves in the USA) rather than the physics laboratory with tables. Certainly this is the image presented by films.[4]

Rather surprisingly, as Kohler has pointed out,[5] there is no general academic history of the laboratory, or even a popular account.[6] There are books about specific laboratories or sets of laboratories, but they are usually about the people who worked in them rather than the laboratories themselves. Numerous biographies (and autobiographies) of the founders of laboratories contain information about their laboratories, but such information is sparse and on closer examination often incorrect (even in the case of otherwise first-rate historians). There have been books on the laboratory as a 'research space' or 'working space', but they are usually of limited interest, being multi-authored, albeit containing individual papers of considerable value.[7] Ernst Homburg's paper on the rise of analytical chemistry and the laboratory of Friedrich Stromeyer does make some of the general points also raised in this volume, such as the early links between the chemical laboratory and the pharmaceutical and factory laboratories, and the transformation of the chemical laboratory in the early nineteenth century.[8]

There is the fundamental question of what a chemical laboratory is.[9] On one level it is a matter of demarcation. When is a space a chemical laboratory rather than a workshop on the one hand, or a physics laboratory on the other? My solution to this question has been entirely pragmatic: a chemical laboratory is a space in which a recognizable chemical activity is taking place, and/or a space that its occupants called a chemical laboratory. On another level, how can we distinguish between a chemical laboratory as a room where experimental work is carried out, and the larger building in which it is located? We all speak about a particular site – be it the Dyson Perrins in Oxford, the Converse at Harvard or one of the Stauffer Laboratories in Stanford – as a 'laboratory', when in fact it is a building containing several laboratories and other rooms such as lecture halls, storerooms and offices. I have distinguished the two by calling the space a 'laboratory', and the larger building containing laboratories a 'laboratory building'.[10]

My aim is to describe how laboratories and laboratory buildings have changed to meet the differing needs of chemistry from 1600 to 2000. I will show how the form of the laboratory has altered over the years to accommodate the functions of chemistry at the time it was constructed. However, it is not just a matter of the laboratory being shaped by chemistry. The new opportunities offered by the most modern laboratories have allowed chemistry to progress; for example, the arrival of coal gas in the laboratory assisted the invention of the Bunsen burner, which in turn enabled the development of spectroscopy. Chemistry pushes the development of the laboratory, and the improved laboratory in turn enables chemistry to move forwards. It is the ambitious chemistry professor who ensures that the two lines of development reinforce each other. As a leading chemist, he (and until recently it was invariably a he) knows what changes need to be made to the laboratory to do cutting-edge chemistry. He also knows what kind of chemistry the new laboratory will make possible. In the absence of such professors the development of the chemical laboratory would have been much slower.

It is clear that competition between individuals and nations is one of the driving forces of laboratory development. At the same time there is much cooperation between chemists (and their architects) across national boundaries. The architect of Bunsen's laboratory designed a laboratory for an American college. France may have been compelled by the German ascendency in chemistry to build a new laboratory building at the Sorbonne in the 1890s, but (August) Wilhelm Hofmann was happy to spend days with Henri-Paul Nénot, the French architect, on the plans for this building. Nonetheless the main aim of these innovators was to do better – and as we shall see, safer – chemistry. I have deliberately eschewed any notion of an *evolution* of the laboratory. Moreover, to avoid any accusations of Whiggism,[11] I have written a chapter on chemical museums to show how certain features of the nineteenth-century laboratory have not survived in any form today.

This is a history of the chemical laboratory as a room or building containing chemists, not as a rhetorical (or actual) space in which various human and non-human actors operate (illus. 2).[12] I have also chosen not to cover the social aspects of the laboratory, interesting though they would be, as I have not found much secondary

2 Chemistry laboratory at Howard University, Washington, DC, c. 1900. The laboratory design created in Germany in the 1850s and '60s had become universal by 1900.

material relating to them; nor are such aspects essential to the telling of my story. I have given some attention to the architect and the architectural details of the laboratory buildings, but I have not found the exact architectural style of the laboratories to be crucial to my core narrative.[13] The external architectural style of the building tends to reflect current fashions (Romanesque or glass plate), without much impact on the internal arrangements. What matters, I would argue, is the sizes and shapes of the various rooms and their relationships to each other.

My main method of analysis has been the close study of pictures of the interiors of the laboratories and their ground plans. Other sources were not readily available. Archival sources are sparse, widely dispersed and too time-consuming for a history on this scale. The descriptions in the literature are useful and have been used, but are no substitute for the actual illustrations. The buildings themselves have been changed over the years (when they have survived at all), and any reconstruction has to be treated with considerable reserve.[14] In passing I would note that several chemistry laboratory buildings have been converted into offices for humanities and social science faculties; for example, the former laboratory of physical chemistry at

Cambridge is now the History and Philosophy of Science Department. As a result this book contains many illustrations of chemical laboratories, possibly the largest number ever gathered together in one volume, which will be useful as a resource for further studies of this important topic.[15] This is not to say that illustrations are unproblematic – they do have to be treated with caution.[16] As far as I can tell, however, most illustrations of laboratories are reasonably accurate and suffice for the purposes of this book.

I was eager for this book to be more than just a history of the chemical laboratory. To provide the context for the laboratories discussed, I have described the backgrounds to their construction, noting national factors where relevant, and sketched the chemical developments that are most pertinent to them. I have also sought to capture the development of chemical practice by having a section on an appropriate technique in each chapter. Acknowledging that the human element adds colour to any narrative and that personality plays an important role in the history of chemistry, I have additionally given a brief biography of a key chemist in every chapter. Unusually in a work of this kind, I have made some autobiographical remarks because my own experience of laboratories seems relevant to understanding what happens in one – something that is very hard to capture solely from documents or pictures.

Choosing the selection of laboratories for this book has been a pragmatic process. For a laboratory to be suitable for this book, I needed clear pictures (the lack of which in a number of cases unfortunately ruled out some otherwise excellent laboratories), good secondary sources because it was impossible time-wise to research every laboratory from original documentation, and laboratories that had some degree of historical significance. It must be stressed that I am not making any claims of historical priority for any of these laboratories, except insofar as I do so specifically in the text. Furthermore, I have chosen these laboratories as exemplars of a larger number of similar laboratories. There is simply no way that this survey can be comprehensive: important laboratories such as those of Isaac Newton and Hermann Kolbe, for example, are absent either due to the lack of good illustrations, or quite simply because of a lack of space to include them. For instance, chapter Six was originally about Kolbe's laboratory in Leipzig, but I eventually decided that it would be more appropriate to discuss Hofmann's laboratory in Berlin. To have covered both laboratories would have made the book much longer and generated significant duplication. Similarly, it was impossible to cover every angle of this topic. To give just one example: apothecaries undoubtedly played an important role in the development of the laboratory, but Apothecaries' Hall had to serve as the sole example for apothecaries in general because of its significance in terms of the introduction of steam to the laboratory.

Of course, there had to be a beginning to the story, and the appropriate starting point was the only alchemical laboratory for which there was documentary evidence for its layout (Weikersheim), supported by the very clear illustrations of a metallurgical workshop in Lazarus Ercker's treatise on assaying. Antoine Lavoisier's laboratory at the Paris Arsenal clearly represented a break with previous chemical practice. The Royal Institution (RI) provided an excellent example of the connection between the laboratory and lecturing. Justus Liebig's laboratory at Giessen is best known as a research school – a 'chemist breeder' in Jack Morrell's evocative phrase – and here we look at the development of the bench and fume cupboard.[17] Robert Bunsen's laboratory at Heidelberg is particularly significant for showing the importance of utilities in laboratories, while Wilhelm Hofmann's laboratories in Bonn and Berlin reveal the development of the large laboratory building. When considering the industrial laboratory it is necessary to look at a number of laboratories, including Apothecaries' Hall in London and the R&D laboratories of Bayer, BASF and Shell. The Government Chemist's Laboratory (LGC) in London's Strand provides a good example of a laboratory that is neither academic nor industrial. The Stauffer Laboratories at Stanford provide a striking parallel to Hofmann's Berlin laboratory a century later, and allow us to examine the extent to which the 'instrumental revolution' affected the laboratory. There were several topics I wanted to cover in the final chapter – pharmaceuticals, health and safety, spin-out companies, the decline of sub-disciplines and the growing importance of separation methods. The central chemical laboratory at Oxford opened in 2004, allowing me to cover all of these themes and conclude the book at the beginning of the twenty-first century.

This book is truly a product of the digital age. Many of the nineteenth-century volumes were available online at Google Books, and I have indicated this in the footnotes with 'GB'. I could not have written this book in the time that I had available without access to Google Books – particularly in the case of the chapter on museums. At the same time, however, Google Books can be frustrating to use. The problems arise mainly with the books in 'snippet view', where you cannot directly check the relevant page, and this also gives rise to doubts about the correct citation. I additionally used Internet Archive whenever possible. This has the advantages of being more stable and having PDF files that are searchable when downloaded, unlike PDFs from Google Books; Internet Archive books are indicated by 'IA'. Furthermore, many Google Books can be read online at the Internet Archive website, including some that are now only available in snippet view on Google's own website. I also bought many twentieth-century books unavailable online using Bookfinder (bookfinder.com). Online library catalogues (OPACs) were an essential aid, especially COPAC (copac.ac.uk). Finally, it would not have been possible to exchange information with my correspondents in fourteen countries without the use of email.

In the primary literature four books stand out, namely Hofmann's masterly account of the erection of the laboratories at Bonn and Berlin;[18] Edward Robert Festing's report on European laboratories;[19] Edward Cookworthy Robins's monograph on laboratory construction;[20] and William Henry Chandler's report on American laboratories.[21] The article in *Strand Magazine* on the Government Chemist's Laboratory was also very useful.[22]

Among the secondary literature the work of Alan Rocke on Kolbe and Adolphe Wurtz is exceptional, although he does not focus on the laboratories as such.[23] This focus is provided by Catherine Jackson's groundbreaking thesis on the organic chemical laboratory in the nineteenth and early twentieth centuries, which covers much of the same ground as this book in that period, but with a greater emphasis on organic synthesis.[24] Even Jackson, however, largely concentrates on what is happening in the laboratory, while making important points about the role of safety in shaping it. Yoshiyuki Kikuchi's work on nineteenth-century Japanese chemistry brilliantly analyses the floor plans of Japanese laboratory buildings, revealing how the arrangements of the rooms show how the balance of influence on Japanese chemistry shifted from Germany to Britain, and also sheds light on contemporary debates about how chemistry should be taught.[25] I would commend the late Jon Eklund's description of the eighteenth-century laboratory, which is also relevant to the late sixteenth and seventeenth centuries. This slim volume with its useful glossary of obsolete chemical terms is now very rare, but fortunately can be downloaded from the Smithsonian website.[26] Peter Hammond and Harold Egan's history of the LGC is the only modern book that is solely concerned with one of the laboratories covered in this volume.[27] Like many similar books, however, it is more about the people who worked in the Laboratory of the Government Chemist and their activities, than about the laboratory as a building. Finally, when I was writing about the techniques used in the laboratory, I almost invariably found that William B. Jensen had written about them first and to good effect, usually in the *Bulletin of the History of Chemistry*.

Another important resource for this book was the 'Sites of Chemistry, 1600–2000' project set up in 2010 by John Perkins (Oxford Brookes University) and Antonio Belmar (University of Alicante), with the help of funding from the Wellcome Trust and Maison Française in Oxford.[28] This project, which was conceived completely independently of this book, has organized a series of conferences on the sites of chemistry from the seventeenth to the twentieth centuries. While the scope of the project is avowedly wider than laboratories, many of the papers are about laboratories and thus have proved to be very useful for the writing of this book.

Almost everyone has been inside a chemical laboratory at some point in their life, even if it was only to take science classes at secondary school. For that reason I have aimed to make this book enjoyable for the general reader by avoiding jargon and academic formalities, such as giving a complete summary of a chapter at the beginning ('this chapter deals with'), and including Socratic dialogue with other scholars ('contrary to what X has claimed'). I have not been able to avoid academic niceties completely, but hope that I have reduced them to a tolerable level. By reading this book, the general reader should gain a better idea of what happens in the chemical laboratory and how it has developed over the last 500 years. Historians of science should find the book intriguing and find new insights into the relationship between chemical knowledge, chemistry as a scientific discipline and the practice of chemistry. I hope that the book will spur the systematic study of chemical laboratories in the future. Above all, I would be delighted if this study of the chemist's training ground and workplace builds bridges between professional chemists and the history of chemistry. If more chemists take up the writing of the history of chemistry as a result, perhaps in the form of histories of their own laboratories, this book will have served its purpose.

I
Birth of the Laboratory
Wolfgang von Hohenlohe and Weikersheim, 1590s

When most people think of the forerunner of the chemical laboratory, they think of a cluttered and dirty alchemical den as painted by artists such as David Teniers the Younger. However, as described in the next section (see page 20), the usual artistic representation of the alchemical laboratory was inaccurate. Nevertheless, there are a few seemingly accurate illustrations, and the reconstruction of an alchemical laboratory based on original documents at Schloss Weikersheim in Germany (see page 27). While the alchemical laboratory may have had some influence on the chemical laboratory of the seventeenth and eighteenth centuries, it was not the only prototype.

The Origins of the Chemical Laboratory

One of the most important forerunners of the chemical laboratory was the pharmacy, and it is perhaps no coincidence that the chemical laboratory first arose at the same time as Paracelsus's concepts of chemical medicine (iatrochemistry). The modern laboratory clearly owes much to the manufacturing pharmacy, which made both plant-based and chemical medicines in line with the teachings of Paracelsus: the pharmaceutical laboratory of Annibal Barlet is discussed in chapter Three, and Apothecaries' Hall in London played an important role in the development of the laboratory. The chemical laboratory copied the pharmacy in many respects, not least in the long counters favoured by pharmacists. Metallurgical workplaces – something between laboratories and factories – were another starting point for later chemical laboratories. This is fortunate as there are more reliable illustrations of the metallurgical workshop than the pharmaceutical laboratory, and the organization of the metallurgical laboratory is discussed in this book in some detail. Finally, there were other industrial sites that were to some extent forerunners of the chemical laboratory. There were the spirit distillers from the early sixteenth century onwards, whose operations would have been on a relatively small scale, and the soap boilers who would have carried out chemical operations of various kinds. Indeed, some of the earliest references to a work laboratory, albeit in the early nineteenth century, place chemical operations in soap-boilers' factories. The industrial laboratory is discussed further in chapter Nine.

How Did Laboratories Begin?

There are several approaches to the issue of the beginnings of the chemical laboratory. In the introduction the chemical laboratory is defined as a room or building *used* more or less exclusively for the practice of chemistry. However, we are mainly interested in rooms or buildings specifically *designed* for the practice of chemistry. For the origins of the laboratory this distinction is an important one. Doubtlessly early pharmacists, metal refiners and alchemists had workshops or at least rooms that were set up for the practice of their craft. These rooms would have contained some kind of furnace and more than likely some type of distillation apparatus. The bringing together of this equipment into a space set aside at least on a temporary basis for the practice of the alchemical craft could be said to constitute the beginning of the laboratory. However, it is hardly possible to date the beginning of this practice, or to even determine in which part of the world it took place, although Egypt perhaps around the beginning of the Christian Era is as strong a candidate as any. A workshop in early Christian Era Egypt may have contained an alembic or two, a water bath and a furnace, among other apparatus. Such an ensemble can fairly be called a laboratory, albeit only with hindsight and a little imagination.

On the other hand, if we seek a building or room specifically designed for the practice of the chemical arts, we have to look much later. While alchemists and early metallurgists probably gave some thought to the positioning of their apparatus and the nature of the room – ventilation was obviously an important factor, for comfort at least – it is unlikely that any medieval alchemists sat down and designed a laboratory, although the metallurgists appear to have been more systematic.[1] The 'chemical house' of Andreas Libavius in 1606 was probably the first laboratory to be consciously designed, although it was never constructed and was in fact developed for purely rhetorical purposes. Ironically, as discussed later in this chapter (see page 33), it was probably not of a good design, if only because it was poorly ventilated.

Alternatively, the issue can be approached as a textual question: when was the word 'laboratory' first used in the sense of a building or room utilized by a practitioner of the chemical arts? It is interesting to note that the term 'laboratory' in Latin and German in the 1560s referred to God's laboratory or nature's laboratory rather than a physical entity.[2] Twenty years later, however, it was used in Latin for the workplace of an alchemist or pharmacist.[3] According to the *Oxford English Dictionary* (OED) the first use of the word in English was by John Dee in *The Compendious Rehearsall* (1592). A few years later Ben Jonson makes its meaning clear – 'A laboratory, or Alchymists workehouse' – in the stage directions of 'Mercury Vindicated from the Alchemists' (1615), which is based on the *Dialogus Mercurii, Alchymistae et Naturae* (1607) by the Polish alchemist Michael Sendivogius.[4] Somewhat surprisingly, 'laboratory' actually predates the more anachronistic word 'elaboratory', which first appeared in 1652, according to the OED, in John Evelyn's *State of France*. The appearance of the word 'laboratory' in the 1590s, followed by Libavius's hypothetical chemical house in 1606, shows how the idea of the laboratory came to the fore at the end of the sixteenth century, culminating in the setting up of a 'public chemical laboratory' by Johannes Hartmann, the first professor of chemical medicine at the University of Marburg, in 1609. So what did these early laboratories contain?

Fact or Fiction?

The few contemporary and unbiased images of early modern laboratories and the similar metallurgical workshops show tidy and clean workplaces. However, some well-known pictures of alchemical workshops were created by the opponents of alchemy, not its supporters. It hardly seems necessary to state the obvious fact that most etchings and paintings of alchemical laboratories were not drawn from life or even during the period they allegedly illustrated, but several well-known historians of alchemy between the 1940s and '60s took them

3 Hans Weiditz, *The Alchemist*, 1520.

at face value.[5] Alchemists themselves tended to produce allegorical pictures that cannot be considered to be factually accurate or representative. Regardless of the subject, filling pictures with still-life objects allowed a painter to show off his skills. Furthermore, an artist had to exaggerate the size of the equipment and tools used by the chemical worker to make them visible in the picture. Hence the scenes portrayed can look unfairly cluttered and disordered, as in the famous picture of a metallurgical workshop by Hans Weiditz in 1520 (illus. 3). This etching raises two important points about such pictures. Representations of metallurgical workshops in the well-known works by Agricola and Ercker were more accurate and more detailed than the few unbiased pictures of alchemical laboratories. The Weiditz etching, however, appeared not in a treatise about metallurgy, but in a volume of moral philosophy, suggesting that most pictures of alchemical laboratories in the sixteenth and seventeenth centuries had a moralizing purpose.

It has to be borne in mind that alchemists were often the subject of moral censure, even if this censure had faded to mild satire by the time of David Teniers the Younger in the late seventeenth century. First and foremost, alchemy was generally considered to be fraudulent.[6] Alchemists who claimed to be able to transmute base metals into gold were regarded as either deluded or con artists and, to be fair, alchemy did attract a fair number of confidence tricksters. Medieval

4 Pieter Bruegel the Elder, *The Alchemist*, c. 1558.

5 David Teniers the Younger, *The Alchemist*, 1650.

Catholics regarded alchemy as a heresy closely allied to Gnosticism. Luther himself approved of alchemy, which he saw as covering metallurgy and pharmacy, as both economically valuable and spiritually symbolic. Other Protestants, who believed in the moral value of hard work, stigmatized alchemy as a way of getting rich quickly without putting in the work, in much the same way that their successors deplore modern celebrity culture. That alchemy, for most of its practitioners, was actually an easy way of becoming poor quickly was neither here nor there. It is more than likely that some alchemists became mentally unbalanced, either because of the effect of working with heavy metals in confined conditions, or due to the stress of trying to find the philosophers' stone with the constant threat of being punished by the state or Church while their funds were running out. Alchemists had to carry out their activities in secret, partly because of official disapproval and because there would be no point in being able to make gold from lead if anyone could copy their process. During the Middle Ages and the Reformation, groups that took part in secret activities were regarded with suspicion and hostility.

The classic satirical image of the alchemist's laboratory was painted by Pieter Bruegel the Elder around 1558, some three decades after Weiditz's moralizing etching (illus. 4). Not only is the satire clear from the picture, which depicts children being neglected and taken into care, and the desperation of the alchemists and their assistants to find the philosophers' stone (their facial expressions are similar to those drawn by Weiditz), but the Latin motto at the base of the painting also mocks the alchemists' activities. Bruegel's approach was adopted a century later by his fellow Netherlandist genre painter David Teniers the Younger (illus. 5). As Teniers is well known among chemists and historians of chemistry for his paintings of alchemy, it has to be stressed that the painting of alchemical laboratories was a very minor activity for Teniers (as it had been for Bruegel), because he only painted 22 alchemical scenes out of a total of 900 paintings. The sheer number of his paintings shows that he was creating paintings en masse for a growing and increasingly wealthy bourgeois audience. The typical Dutch burgher would have disapproved of alchemy as morally dubious 'get-rich-quick' superstition, in common with the German Lutheran bourgeois schools' inspector Libavius, whose ideal chemical laboratory was a typical burgher house (see below). At the same time, alchemy was already receding into the past, at least in bourgeois Dutch circles, and Teniers's satire is gentle whimsy rather than the severe moral censure of Bruegel. His alchemists are kindly old men (not unlike Santa Claus), who are perhaps forgetful and deluded, but not evil or even untidy – with one or two exceptions.

It might be thought that a more accurate picture of the early-modern laboratory could be obtained from archaeological excavations, notably the ongoing research on the laboratory at Oberstockstall in Austria. The findings from Oberstockstall and other sites, such as the late seventeenth-century chemical laboratory in Oxford, shed light on what apparatus early chemical workers used and hence what they were doing, but they do not tell us much about the space in which this work was done.[7]

Medieval Assaying Laboratory

The best (and as far as we can tell, the most accurate) illustrations of a medieval chemical working space are of the metallurgical workshop or the assaying laboratory – if we can call it that – in the woodcuts in Georgius Agricola's *De re metallica* (On the Nature of Metals) of 1556 and Lazarus Ercker's *Beschreibung aller furnemisten mineralischen Ertzt vnnd Bergtwercks arten* (Description of the Most Important Mineral Ores and Methods of Mining) of 1574.[8] The illustrations of the laboratory in Ercker provide an immediate impression of tidiness and order, of careful arrangement and space. The retorts are kept in a rack on the wall and the furnaces are neatly lined up. The work is done in the clear space in the centre of the workshop away from the furnaces. The workshop is clearly arranged for assaying, testing and experimental work such as the separation of gold from mercury, and not for the large-scale production of metals. However, the workshop did produce the strong acids it required by distillation, so that the scales of the laboratory in general and the furnaces in particular are no larger than in the later chemical laboratory. The most striking feature of the workshop is the prevalence of distillation, an operation that would not normally be associated with assaying or metals (apart from mercury). This shows how significant furnaces and distillation were to the medieval and early-modern chemical arts.

The first image (illus. 6) forms the frontispiece of Ercker's treatise and it may be intended to be a compilation of the techniques described in the book rather than any attempt at a realistic depiction.[9]

Birth of the Laboratory

6 The assaying laboratory, from Ercker's *Beschreibung* of 1574.

However, it does show how a laboratory of the period could bring together different techniques in an ordered manner. At the back of the laboratory there is a cementation furnace used to increase the purity of precious metals, especially gold.[10] The man next to the furnace is filling the 'Piger Henricus', or philosophical furnace (see the section on furnaces, page 33). On the right at the back are wind furnaces, used in the sulphur process to separate gold from silver.[11] The pot circled by burning coals in front of it is preheating the granules used in this process. On the near right a man is carrying out the fusion assay of copper using an early steam-driven bellows that looks like a retort, but is actually a version of Hero's æolipile.[12] The fuming pot in the middle of the back wall is holding the flux for this process. In the centre of the foreground another man is distilling 'parting acids', strong acids used for separating gold from silver.[13] Through the doorway on the right can be seen a standard assay furnace.[14] The most unusual feature is the sunken floor at the front. This may have actually existed – in another picture it houses a long furnace and it seems to have been used to store fuel and spare apparatus – but it may just be an artistic convention to allow more objects to be displayed in the picture. The broken cucurbit at the front adds a touch of realism, as the delicate glassware would have often been broken. The impression provided by the picture is that metallurgical workshops were similar in arrangement

7 Amalgamation of gold and recovery of mercury.

and scale to other chemical laboratories, with the necessary addition of the equipment that they needed to carry out metallurgical operations – in this case the refining of gold and assaying.

The second image (illus. 7) shows the refining of gold using amalgamation with mercury; the main difference from the ordinary laboratory is the large furnace with its bellows at the back of the picture, which was used to melt the gold. The mortar in the centre is being used to pound the impure gold and mercury together to form the amalgam. The Piger Henricus in the middle of the room is being used to distil mercury from an earthenware alembic and cucurbit into an earthenware receiver. Mercury was usually stored in pottery bottles well into the twentieth century. The master is cleaning the mercury by squeezing it through a damp chamois leather bag (an empty bag can be seen lying on the floor), with the clean mercury coming out as droplets and falling into the basin. Nowadays this process is still used by freelance gold panners and considered to be very hazardous. In this illustration the master is exposed to mercury through the skin of his hands (although he is clearly being careful to only squeeze the bag at the top away from the mercury), and more importantly by inhaling mercury vapour – although this would have been mitigated if there was a draught passing through the open doors of the laboratory.

The Alchemical Laboratory in a Castle

By contrast, there are no reliable contemporary illustrations of a sixteenth-century alchemical laboratory that show any detail. There is, however, the scholarly reconstruction of Wolfgang von Hohenlohe's laboratory in Schloss Weikersheim by Jost Weyer.[15] It must be emphasized that this sketch (illus. 8) is not a drawing from the time

8 Drawing of the interior of the reconstructed alchemical laboratory at Schloss Weikersheim, Weikersheim, Baden-Württemberg, Germany.

or a room that has been left untouched since its owner died in 1610, but it is based on the detailed records left behind by Hohenlohe and has to be given a certain cautious respect. As might be expected from trustworthy contemporary pictures (see illus. 9 and 10), it is relatively simple. It looks like a cross between a blacksmith's forge and a manufacturing pharmacy. It shows two furnaces and the central hearth, which could be used for a variety of purposes, including heating the room. A large fixed bellows is attached to the hearth, providing the necessary draught as in a forge. There is a strong similarity between this furnace and the one shown in Ercker's *Treatise on Ores and Assaying* (see illus. 7). Alongside it is a smaller furnace specifically for distillation and sublimation. Finally, there is an assaying furnace (not shown in illus. 8) showing the prevalence of assaying in such laboratories, as in the case of the similar laboratory at Oberstockstall in Austria. As in Ercker's illustrations, there are short shelves and racks for alembics, retorts and other glassware. Retorts are also hung directly on the wall, as shown more dramatically in an unreliable painting of a German laboratory in 1638.[16] Heavier apparatus, a pan and a brass mortar rest on a stool. The open flue and hood would have acted as a crude fume hood drawing smoke and fumes away from the laboratory. As described in chapter Four, such

9 Alchemist at the furnace, 1503.

10 Distiller preparing flavoured brandies, 1478.

hoods were the forerunners of the modern fume cupboard. There is also a large window that would have assisted the draught when necessary; it shows the concern with ventilation, even in this early period. Water is provided by a stone sink and barrel, and ladled out with a stoop. The waste water ran into a drain, which emptied into the castle moat. The apparatus was stored in four large cupboards. The Oberstockstall excavation has shed light on the apparatus used in this period, with a predominance of distillation apparatus, mostly but not entirely made of glass, and triangular crucibles (known as Hessian crucibles), made from refractory clay.[17] Crucibles and distillation vessels were also found in the archaeological excavation of the later seventeenth-century laboratory at Oxford.[18] Another common item at Oberstockstall was the scorifier, a shallow, fired clay plate or bowl used to burn off lead to obtain the precious metals it contained.

There are two simple and probably trustworthy contemporary images of laboratories that are strikingly similar. One (illus. 9) shows an alchemical laboratory, while the other (illus. 10) features a distiller making flavoured brandies for medical use, utilizing the plant material strewn on the floor in front of the furnace. They both contain

Locus Occidentale & Septentrionale.

11 Libavius's ideal laboratory from the rear, 1606.

a central furnace that is being heated up by a hand bellows, looking much the same as the fireplace bellows commonly used until the mid-twentieth century. The distiller's alembic is being heated by the central furnace, but the alchemist's is being heated separately by a small Piger Henricus. Other furnaces and apparatus are not much in evidence, except for some flasks on a shelf in one picture. Thus both pictures generally support the image of the laboratory given by Weikersheim, but suggest that the average alchemical laboratory – if one can speak of such a thing – was probably smaller and less lavish than Wolfgang von Hohenlohe's laboratory, which is hardly surprising. Yet the theoretical laboratory postulated by Libavius was even more elaborate.

Laboratory of the Future

Wolfgang von Hohenlohe was a German aristocrat who was born in 1546, the son of Count Ludwig Casimir of Hohenlohe-Waldenburg in Waldenburg, Württemberg. He married the sister of the Dutch stadtholder William the Silent in 1567, and restored the family castle at nearby Weikersheim, which he inherited, into a Renaissance palace with the help of a Dutch architect. He died at Weikersheim in 1610. By contrast Andreas Libavius (Andreas Libau) was born in 1540 in Halle, the son of a linen weaver. Despite his humble background he trained as a physician at Jena. He was a schools' inspector and physician for several years in Rothenburg ob der Tauber, in Franconia, which remains much as it was in his day. He was also interested in history and became professor of history and poetics at Jena in 1588, the same year he obtained his MD from Basle. He died in Coburg, where he was rector of the Casimirianum Gymnasium (grammar school), in 1616.

In 1597 Libavius published *Alchemia*, one of the first attempts to systemically teach 'chymistry',[19] without the spiritual dimensions of alchemy, which as a devout Lutheran he rejected, and the doctrines of Paracelsus, which he detested. When he published the second edition in 1606, he attached a commentary on the 'chemical house' that was a polemical attack on the Danish astronomer Tycho Brahe, and the aristocratic, aloof, secretive and contemplative 'castle-science' that his observatory on the island of Hven and Hohenlohe's Schloss Weikersheim represented.[20] His design of the chemical house has also been criticized for its lack of understanding of classical architecture. However, leaving aside its architectural weaknesses and its polemical intentions, it was a remarkable plan that foreshadowed the later

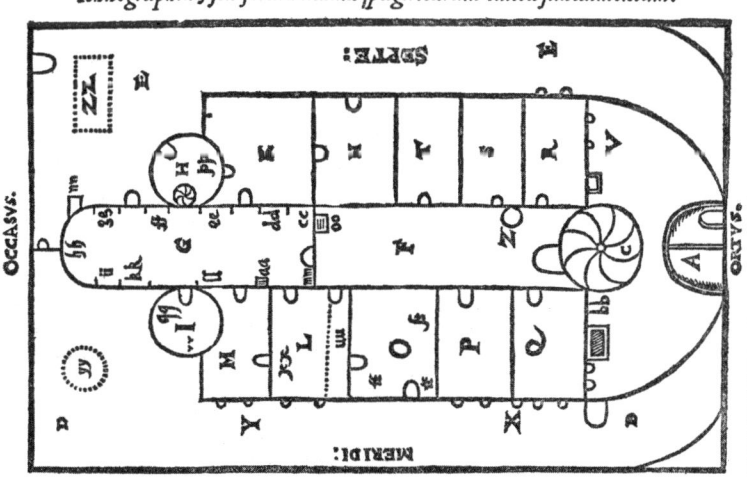

12 Plan of Libavius's ideal laboratory, 1606.

development of chemical laboratories. It cannot be considered as a real laboratory in the same way that we consider laboratories in subsequent chapters, but nonetheless it is comparable. One must also wonder if there was a certain amount of wish-fulfilment in Libavius's plans, and perhaps he would have constructed a laboratory like this if he had enjoyed the same royal patronage that Brahe enjoyed, or the wealth of Hohenlohe.

The exterior of the house is solidly German bourgeois style – perhaps a merchant's house (which would contain his working quarters), a warehouse or an inn. In contrast to Brahe's Hven, it stands in the centre of a town and is open to public gaze, if not actually accessible to the general public (illus. 11). It is thus the diametrical opposite of the secrecy of alchemy and aristocratic science. The most noteworthy aspect of the chemical house's interior is the side rooms; it is not just a single-room laboratory (illus. 12). Specialized rooms away from the main laboratory became and have remained an important aspect of the chemical laboratory. There were various storerooms, including a general store for chemicals and apparatus (N in illus. 12), and specific store rooms for wood (P), wine (X), fruit (R) and vegetables (V). The latter rooms were presumably partly for the production of plant-based medicines, but were also a result of the house being the residence of the chemist and his assistant, which became a tradition in German academic chemistry that lasted up to the twentieth century. The chemist's private laboratory (H), with its Piger Henricus (pp), was linked directly to his study and living quarters. It has been suggested that the private laboratory was deliberately set aside from the main laboratory in order to prevent visitors from discovering the chemist's alchemical secrets.[21] The private laboratory strikes me as being exactly like the professor's personal laboratory in a nineteenth-century university, and not at all secret in nature. There was also a changing room/ bathroom (S) and an aphodeuterium or toilet (T). Even more striking are the rooms set aside for specific operations, such as the analytical laboratory (J) with its assay furnaces (qq) and balances in cases (rr), the coagulatotorium or crystallization room (O) with its tubs and vats (rr), and tables for vessels (tt), a preparation room (L) and a pharmacy for the making of medicines (K). Libavius made the ordinary hearth (hh) the central feature of the main laboratory, situated in the middle of the apse opposite the entrance, but this was mostly for heating the room.

Around the walls, specialized furnaces were placed in a carefully thought-out arrangement, in contrast to the random ordering of the furnaces of many alchemical (and chemical) laboratories, including – crucially – Brahe's laboratory at Hven. The steam bath (cc) and ash bath (dd) were near the entrance, followed by the water bath (ee), the apparatus for upwards distillation (ff) and the sublimation apparatus (gg). On the other side of the hearth were the reverberatory furnace (ii), the ordinary downwards distillation apparatus (kk), and the distillation apparatus with the spiral condenser or worm (ll). The dung bath (mm), used for gentle heating, was placed discreetly near the main entrance to the laboratory and the stairs leading down to the laboratory cellar (aa).

It is difficult to say how much influence the chemical house had on laboratory design during the seventeenth century. The laboratory at Altdorf in 1682 was rather similar, with different furnaces arranged around the walls in a specific order with spaces between them, rather than being conjoined. It is interesting that these spaces in Altdorf were occupied by semicircular tables, and this may have been Libavius's intention as well. A table or tables in the centre would have been a logical feature of Libavius's dream laboratory, but would not have featured on a ground plan. As there are no internal drawings of the chemical house, we cannot tell if Libavius would have hung retorts and alembics on the walls as at Weikersheim. I suspect they would have offended his sense of elegance. He probably would have had bare walls, as in Altdorf. Indeed, one drawback of Libavius's design is the lack of ventilation except from the hearth (which may have been deliberately designed to act as a crude fume hood drawing smoke and gases away from the laboratory). The laboratory is partly inside the building as it is surrounded by the specialized rooms, and the few windows at the end are small. It would therefore seem that if it had ever been constructed, it would have been very smoky and probably very warm for much of the year, even in Germany.

Furnaces

The furnace was the key feature of any early chemical laboratory and it is practically impossible to find a picture of such a laboratory – alchemical, pharmaceutical or metallurgical – without one.[22] The alembic may have been the icon of alchemy, but it was not essential

13 Assembling a Piger Henricus, 1580.

in the same way as the furnace. One important function of the furnace is often overlooked – it helped to keep the laboratory (and its occupants) warm, although the heat must have been almost unbearable in the laboratory during hot summers. It could also provide a modicum of light at a time when artificial light was not easy to come by, and it is interesting that pictures of early laboratories rarely show any other means of lighting such as candles or rushes. Given the need for light and also for ventilation, it is surprising that the windows in alchemical laboratories, while usually present, are often small. It is likely that the door was left open in good weather.

These practical considerations do not explain why the furnace was so central to the laboratory up to the early nineteenth century. Most modern chemists try to carry out reactions at moderate temperatures, and even at room temperature if possible. Yet before powerful reagents on one hand and highly selective reagents on the other were introduced in the twentieth century, heat was the predominant agent of chemical change.[23] Apart from the strong mineral acids, the only other important agent was time, as demonstrated by the industrial manufacture of copperas and white lead over a period of weeks and months. If 'philosophers' wanted to find out what something was made of, they heated it. If they wanted two compounds to react, they mixed them and heated them. If they wanted to separate a mixture, they heated it to remove the more volatile components by distillation or sublimation. This was a very limited repertoire of techniques and explains why sublimation as a process was so highly regarded (apart from its spiritual overtones), and why it was desirable to find new ways of doing alchemy as symbolized by the philosophers' stone and the universal solvent, the alkahest.

By the sixteenth century several types of furnace were available. One of the most important was the self-feeding furnace, called the athanor, Piger Henricus, faule Heinz, lazy Heinz or slow Harry. It could be left unattended for long periods, which made it popular with alchemists and early chemists.[24] Ercker shows clearly how the Piger Henricus was constructed (illus. 13).[25] It was built of bricks and could be either square or round. The central furnace was tall and was closed with a metal cap. It was charged with a day's supply of charcoal. If necessary the cap could be removed and the furnace refilled without affecting the heating operations. At the bottom there was an iron grate with air vents above and below. The central tower fed into smaller circular side furnaces by means of large holes between them. It is important to note that there was no fire in the central tower; it was simply a silo for the side furnaces. The charcoal fell by gravity into the two side furnaces where the fire was actually burning. The key feature of the Piger Henricus was the good control of heat. The fire was controlled by air holes that could be closed when necessary by pottery plugs, and the heat in the side furnace was controlled by pottery slides (dampers) between the two furnaces.[26] The amount of heat given off by the furnace was not considerable, but it was sufficient for most distillations or digestions. There was also the portable furnace, made of terracotta, which came apart and could be reassembled elsewhere, and was not dissimilar to a modern chimenea. Other forms of heating included the

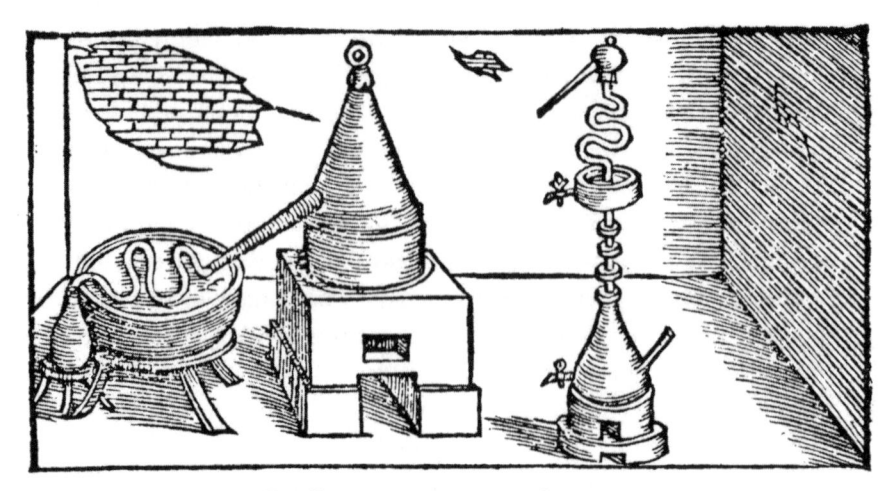

14 Distillation using a worm condenser, 1540.

ash or sand bath, a box of ashes or sand, heated underneath to produce a uniform indirect heat. To produce a stronger heat, the sand could be replaced by iron filings. For intermediate heating there was the water bath, the bain-marie of the kitchen, which is still in common use in the modern laboratory. Conversely, to produce a gentle heat of about 40°C, for example, for delicate digestions, the alchemist would have used the dung bath, where the heat was produced by the fermentation of animal dung.

Distillation and Condensation

The distillation of liquids and solids with the condensation of the resulting vapours was one of the most prominent activities in medieval laboratories.[27] The alembic is usually associated with alchemy, but distillation was actually more important in medicine, as plant materials were distilled to produce medicinal products such as essential oils. Modern spirit beverages also arose out of this medicinal distillation and were originally used as medicines. Indeed, brandy (and whisky in the form of a toddy) remained a standby of household medicine until modern times, when physicians began to frown on the practice of giving injured or sick people spirits. For this reason, fragments of pottery and glass-distilling apparatus, including alembics and cucurbits, are often found in the grounds of medieval monasteries, for example Pontefract Priory in Yorkshire and Selborne Priory in Hampshire.[28] The alembic was used for almost the entire

Christian Era and only fell out of general use in the laboratory during the nineteenth century.[29] It has two parts: the alembic itself with its dome-shaped still-head and spout, and the round-bottomed cucurbit, which contains the material to be distilled. Alembics were often used with the Piger Henricus, as this allowed distillation for long periods unattended (see illus. 29). The cucurbit was placed inside the side furnace, supported by two semicircular plates, and the space between the cucurbit and the furnace was sealed up, a process called luting. The alembic was then placed on top of the cucurbit, and could be removed without having to free the cucurbit from the furnace. Alembics were both expensive and delicate, which is why few medieval ones have survived intact, and the simpler retort was often used in their stead.

The retort was particularly useful for the destructive distillation (pyrolysis) of solids, and hence was an important tool for organic analysis, which relied largely on pyrolysis up to the mid-nineteenth century. It continued to be used until the mid-twentieth century

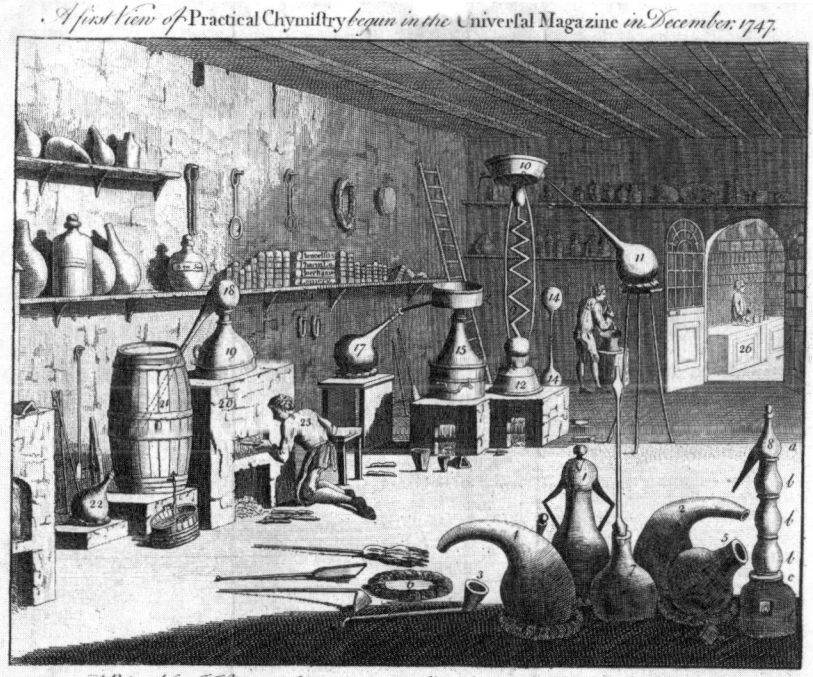

15 Pharmaceutical laboratory, 1747. Although later in date, it is similar to pharmaceutical laboratories of the 17th century.

(I bought a glass retort in my local department store around 1970). The vapours from the alembic and the retort were usually condensed by the cooling of their spouts by air, and the spout of a retort can be extended with a glass adapter, producing a larger cooling surface. A long spout was sometimes passed through a barrel of cold water to improve the yield. The 'worm', a spiral condenser often made of copper, which could be cooled by air or a water bath, was first described by Taddeo Alderotti in the thirteenth century (illus. 14). It was often used from the fifteenth century onwards for the production of distilled water.[30]

The Roots of the Modern Chemical Laboratory

Like chemistry itself the chemical laboratory has several origins; it is not simply a successor of the alchemical laboratory. It has its roots in various areas of activity, all of which involved using chemical change and could be considered to come under the umbrella of alchemy, as Luther stated in his *Tischrede*.[31] The most important of these fields was pharmacy, and chemistry struggled to distinguish itself from pharmacy (and medicine more generally) until well into the nineteenth century (illus. 15).[32] Early metallurgy and assaying was another significant model for early chemical laboratories. By the early seventeenth century Libavius could present a plan for a 'chemical house' for purely rhetorical purposes, which nonetheless reflected current practice and even foreshadowed some aspects of the nineteenth-century laboratory. In its details, however, the chemical house was at least partly a pharmacy. The main features of all these laboratories were the furnace, especially the Piger Henricus, and the distillation vessels, usually alembics and cucurbits, even in the metallurgical workshops. They were to remain important elements of the chemical laboratory until the early nineteenth century. However, the modern laboratory could not take shape until the furnace was removed. The gradual displacement of the furnace is a thread running through the next four chapters. The next chapter describes how the rise of a new field of chemistry, the study of gases, began this process in the 1770s.

2
Form and Function
Antoine Lavoisier and Paris, 1780s

Most eighteenth-century laboratories were dominated by their furnaces. Many were effectively workshops, places for making things, albeit chemical products rather than simply physical objects. In the case of pharmacists, glass and porcelain manufacturers, and metal workers this production was commercial, but consultants and universities, including Oxford and Padua, had similar laboratories. Furnaces remained the focal points of laboratories both visually and in terms of their functional importance until the early nineteenth century.[1] The laboratories of Michael Faraday and William Pepys still contained furnaces in the early 1800s (see chapter Three).

A Decisive Shift

Once the focus of chemistry began to shift from making to investigating and from quantity to precision, the days of the multi-furnace laboratory with all its dirt and clutter were numbered. The pneumatic chemistry (study of gases) of the late eighteenth century was an important driving force in this transition.[2] In this context there was a decisive break when Antoine Lavoisier set up his new laboratory at the Paris Arsenal in 1777. Furnaces were inconspicuous in drawings of this laboratory, which was dominated by large tables and table-like pneumatic equipment. Rather than featuring rows of retorts, the walls were lined with large pigeonholes for the storage of apparatus, which was surprisingly unusual for this period. In his use of tables Lavoisier was linked to a distinctively French line of laboratory development that progressed in parallel with the development of physics and engineering laboratories rather than the specifically *chemical* laboratory which arose later in Germany.

The Furnace Endures

Zacharias Conrad von Uffenbach described the basement *Officina Chemica* of the Ashmolean Museum in Oxford in 1710 in the following terms:

> But to return to the Laboratory. I must say it is right well built.... It is completely vaulted over and is provided with many kinds of furnaces, some quite remarkable, all of which are decorated in the most costly manner with architectural decorations and the like.[3]

It is striking how he assessed the value of the laboratory in terms of its furnaces, just as we might now assess a laboratory's standing by the presence of FT-NMR machines or the most up-to-date mass spectrometers. He lamented the neglect of these furnaces by the 'good-for-nothing' laboratory technician ('operator') and went on to ask, 'Is it credible that so little attention should be given to so costly and beautiful a work?'

16 The pharmaceutical (east) side of Ambrose Godfrey's laboratory,
'The Golden Phoenix', Covent Garden, London, *c.* 1728.
The other side manufactured phosphorus.

The importance of the chemical furnace is also shown by the creation of a constellation in its honour. The French astronomer Abbé Nicolas de Lacaille is often cursed by his modern successors for creating numerous small constellations in the southern sky.[4] A good example of these is the rather shapeless constellation that he named Fornax (or rather Fourneau, as he used the French name) in honour of chemistry in 1752.[5] It is sometimes claimed that he named the constellation in honour of Lavoisier, despite the fact that Lavoisier only became a pupil of Lacaille in 1760 at the age of seventeen, eight years after Fourneau appeared in Lacaille's planisphere.[6] I suspect that the confusion arises because Johann Bode renamed the constellation 'Apparatus chemicus' in 1782 to honour Lavoisier, but this name was never widely adopted.[7] Having paid tribute to astronomy and navigation (Telescopium, Octans and Pyxis), clockmaking (Horologium), mathematics (Circinus), the fine arts (Sculptor and Pictor) and even carpentry (Norma and Caelum), Lacaille clearly felt the need

17 William Lewis's Laboratory, Kingston, Surrey, 1763.

to acknowledge chemistry. That he recognized chemistry in this way tells us something about its status in mid-eighteenth-century France, a point that is explored further below. It is also interesting that he chose the chemical furnace as the symbol of chemistry, in the same way that the telescope had come to represent astronomy.

As already noted, the laboratories of alchemists and assayers were centred around the main furnace. This was also true of the laboratories of the chemical physicians such as Annibal Barlet. It remained the case in the early eighteenth century, a period in which many chemical laboratories were remarkably similar in their fittings and layout. Whereas the alchemists had been mostly content with one furnace, or perhaps one large furnace and a smaller portable one, some eighteenth-century laboratories had several furnaces of different sizes. Examples of such systems can clearly be seen in the drawing of the laboratory of Ambrose Godfrey in Southampton Row, Covent Garden, at the beginning of the eighteenth century (illus. 16), and in the laboratory of William Lewis at Kingston upon Thames in the 1760s (illus. 17). As William Gibbs remarked of Lewis's laboratory, 'This was one of the first laboratories specifically designed for research in applied chemistry and physics'.[8] There are strong similarities with Godfrey's laboratory, with glassware being stored on shelves above the furnaces, and *several* furnaces raher than just one large example. There is an impression of what one might call 'tidy clutter', with considerable amounts of glassware and multiple furnaces in rows, and with other pieces of apparatus lying around, such as the rotary drum evaporator with some cloth draped on it, presumably having been dyed, and a ball mill for grinding materials at the front of the picture.[9] These items reveal that this was not a natural philosopher's room, but a consultant's laboratory with strong industrial connections – as shown by the title of Lewis's book, the *Commercium Philosophico-Technicum*. As Gibbs noted, Lewis put considerable effort into developing a small furnace made of blacklead (graphite) that was suitable for the laboratory. Gibbs believed that the laboratory was organized in a way that put chemistry on the left and physics on the right. It is more likely that the barometer and balances were placed at the window so that they could be read easily in a period when interior lighting was practically non-existent. Whether they would have been stored on the windowsill is a moot point. The direct sunlight would clearly have not been good for them, so it is likely they were put in a store cupboard when not in use (or on display for the purposes of the illustration).

Gas Manipulation

Since today we can handle gases with ease, and weigh them with modern apparatus to a millionth of a gramme if necessary, the problems of manipulating gases are not obvious. Furthermore, gases are surprisingly easy to generate. We produce a chemically distinctive gas every time we breathe or burn something. Stagnant ponds and marshes also produce gases, as does putrefying matter. Chemically interesting gases can be produced by dropping liquids onto solids (hydrochloric acid on marble) or heating solids (nitrous oxide from ammonium nitrate). The problem lies in collecting the gases and storing them.[10] Chemists had never handled individual gases before, and while some basic apparatus that had been used with air could be utilized (such as a pig's bladder), they were fairly limited. To make matters worse the available technology (for example glass making) was barely able to meet the more stringent demands of this new field. There was always the danger of leaks or explosions. The founder of modern gas chemistry, Joseph Black, was very wise, in the light of the poor apparatus available to him, to sidestep the manipulation of gases altogether. He studied the properties of fixed air (carbon dioxide) by only weighing solid substances and calculating the weight of

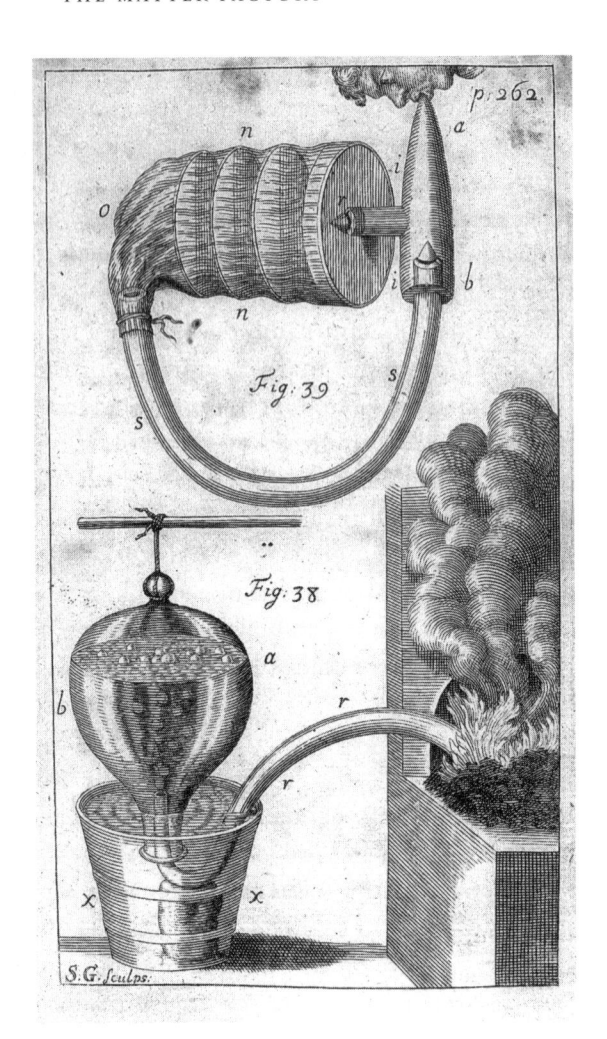

18 Stephen Hales's gas-collecting apparatus, 1721.

fixed air by difference. Later chemists continued to use the absorption of gases in liquids or on solids as the most accurate way of weighing very light gases – a litre of air at room temperature weighs just over a gramme.

However, if gases were to be studied properly their manipulation had to be mastered. In the 1650s Jan Baptist van Helmont tried stoppering the containers in which the gases were generated, with explosive results. During his work on plants in the 1720s, Stephen Hales introduced the collection of gases over water in an upturned container (illus. 18). His main aim in doing so was to remove sulphurous fumes from the gas produced in a retort, a function of the pneumatic trough that remained important for the next two centuries.

Hales did not, however, recognize the chemical distinctiveness of the gases he collected and was only interested in the volumes of gas produced. Henry Cavendish used his apparatus to collect hydrogen for the first time in 1766. Hales deserves credit for the idea of collecting gases over water, but it would be going too far to call Hales's crude apparatus a pneumatic trough. A pneumatic trough implies three pieces of apparatus: a trough to hold the water, a stand to hold the gas jar above the bottom of the trough and the gas jar itself. William Brownrigg made the first apparatus to resemble this set-up in 1765, but he only used it to transfer gases from a bladder to an upturned container, and his shelf was above the water (as was the later shelf made by Torbern Bergman). The complete pneumatic trough as used in schools was first made in the early 1770s by Joseph Priestley, and has changed little since (illus. 19). Using this apparatus, Priestley transformed our knowledge of gases. Between 1772 and 1780 he discovered eight gases, including nitrous oxide, oxygen, nitric oxide, ammonia gas, hydrogen chloride and sulphur dioxide.[11] The nearest parallel to his work is the transformation of astronomy by Galileo's use of the telescope, although like Priestley and the trough, he did not invent the basic instrument.

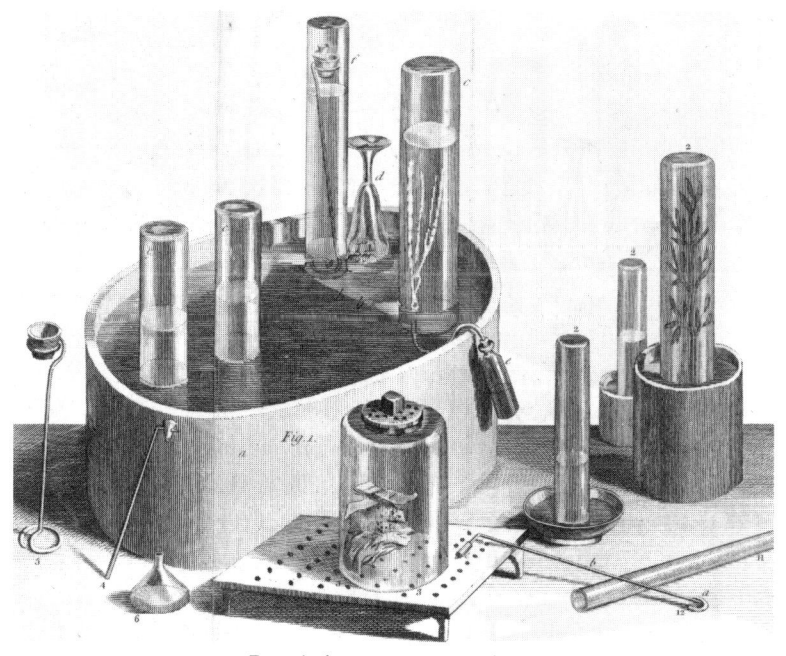

19 Priestley's pneumatic trough, 1774.

It was all very well for Hales to remove acid fumes by collecting the gas over water, but this is not much use if you actually want to study sulphur dioxide or nitrogen dioxide. Fortunately (at least in the days before the health and safety people took over), there was a simple solution; you could replace the water with mercury. Cavendish was the first person to use mercury to collect and study fixed air in 1766. Once again it was Priestley who used this idea to its full extent to study hydrogen chloride, ammonia and sulphur dioxide, which are all water soluble. Sometimes, however, you want to wash out a component of a gaseous mixture, either to purify it or to weigh that component more easily. For example, you might want to remove acidic fumes by bubbling the gas through a solution of caustic soda, or to dry it by passing it through concentrated sulphuric acid. The eccentric Irish chemist Peter Woulfe created a way of doing this in 1767 by passing a gas through a jar fitted with two necks and partly filled with water to remove acidic fumes that would otherwise be 'very hurtful to the lungs'.[12] His main problem was fitting the glass tubes into the corks. He cut the corks half open, then sealed the joints with so-called lute – clay or cheese being favoured for this purpose. The problem was eventually solved by the invention of the cork borer by (Karl) Friedrich Mohr in 1837.[13]

An Exact Chemist

By the middle of the 1770s much progress had been made in less than a decade in the field of gas chemistry, largely due to the work of Priestley and Cavendish. However, another chemist was soon to dominate the scene. Lavoisier was a tax farmer, a businessman and a government official.[14] In our era he would probably have been an accountant. He certainly applied the accountant's mindset to chemistry. Born in 1743, Lavoisier was the son of a wealthy lawyer and took a law degree to comply with his father's wish that he maintain the family link with the law. However, Lavoisier was attracted to science, partly as a result of being taught by Lacaille at the Collège Mazarin after the astronomer's return from the Cape of Good Hope in 1754. Lavoisier initially pursued geology under the influence of Jean Etienne Guettard, and he was elected to the French Academy of Sciences in 1768 (when he was only 25) on the strength of his geological research. The Academy remained the main arena for his scientific work for the rest of his life. At the same time he had to find an income and joined the Ferme Générale, a private organization that had the task of collecting taxes for the French government, the eighteenth-century French equivalent of joining a Lloyds insurance syndicate. The job was essentially an administrative one, involving the supervision of the numerous tax collectors spread across France. This choice of career had two momentous consequences for the young man: Lavoisier married the daughter of a fellow tax farmer, Marie-Anne Pierrette Paulze, in 1771 (illus. 20), and as one of the hated tax farmers he was guillotined in 1794.

During his brief spell in office as controller-general, the reformer Anne-Robert-Jacques Turgot appointed Lavoisier as one of the four directors of the new Gunpowder Administration in 1775. Consequently Lavoisier moved to the Paris Arsenal in April 1776 and set up his laboratory there. At a time of famines in France, Turgot sought to reform the Ferme Générale – the creation of the Gunpowder Administration was one of his reforms – and his secretary Pierre Samuel du Pont believed that it should concern itself with the development of agriculture. Du Pont was dismissed in 1776 along with his master, but returned to the finance ministry two years later. In the same year his friend Lavoisier bought an estate in the Loire mainly to carry out agricultural experiments. Central to the improvement of both gunpowder manufacture and agriculture was the study of saltpetre production, to which Lavoisier now devoted himself. One of his assistants at the Arsenal was du Pont's son Eleuthère Irénée du Pont, who also ran a saltpetre refinery. He emigrated to the USA in 1799 and founded the eponymous gunpowder mills in Delaware, the forerunner of the modern chemical firm DuPont.[13]

As an ambitious and increasingly wealthy amateur scientist, Lavoisier was attracted to the new field of gas chemistry. He did not discover any new gases or elements himself, but due to his clear thinking and precise way of working he brought about a revolution in chemistry. He saw that this new field was throwing up problems that its pioneers – Priestley, Cavendish and Carl Wilhelm Scheele – were struggling to conceptualize and solve. Unlike Priestley and Scheele, he could afford to buy the very best apparatus for his work. By redefining issues such as the nature of combustion and the composition of water into physical problems to which basic principles such as the conservation of weight could be applied, Lavoisier hoped to

20 Antoine and Marie-Anne Lavoisier, after the oil painting
by Jacques-Louis David, 1788.

clarify the phenomena thrown up by gas chemistry. He found that he was only able to do this if he abandoned the existing theoretical framework, Kuhn's well-known 'paradigm',[16] of phlogiston accepted by Priestley and Scheele, and created a new one based around his newly discovered element, oxygen (*oxygène*). Effectively he turned the chemistry of combustion on its head. Instead of combustion occurring because of the presence of the 'principle of inflammability' (called sulphur, phlogiston or Feuerstoff), according to Lavoisier it took place only when oxygen (Priestley's dephlogisticated air) was present. Instead of phlogiston leaving the burning substance, oxygen entered it to form a compound, an oxide. As long as it could react with oxygen, the exact chemical composition of the inflammable substance was irrelevant.

Lavoisier's most crucial action did not, however, require carefully designed experiments or precise measurements. By the mid-1780s chemists either had to accept the existence of oxygen or argue that phlogiston had a negative weight.[17] While most chemists reluctantly agreed that oxygen was more feasible than negative weight, they were hesitant to abandon phlogiston, partly because it had greater explanatory power. Lavoisier forced their hand by the Orwellian tactic of creating a new nomenclature based on his reforms. Either you accepted the oxygen theory and used the new language of chemistry, or else you faced being locked out of chemical discourse. There was a third alternative – to keep phlogiston but also accept oxygen, which is in fact the closest to modern chemistry if one equates phlogiston with the electron,[18] but this never caught on because of Lavoisier's tactics. Lavoisier also proposed the first clear definition of an element and produced a list in his *Traité élémentaire* published in 1789. However, not all of Lavoisier's ideas were correct; he believed that all acids contained oxygen and hence that chlorine was the oxide of an unknown element. He also thought that heat was an element, which he called 'caloric'.

As a supporter of the reformers such as Turgot and du Pont during the Monarchy, Lavoisier also supported the French Revolution in its early, more moderate phase, and proposed various progressive measures, including the reform of the Academy of Sciences and the introduction of the metric system. He was dismissed as a director of the Gunpowder Administration in the summer of 1793, and was put under arrest as a member of the Ferme Générale along with his father-in-law in November. The former fermiers were put on trial in May 1794 at the height of the Terror – Danton had been executed only a month earlier – and they were all guillotined on 8 May. They were only a few weeks from safety – Robespierre, the architect of the Terror, fell from power on 27 July 1794. By contrast, Lavoisier's friend Pierre du Pont had also been condemned to death, but having survived to see the fall of Robespierre, he was released.

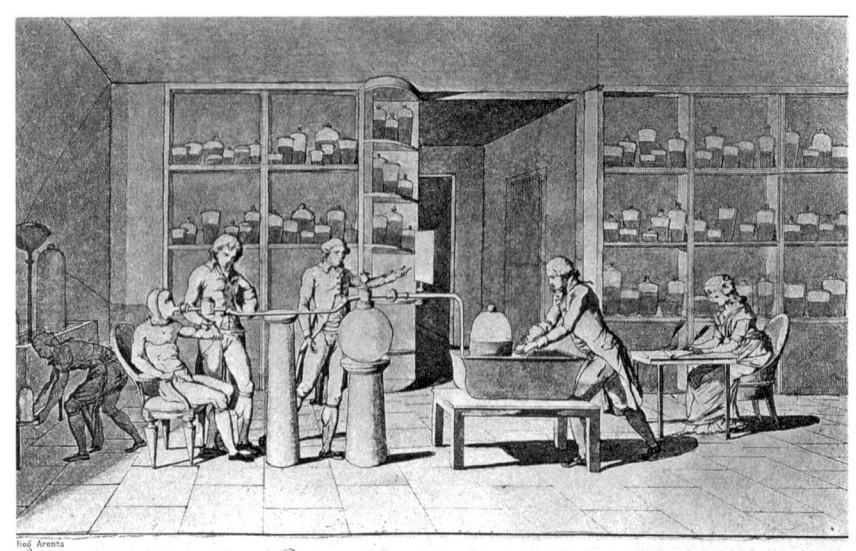

Lavoisier dans son laboratoire
Expériences sur la respiration de l'homme au repos

21 A view of Lavoisier's laboratory in the Paris Arsenal, *c.* 1790.

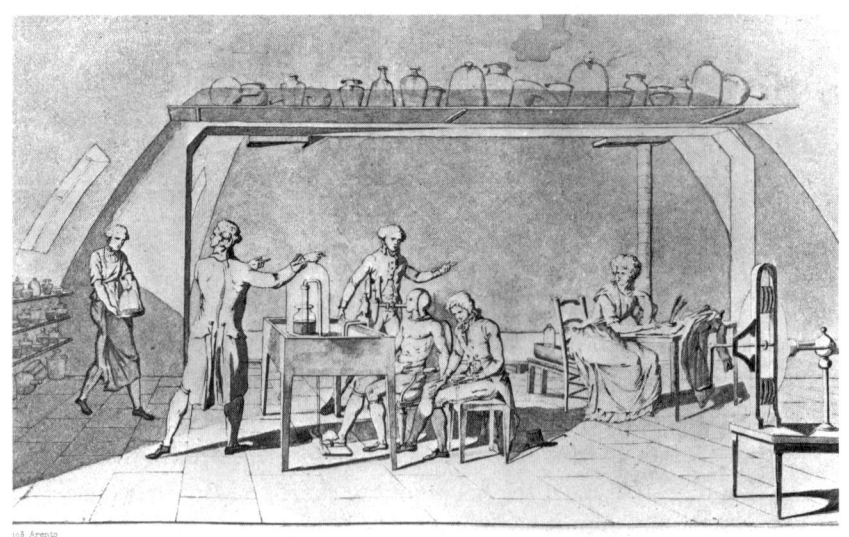

Lavoisier dans son laboratoire
Expériences sur la respiration de l'homme exécutant un travail

22 Another view of Lavoisier's laboratory, *c.* 1790.

A Functional Laboratory

Marie-Anne Lavoisier's famous drawings of her husband's laboratory depict the experiments on human respiration carried out on his assistant after the publication of the *Traité* in 1789 (illus. 21, 22).[19] From our perspective what matters is not the precise details of the experiments or even their outcomes, but what Madame Lavoisier's drawings tell us about Lavoisier's laboratory at the Arsenal. While doubts have been raised about the accuracy of the pictures as far as the experiments are concerned, it seems reasonable to assume that the overall situation depicted is fairly true to life, as Marco Beretta has shown conclusively that the drawings were made for publication in a monograph, probably Lavoisier's *Mémoires de physique et de chimie*.[20] Doubtless the laboratory is tidier than it might have actually been, and the bottles on the shelves seem stereotypical rather than accurate representations, but this is a problem with all drawings of laboratories.

What stands out is the functionality of the laboratory and the people working in it. Tables of different heights and size are used as required, and two tables are effectively water tanks on legs, presumably designed by Lavoisier for the manipulation of gases. Bottles or carboys of chemicals are stored on a proliferation of shelves, which are effectively large pigeonholes. By contrast, the apparatus is stored on a shelf above an archway and on a side table, reflecting earlier laboratory practice. Everything is in its place and nothing is stored on the floor – the clutter of Lewis's laboratory is absent. Furthermore, the people in the prints all have a definite function, as Beretta has pointed out.[21] Lavoisier directs the experiments, Armand Séguin is the subject – one assistant is taking his pulse and another one seems to be counting the time. Madame Lavoisier is recording the details seated at a table. Finally, the laboratory assistant brings, and presumably sets up, the apparatus.

The use of tables for experiments is noteworthy. Just as tables have always existed in kitchens, so too were they an incidental feature of alchemical and metallurgical workshops. They were used for putting down tools between operations, or perhaps for particularly delicate procedures such as weighing, as in the picture of Johann Barchusen's laboratory (illus. 23). However, the table remains insignificant relative to the furnace or even the windowsill. There are no tables to be seen in the pictures of the laboratory portrayed in the

Universal Magazine of 1748, or in Lewis's laboratory in 1765 (see illus. 17).[22] When tables are present, they are not there for chemical experimentation: lecturers such as Barlet used a table for lecture-demonstrations (see illus. 32), but Barlet's table appears to be a trestle table that was probably put away between lectures, and the prominent table in the etching of the Altdorf laboratory of 1682 (see illus. 33) was only there to be used by the students for taking notes.

The first clear example of a table being used for chemical operations was in the drawing of a laboratory in the famous *Encyclopédie* in 1763 (illus. 24). This picture has to be treated with some caution as it represents an idealized laboratory rather than a real one of the time. Coming to chemistry from a natural philosophy background and shaped by the French Academy of Sciences, Lavoisier was probably influenced by the then popular *cabinet de physique*, a room containing shelves of physical apparatus and a central table for experiments, or perhaps more often demonstrations to visitors. Abbé Nollet's famous cabinet does resemble Lavoisier's laboratory in several respects.[23] Most important of all, however, is that a table is much more suitable for gas manipulation than a furnace or a narrow bench. Pneumatic troughs, bell jars, gas jars, bladders, flasks, Woulfe bottles and eudiometers require a level and stable space, but little or no heat. While some methods of gas generation require heat, the

23 Johann Barchusen's laboratory, 1718.

24 Idealized laboratory from the *Encyclopédie, ou dictionnaire raisonné des sciences, des arts et des metiers* (1763).

amount needed is usually small and can be provided by a spirit lamp or burning glass. Priestley also used a table for his experiments in the sitting room of his house on the green at Calne, Wiltshire, near Bowood House where he worked as the librarian, a conclusion I reached after comparing the present-day sitting room and fireplace in the house with his drawings of his experiments. We know he had six tables in his house and laboratory in Birmingham, as he listed them in a claim he made after they were destroyed by a loyalist crowd in July 1791.[24] He also had a carpenter's table, which shows that they were not all used for pneumatic chemistry. It is very likely that he continued to use a table at Northumberland, Pennsylvania, but direct evidence for this is lacking.

Why France?

In the Netherlands and Scotland in the early to mid-eighteenth century, chemistry became increasingly important, but it was usually subordinate to medicine. Herman Boerhaave in the Netherlands,[25] and William Cullen and Joseph Black in Scotland, were all first and foremost professors of medicine and practising physicians.[26] In Oxford chemistry was part of anatomy and was taught alongside it.[27] Similarly in Germany it came largely under pharmacy.[28] In Sweden it was tied to mineralogy, which was also partly the case in Scotland.[29] Only in France did chemistry develop an autonomous and distinctive position from the mid-eighteenth century.[30] This was in part due to the increasing demands of the state, nationally and locally, for scientific advice on everything from industrial processes to public health, issues that were increasingly seen as chemical. At the same time in Paris and in the many other large cities in France, an expanding educated public came to see chemistry as the enlightened science par excellence; rational, progressive, practical and useful.

Furthermore, in France chemistry was seen as part of natural philosophy rather than natural history or medicine. There had been a strong movement in early eighteenth-century England to develop a system of chemistry based on Newton's mechanistic and atomistic ideas, which he outlined in his *Opticks* of 1704, but this had died out by the 1740s and had been replaced by a largely medical approach.[31] In France, however, chemistry emerged as a very broad church.[32] Having established its autonomy, French chemists concerned themselves with the systematic investigation of problems that ranged almost seamlessly from the practical (mineral water analysis, dyestuffs and the improvement of gunpowder manufacture) to the theoretical (the nature of acidity, affinity, and the role of gases in composition and reactions).

What is most striking about chemistry in France in this period is how popular it was and how eager people were to learn it.[33] In mid-eighteenth-century France there was an extraordinary explosion of public interest in learning about chemistry and even taking formal courses in the subject. John Perkins has shown that several thousand people took chemistry courses each year in Paris alone. Even allowing for the size and importance of Paris, these are remarkable figures. It would not be unreasonable to say that more people took chemistry courses in Paris in, say, 1785 than during the entire eighteenth century in Oxford; indeed, possibly the whole of England.[34] These three aspects of French chemistry made it an attractive subject for the ambitious and increasingly wealthy Lavoisier.

The Laboratory Table

Lavoisier's introduction of the table into chemistry did not lead to an immediate increase in the use of tables by chemists. This was partly because an alternative arose in the same period, which was a forerunner of the chemical bench – a freestanding counter, not unlike a shop counter, made of brick or stone. Such counters probably arose in pharmacies, as shown in the wonderfully elaborate pharmacy in

Pressburg (now Bratislava) drawn by Salomon Kleiner in 1751 (illus. 25).[35] They were eminently suitable for chemical experiments that generated heat. This point is nicely illustrated by a nineteenth-century illustration showing a dramatic demonstration by Guillaume-François Rouelle (illus. 26). Similar counters can be seen being used by Michael Faraday at the Royal Institution (see illus. 30), and at the University of Zurich in the 1840s (see illus. 48). However, Harriet Moore's painting of Faraday shows the drawback of the counter. It was usually rather narrow and did not have cupboards. Indeed, Faraday was forced to keep some of his apparatus on the floor, like Lewis nearly a century earlier.

The closest example of a laboratory similar to Lavoisier's in the early nineteenth century was the laboratory at the University of Leiden set up in 1831 by Antonius Henricus van der Boon Mesch to teach technical chemistry, which had not only the table but also the pneumatic trough on legs that Lavoisier used (illus. 27). Despite its modernity in these respects, the retorts on the back wall hark back to Ercker's metallurgical workshop (see illus. 29). William Pepys at the London Institution also had a very traditional laboratory with a furnace and rows of retorts on the wall, and he used a sturdy table (see illus. 39). The work in Berzelius's laboratory was carried out

25 Salomon Kleiner, *Interior of a Pharmacy*, 1751.

26 Nineteenth-century representation of Guillaume-François Rouelle giving a lecture demonstration in the 1750s at the Jardin du Roi, Paris, 1874.

on two pinewood tables, one of which was for his use and the other for his assistant.[36] They were simply kitchen tables, and in fact the kitchen was in the next room. Liebig's laboratory at Giessen had tables as well as benches against the walls, in contrast to the counters at Zurich. What is most striking, however, is the continuing use of tables remarkably similar to Lavoisier's by French chemists well into the nineteenth century. The French chemist and politician Marcellin Berthelot became professor of chemistry at the Collège de France in Paris in 1865 (illus. 28). It appears that the laboratories of the Collège de France had not changed very much since the 1830s. They contained long rooms fitted with glass-fronted cupboards (like Heidelberg). In the open space in the middle there were tables for experiments and a desk for the professor. This description fits both extant photographs of Berthelot and the well-known painting of Louis Pasteur's laboratory that was located in the same building. The chemical laboratory at Lisbon Polytechnic School used tables and stored chemicals in glass-fronted cupboards (one of which looks like a bookcase) and a chemical desk before the renovation of 1888–90,[37] when the tables were replaced by classical benches with the usual bottle racks.[38] This perhaps shows a shift from French to German influence in Portuguese chemistry.

In modern terms Pasteur was as much a microbiologist as a chemist, and Berthelot's later research at the Collège de France straddled the boundary between chemistry and physics. The classical bench is specific to chemistry (and the related biochemistry). Other sciences, notably physics and biology, are better served by tables. This is also true for engineering. When William Thomson (later Baron Kelvin) took up his chair at the University of Glasgow in 1846, the concept of a dedicated physics laboratory for research or teaching was almost unknown.[39] The university gave him some new apparatus, but this was not directly linked to a laboratory, new or otherwise. Thomson used classrooms as a research space for himself and his student assistants. By 1855 his work had spilled over into other classrooms, leading to a complaint from an older colleague. The faculty resolved this complaint in 1857 by giving Thomson several cellar rooms beneath his classroom, thus, perhaps unwittingly, creating one of the first academic physics laboratories in Britain. Perhaps crucially Thomson was used to chemical laboratories, as he had been taught chemistry by Glasgow's professor of chemistry, Thomas Thomson.[40] As seen in later photographs, Thomson's laboratory was a large, open room with tables for equipment and a desk for the professor or demonstrator – again almost identical to Lavoisier's arrangement.[41]

27 Laboratory for teaching technical chemistry, University of Leiden, 1831. Note the retort racks, which are similar to the ones in the Ercker etching 250 years earlier (see illus. 29).

28 Marcellin Berthelot in his laboratory at the Collège de France, 1901.
Photograph by Paul Dornac.

Putting it Away

Lavoisier's laboratory also represented a major advance in the storage of chemical apparatus. From its inception chemistry involved the use of bulky equipment such as alembics, retorts, crucibles, reaction vessels, small furnaces, braziers, mortars and filters. Clearly storage space was always an issue. The more expensive apparatus might be displayed on side tables, as at Altdorf (see illus. 33) – just as one might display an expensive household object – but in many respects the chemical laboratory's need for storage was similar to that of the kitchen. However, many pictures of laboratories between the Middle Ages and the mid-eighteenth century do not show any storage space apart from a single simple shelf, which could be quite broad, and the furnace mantelpiece (illus. 29). At Altdorf and in Lewis's laboratory, apparatus was also seemingly stored on the windowsill. As Lewis's laboratory illustrates, even expensive equipment such as balances were kept ready for use rather than being stored away or used in a separate room, despite the potentially damaging effects of light, smoke and fumes. However, such a configuration required space and not all laboratories were large enough. It would be logical to assume that many laboratories, even in the seventeenth century, had a separate storage room for both equipment and chemicals, like a pantry in a kitchen, as shown in the hypothetical 'chemical house' of Libavius (see chapter One). Cupboards under the bench of the type found in modern laboratories were rare before the mid-nineteenth century.

The Eudiometer

This is a graduated vessel that is used to measure the volumes of gas that take part in a reaction, usually followed by a diminution of the

29 Distillation of nitric acid (parting acid) in the medieval laboratory, showing the use of simple shelves for storage and special racks for alembics and retorts, 1580.

volume. After Priestley discovered dephlogisticated air (which we now call oxygen), there was feverish activity to measure the amount of this new 'air' in common air.[42] This was because it was assumed, not unreasonably but incorrectly, that there was a connection between the 'goodness' of air and the amount of dephlogisticated air present. It was thought, for example, that seaside air was best because it contained more dephlogisticated air.[43] The disease malaria was given its name because of an assumed link with bad air – in reality the putrid smell came from the Roman marshes, which also harboured the parasite-carrying mosquitoes. We now know that the amount of oxygen at sea level remains remarkably constant, and hence the healthiness of the air is related to other factors. It is a different matter at higher altitudes, but this effect was not the prime interest of the early investigators, although they did discover that oxygen levels drop with increasing elevation. This is actually a result of the lower air pressure; the percentage of oxygen in the atmosphere remains constant at 21 per cent. The supporters of dephlogisticated air and oxygen were equally interested in the goodness of air as they both linked the gas with respiration; thus the development of the eudiometer was not hampered by the increasingly deep theoretical divide.

When Priestley first made nitrous air (nitric oxide) in 1772, he noticed that it reduced the volume of common air stored above water. Two years later, when he discovered dephlogisticated air, he found that a much larger contraction took place and realized that the nitric oxide must be reacting with oxygen to produce a soluble product (in fact nitrogen dioxide). Thus the reduction in the volume of air above the water when nitric oxide was bubbled into the air showed how much oxygen was present. The first eudiometer was invented by the Italian aristocrat Marsilio Landriani in 1775. He was convinced that there was a relationship between the amount of de - phlogisticated air in common air and salubrity. He called his new instrument the 'eudiometer' from *eúdios* (ancient Greek for 'clear weather', but also with the implication of good air, because *-dios*, stemming from Zeus, can mean 'weather' or 'air').

Landriani vigorously promoted the eudiometer, but his friend Alessandro Volta was sceptical about any relationship between salubrity and dephlogisticated air, making the distinction between the healthiness of air and its respirability. Volta also felt that it was a complicated way of carrying out a simple test. He was studying inflammable air (hydrogen) at the time, and in 1777 he developed an electric pistol that produced a report by igniting the explosion of hydrogen and air with an electric spark. Volta's hopes that his pistol could replace gunpowder were not fulfilled, but it became a well-known demonstration apparatus in the nineteenth century. Carrying out experiments with his new pistol, he then realized that it made a better eudiometer as the water produced effectively disappeared, whereas the nitrogen dioxide formed in Landriani's eudiometer only dissolved imperfectly in water. Volta then converted his electric pistol into the Volta eudiometer, which became the standard instrument. He thereby widened the use of the eudiometer to the study of all gas reactions involving a change of volume rather than just the respirability of air. Although the study of the 'goodness' of air continued into the nineteenth century, this extension of the uses of the eudiometer ensured its long-term survival. In 1783 Cavendish introduced a brass and glass eudiometer. His eudiometer, based on Landriani's original design, used nitric oxide and has no connection with the later so-called Cavendish eudiometer, which is in fact a pear-shaped variant of Volta's electric eudiometer.

A Path Not Followed

The eighteenth century opened with the furnace still firmly entrenched in the chemical laboratory. While the furnace continued to feature in some chemical laboratories up to and even beyond the 1830s, a shift to a new kind of laboratory got underway by the mid-eighteenth century. At first it was a matter of new apparatus intruding into the traditional laboratory, often related to natural philosophy (physics) as in Lewis's laboratory, which was awkwardly placed on windowsills, shelves and even on the floor. The pace quickened with the rise of pneumatic chemistry, which did not require furnaces, but did need large, flat spaces for the new pneumatic apparatus. Priestley just used the ordinary tables that were in his house, but this was clearly unsatisfactory in the long term. Lavoisier developed a new type of laboratory geared to the new kinds of chemical and biological investigations he had in mind. It contained tables, free-standing pneumatic troughs (as an improvement on tables) and organized shelving. Above all, it was highly functional and clear of clutter. It might thus be seen as the physical counterpart to Lavoisier's precise ways of working on paper – it was a laboratory for a natural philosopher or even an accountant.

Lavoisier's laboratory hence marked a decisive break with both the furnace-dominated workshop-laboratory and the old amateur chemistry carried out in the home, as in Priestley's early experiments. While William Hyde Wollaston may have carried out experiments in his drawing room some three decades later for visitors, he actually did his research in a laboratory out at the back of the house.[44] However, later chemists, especially in Germany, did not adopt the table-based style of the natural philosophers and physicists, but created their own bench-centred design, which was better suited for inorganic analysis and organic synthesis.

3

Laboratory versus Lecture Hall
Michael Faraday and London, 1820s

In the first two chapters the focus was on chemical workers who were operating in industry as consultants or as private researchers. The teaching of chemistry has not hitherto featured very prominently in the laboratories that have been examined, but such teaching to students or private individuals was and remains an important function of chemistry.

A Tense Relationship?

If there was no chemistry teaching, there would be hardly any chemists, and there can be no doubt that the lecture hall is central to modern chemistry. The famous auditorium in the Royal Institution (RI) in London, hallowed by the lectures of Humphry Davy and Michael Faraday, and the large lecture halls of the leading nineteenth-century chemistry departments such as those in Berlin, Munich and Leipzig are evidence for this assertion.[1] Apart from being in the same building, however, what do these lecture halls have in common with laboratories? Nowadays the connection is far from obvious. At the RI the demonstrations are brought into the auditorium on a trolley. In chemistry departments the demonstrations are set up on a bench in the lecture hall beforehand, or are brought through from another room. In any event, the use of live demonstrations during lectures is declining, often replaced by videos or computer graphics. This hides the long association between the laboratory and lecturing. Were lectures once held in laboratories and how did the connection become broken to the extent that we now seem embarrassed at any sign of such a connection? Some might be surprised to see the words 'tense relationship' being used in this context. Yet the development

30 Harriet Jane Moore, *Michael Faraday (1791–1867) in his Basement Laboratory*, 1852.

31 Engraving of Faraday's laboratory from H. Bence Jones, *The Life and Letters of Faraday* (1870).

of the lecture hall over the last four centuries has seen the laboratory being pushed out of the lecture hall. This tension can be seen in Harriet Moore's famous painting of Faraday's basement laboratory (illus. 30).[2] It is not often realized that this laboratory was built to serve an adjoining lecture room rather than as a research laboratory. Moore, with her focus on Faraday and his research, does little to enlighten the viewer. The archway to the lecture room can be seen on the far right, but it is barely visible. Curiously, an etching of this painting made for Henry Bence Jones's *The Life and Letters of Faraday* shows the link much more clearly (illus. 31).[3] As Frank James has pointed out, the etching is a poor copy of the painting – its representation of Faraday is almost a caricature – but it does show how the lecture room is in the same space as the laboratory.[4]

This chapter describes how the lecture space gradually moved away from the laboratory space and in the mid-nineteenth century became the lecture theatre. This lecture theatre still required the services of a laboratory, albeit often a rudimentary one, to provide the materials required for lecture-demonstrations. The removal of lecturing from the laboratory thus implies the rise of the preparation room, usually linked directly to the hall but hidden from the audience. The large lecture hall with its preparation room became a central feature of the palatial laboratories of the mid-nineteenth century. Just as Wilhelm Hofmann was a leading figure in the development of the classical laboratory, he also encouraged the production of demonstration apparatus for the lecture hall. The preparation room declined in the late twentieth century as potentially dangerous demonstrations were replaced by videos and animations.

The Lecture-demonstration

How do you teach chemistry? On one level this may seem a simple enough question, but on closer examination it is a very complex issue. If chemistry is regarded as a craft, the traditional route would be the apprenticeship – young chemists learning chemistry by practical work under the instruction of a master. This was how the teaching of chemistry began in the early seventeenth century, and for this purpose one clearly needs a workspace; in other words a laboratory. The apprenticeship route can still be found in the form of modern postgraduate training in the laboratory under a supervisor. However, in

32 Annibal Barlet lecturing in his laboratory, 1657.

the middle of the seventeenth century a new way of teaching chemistry arose: the lecture-demonstration, based in part on similar medical-anatomical lectures. Lecture-demonstrations were more than just conventional lectures with a few 'magic tricks' tacked on; they were a series of set experiments shown to the audience. The demonstrations fulfilled a dual purpose. They presented the tangible evidence for the claims made by the lecturer, and at the same time often showed students how chemistry should be done. If they also entertained the students, retained their interest and thereby ensured that they enrolled for the next course, so much the better. The lecture-demonstration had the advantage of teaching many students at the same time and presenting them with best practice, as opposed to the hit-and-miss method of learning to do standard procedures on the job. For the lecturer who was paid by the student, the lecture-demonstration had the additional advantages of using less material and attracting students who might not wish to engage in practical work.

One of the earliest examples of the chemical lecture-demonstration can be seen in a series of illustrations of a lecture by the physician Annibal Barlet in mid-seventeenth-century Paris (illus. 32).[5] These pictures show three key elements of the lecture-demonstration: the use of a long table or lecture bench for the demonstrations, an assistant to help with the demonstrations and the use of specialized equipment. Suitable equipment for lecture-demonstrations has to be of the right scale for the table, and here we can see the use of a small furnace – with a small alembic and cucurbit – and a charcoal heater. These small heating devices were in general use, but chemists soon developed specialized equipment that enabled them to illustrate the points they were making in their lectures. In this case the lecture table is still in the laboratory, although it is set apart from the working area with its distillation apparatus and larger furnaces. If we look at the illustration of the laboratory at Altdorf in 1682 (illus. 33), we can see that the lecturer gave his lectures from a pulpit-like podium similar to the podiums found in anatomy theatres and churches. Presumably there were no demonstrations unless they were carried out by an assistant, or the lecturer moved over to one of the furnaces that surrounded the central table. It has to be emphasized that the table was used by the students for taking notes and was not employed for demonstrations or experimental work. When the Ashmolean Museum opened in Oxford a year later, chemistry lectures were apparently given on the ground floor.[6] As Anthony Simcock has pointed out, the *Officina Chimica* in the basement, which contained the furnaces, must have looked strikingly similar to the Altdorf laboratory, and there must be a possibility that lecture-demonstrations were given there.[7] We do not know, however, if the lecturer in Oxford used a table for demonstrations as did Barlet, or a podium similar to that at Altdorf. Regardless of whether the lectures took place on the ground floor or in the basement, the fittings must have been movable. By coincidence, in the earliest photograph of the *Officina Chimica*, long after it had ceased to be a laboratory, it contains a heavy wooden lectern that would have been very suitable.[8]

The lecture-demonstration did create new problems. Where should it take place? If the lecturer wished to show practical procedures or simply eye-catching experiments, there had to be chemicals and apparatus on hand. A conventional lecture room would not be enough; there needed to be a laboratory nearby. In the case of Barlet, the lectures took place at one end of the laboratory, and his fellow French chemist Guilluame-François Rouelle also lectured in his laboratory at the Jardin du Roi a century later.[9] However, most laboratories lack suitable space for seating, lectures disrupt the work of the laboratory and the room's atmosphere would often not be very

33 Laboratory and lecture room at the University of Altdorf, 1682.

34 Lee Building, Christ Church, Oxford, early 19th century.

pleasant. Furthermore, many lecturers would wish to keep the work being done in their laboratories from prying eyes and thieving hands.

Johan Gottschalk Wallerius became professor of chemistry in Uppsala, Sweden, in 1750. He designed a teaching laboratory building that was completed in 1754 and is still standing today, although it is no longer used as a laboratory.[10] The style of an early to mid-eighteenth-century science laboratory was very much like that of a gentleman's house, as illustrated by Dr Lee's Laboratory at Christ Church, Oxford, erected in 1766–7, which looks like an elegant townhouse and is now used mainly for functions (illus. 34). While he would live in the new chemistry building, Wallerius was anxious to obtain a building that looked more institutional and imposing to show the importance of chemistry as an up-and-coming subject, all the more so since it would be the first university building seen by visitors coming from the south. Ideally he wanted a building similar to the University Observatory, with its central tower, but he had to make do with a single-storey building that was substantial but prosaic. Crucially, Wallerius made the auditorium and laboratory into separate rooms. It has been suggested by Hjalmar Fors that this shows a greater openness for chemistry in the Enlightenment, as there were several doors into the laboratory and front windows overlooking the street. Fors admits that the windows were needed for ventilation, but this could have been achieved by high windows that did not allow the public to look in. Given that the chemical operations were clearly visible to students attending Barlet's lectures, it could be argued that Wallerius's design actually separated the chemical work from the public lecturing. Indeed, you could even argue that far from showing a movement away from alchemical secretiveness, this separation reveals Wallerius's alchemical roots in contrast to Barlet's openness as a medical practitioner. As described in chapter Eight, Wallerius's successor Torbern Bergman converted the auditorium into a museum and moved the teaching back into the laboratory.

As the professor of chemistry at Edinburgh in the second half of the eighteenth century, Joseph Black gave demonstration-lectures, as can be seen in the famous caricature by John Kay (illus. 35).[11] He was not given a stipend as a professor and was thus dependent on the fee income from giving lectures. This is one of the reasons why Black never built up a reputation as a researcher – he put all his efforts into his lectures, and in particular the demonstrations. As Robert Anderson has remarked, 'Black transformed the chemistry demonstration into a *tour de force*'.[12] The preparations for these demonstrations took up much of his time. In 1772 he wrote to his friend James Watt:

> my attempts at chemistry at present are chiefly directed to the exhibition of Processes and experiments for my Lectures, which require more time and trouble than one would imagine.[13]

Asking for a house on the premises when the university was being rebuilt in 1789, he emphasized the time he had to spend preparing his lectures:

> he has it is true one hour only of teaching but he must spend several hours every day in his laboratory in prepareing for the experiments & operations of the next lecture or in finishing those already begun and as these operation often last ten twelve or twenty four hours, or some of them several days he is under the necessity of looking into it frequently during the day & occasionally must be there early in the morning and late at night.[14]

The place where these labours took place and the subsequent lectures delivered changed during his time at Edinburgh and were

35 John Kay, *Joseph Black Lecturing*, 1787. The lecture is clearly about combustion and fixed air, the subject of his famous MD thesis.

rarely satisfactory. When he first arrived at the end of 1766, Black probably taught in the house and laboratory of James Scott, a manufacturing druggist. In 1777 Black was given a new classroom in a building containing the natural history and geological collections. In 1781, after this building had become crowded with the arrival of John Walker's mineralogical collections, Black was provided with a proper laboratory, which was used mainly for lecture preparations rather than for chemical research. As this was also described as a classroom, it is likely that Black lectured in his laboratory at least after 1777; and possibly from the outset of his time in Edinburgh, if he used Scott's laboratory for lecturing. This laboratory was demolished in 1790 during the construction of the new university buildings designed by Robert Adam, and as the completion of the new buildings was delayed, Black carried on his teaching without the benefit of a laboratory until his death in 1799.

A Religious Chemist

Michael Faraday was born in Newington Butts on the south side of the Thames, an area now usually called the Elephant and Castle, in 1791, soon after his family moved to London from Westmorland (illus. 36).[15] The family then moved to mews accommodation in Manchester Square, Marylebone. Faraday had the good fortune to become an apprentice to George Riebau, a bookbinder in nearby Blandford Street. Not only was Riebau a kindly employer, but he allowed his apprentices to read the books they were binding. Having honed his mind by reading Isaac Watts's *The Improvement of the Mind*, Faraday learned about electricity from the *Encyclopædia Britannica*. He then studied chemistry using Jane Marcet's *Conversations on Chemistry* (he was thereafter eternally grateful to Mrs Marcet) and the lectures of Humphry Davy at the RI.

Deciding that he did not wish to become a bookbinder, he approached Davy in 1812 about a position at the RI. Davy told him to stick to bookbinding, but fate played a hand. Soon afterwards Davy injured his eyes in a laboratory explosion involving nitrogen trichloride, and he employed Faraday as an amanuensis. Then, in February 1813, the laboratory assistant William Payne got into a fight with the instrument maker John Newman and was subsequently dismissed. Davy offered the now-vacant post to Faraday. Afterwards the two men went on a tour of Europe during which Faraday served as Davy's valet. He then assisted Davy with his research until Davy ceased active research in 1826. Faraday was made superintendent of the RI in 1821 and director of the laboratory four years later. Davy's successor William Brande freed Faraday from his duty to assist the lecturers a year later. At this time Faraday established the now-famous Friday evening discourses, which he then delivered regularly for nearly four decades. In 1833 his friend, the eccentric philanthropist John 'Mad Jack' Fuller, endowed the Fullerian chair of chemistry, which freed Faraday to concentrate on research, although in practice he continued to lecture.[16]

36 Michael Faraday, c. 1830.

Faraday's research on electricity, notably his discovery of the relationship between a moving magnet and an electric wire (which led to both the electric motor and the electric dynamo), is well known. His contributions to chemistry, however, are often overlooked, partly because he did not develop his findings into a complete research programme. He trained as a chemist and did not see any distinction between his chemical research and his work on electricity. While working with Davy on chlorine in 1823, Faraday discovered by accident that chlorine gas could be liquefied under pressure, and with Davy he went on to liquefy other gases. During this research on chlorine, Faraday also prepared hexachlorobenzene and tetrachloroethylene, but their theoretical significance for organic chemistry was not appreciated for many years. In 1825 he discovered benzene (which he called 'bicarburet of hydrogen') in illuminating gas, which had been made from whale oil, and achieved a remarkably high purity of 99.7 per cent.[17] However, it was not until benzene was found in large amounts in coal tar in the 1840s that it became important industrially, first as a solvent then as a raw material for synthetic dyes.[18] During his research on the equivalence of the different forms of electricity, Faraday established the framework for the electrochemistry of solutions with his laws of electrolysis. He had carried out the analysis of water from different sources when he first started working for Davy, and in the 1850s he became concerned about the highly polluted state of the River Thames. In 1827 Faraday published *Chemical Manipulation*, which was the first manual of chemical practice and was very influential.

Faraday was a devout Christian, which was not uncommon for scientists at the time, but he was also a member of a denomination that was based on the inerrancy of the Bible.[19] He made his confession as a member of the Sandemanians – a small Protestant sect founded in Scotland in the 1720s – in 1821 and stood in high regard in his congregation. For a short time in 1844 he was excluded from the congregation, but this was a result of internal divisions, not any disagreement on Faraday's part with the doctrines of the Sandemanians nor – as the popular myth has it – because he allegedly visited Queen Victoria on the Sabbath. It is clear that Faraday regarded his faith and his science as two separate spheres, but – whether by design or not – he avoided researching areas of science where the latest developments could have posed problems for his faith. Certainly, his Christian faith was reflected in his scientific work; for example his belief in the unity of science and the importance of force and power. Having turned down the presidency of the Royal Society in 1857, mainly for health reasons, Faraday retired from the RI a year later and moved to a house given to him by the Royal Family (a 'grace and favour residence') at Hampton Court to the southwest of London, where he died in 1867.

The Royal Institution

If it was problematic to give lectures in laboratories, could purpose-built lecture halls incorporate laboratories? This was the case at Robert Hare's lecture hall in the University of Pennsylvania Medical

School in the 1830s (illus. 37). The space is clearly a large lecture hall with a demonstration bench (or 'counter'; see chapter Two), but there is a complete laboratory behind it that was used for ordinary experimental work as well as preparations for lectures.[20] This was a very logical arrangement, but there was clearly an uneasiness about putting the laboratory on full display in this way as it was not widely imitated.

A different way of joining the laboratory and lecture hall had been developed a couple of decades earlier in London. The Royal Institution of Great Britain was established in March 1799 with the aim of spreading knowledge about science and technology in ways that the Royal Society was not doing at that time. This was to be achieved by a mixture of public lectures and the teaching of artisans. The moving force behind the new institution was a loyalist American army officer and schoolteacher, Benjamin Thompson, Count Rumford, who departed in 1802 under a cloud, but left an indelible mark on the infant organization. He quickly acquired a house in Albemarle Street off Piccadilly, and designed its conversion into a scientific establishment with the help of the Orcadian clerk of works, Thomas Webster.[21] Webster was a trained architect but he was also interested in geology.[22] He became curator and librarian of the Geological Society's collections, and eventually professor of geology at University College London three years before his death in 1844. After leaving the RI in 1802, he also made a living painting watercolour landscapes for books. As a private institution the RI needed the income it could make from public lectures, so the famous horseshoe-shaped auditorium on the

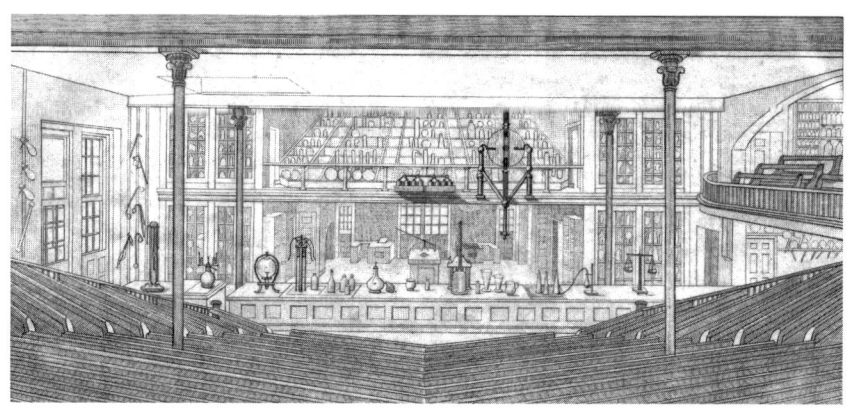

37 Robert Hare's laboratory and lecture hall at the University of Pennsylvania Medical School, *c.* 1830.

38 Royal Institution Laboratory and lecture room, 1819; the sand bath on the right was moved to the middle of the laboratory at around that time.

first floor was one of its earliest features, the planned (and current) library having been turned into a temporary lecture space in the meantime.

The laboratory in the basement, with an adjoining lecture space for teaching rather than public lectures, was installed and probably designed by Davy after Rumford's departure in 1804.[23] He effectively created a halfway house between a completely separate preparation room and a laboratory-lecture hall. Although the laboratory was originally intended as a lecture-preparation room, Davy was soon using it for research between lectures. It was connected to the lecture room via an open proscenium archway, but much of the laboratory was not directly visible to the audience, thus giving the convenience of an open arrangement without making the whole laboratory embarrassingly visible. Looking at the laboratory (illus. 38), it can be seen to be very much a working laboratory with equipment and glassware scattered around, and chemicals on bookcase-like shelves. The main working area is simply a table, and the room also contains a small stove and furnace. The brickwork sand bath is at the back of the archway in the picture, but was moved to the middle of the laboratory by 1819 (as seen in the painting by Moore). The equipment is an odd mixture of electrical apparatus and rather old-fashioned chemical apparatus with a couple of bellows. The large bellows hanging from the ceiling had been installed for metallurgical experiments. Given the laboratory's role as a preparation room, surprisingly there is hardly any equipment specifically intended for lecture-demonstrations.[24]

In general, the laboratory looks very similar to William Pepys's laboratory at the rival London Institution at Finsbury Circus, which was fitted out in 1819 (illus. 39).[25] Pepys's laboratory had a hood over the furnaces, which was more typical of laboratories of the period such as Priestley's laboratory in Northumberland, Pennsylvania. Despite its simplicity, Brande remarked that 'in its completeness and convenience; [the RI laboratory] comprises all that is required in the pursuit of experimental chemistry'. The lectures were delivered from the archway itself from a lecture bench that looks very much like the one in the main auditorium, with a semicircular recess for the lecturer. This lecturer's desk was, however, only a shadow of the later lecturing benches, which were fitted with sinks and all the facilities needed to carry out demonstrations. This level of facilities of course required the presence of running water and gas, which were not available when the RI was established (see chapter Five).

The Rise and Fall of the Prep Room

In the main auditorium of the RI, Rumford and Webster had put an anteroom next to the lecture hall with a doorway between them. This contained the apparatus used in the lectures, as can be seen in the caricature by James Gillray (illus. 40). It does not appear to contain chemicals and was probably not used for the preparation of the lectures beyond the simple setting up of apparatus. The next step was to put the main laboratory next to the main lecture theatre, linked by a doorway, rather than just an anteroom. John Webster appears to have been one of the first chemists to plan this kind of arrangement, at Harvard in around 1826. As William Jensen has remarked, Webster was famous for a spectacular demonstration:

> Webster . . . is known in Harvard chemical lore for his famous volcano demonstration – a large plaster mountain filled with potassium perchlorate and sugar which Webster

Overleaf: 39 William Haseldine Pepys's laboratory at the London Institution, Finsbury Circus, London, 1822.

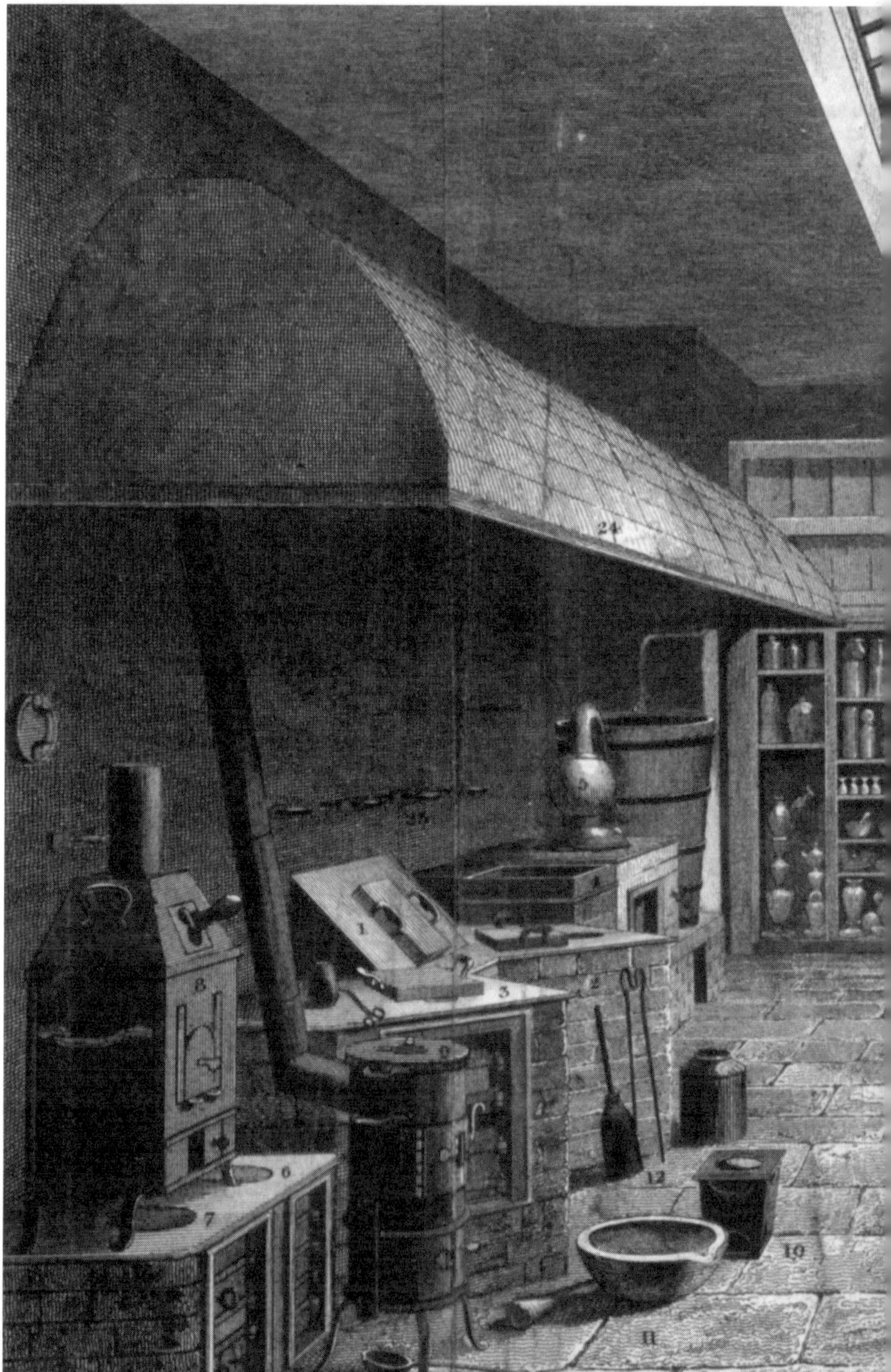

24
1
2
3
6
7
8
10
11
12

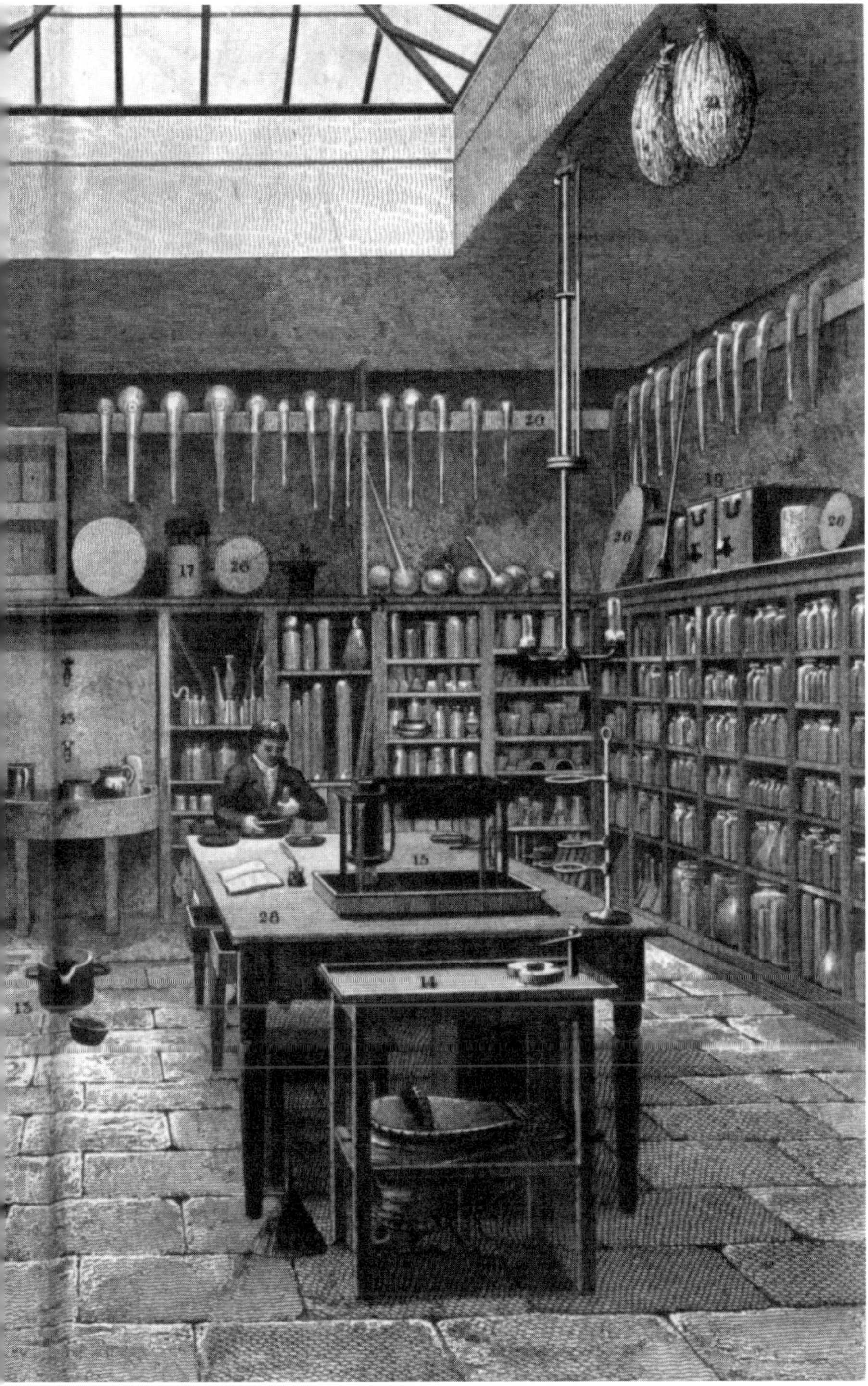

20
17
26
15
28
14

would ignite without warning using a drop of sulfuric acid and then dash for the nearest door, leaving the students to fend for themselves.[26]

Similarly, Justus Liebig's newly expanded laboratory at Giessen, which opened in 1839, had a communicating door between the student laboratory cum preparation room and the lecture room.

As laboratories became more specialized, it was logical to convert the laboratory next to the lecture hall into a dedicated preparation room for lecture-demonstrations, although in many cases it was also doubtless used by the laboratory technicians for the making up of chemicals for student work and similar routine chemical work. It is not clear where the first such preparation room was located, not least because of the difficulty of distinguishing it from an apparatus store (as at the RI), or a laboratory used for other purposes (as at Giessen). The new laboratory in Christiania (now Oslo) in 1852 had a preparation room in the lower part of the lecture hall linked by a door and an apparatus store under the raised part of the hall.[27] Liebig's lecture hall at the Bavarian Academy of Sciences in Munich (illus. 41), which was opened in the same year, had a preparation laboratory despite the fact that as a former laboratory, there were facilities to undertake preparations in the lecture hall itself, but the preparation laboratory was

40 Caricature of a lecture at the then new Royal Institution by James Gillray, 1802.

41 The Lecture Hall in the chemistry laboratories at the Academy of Sciences, Munich, 1852.

not directly connected to the auditorium.[28] There was a specialized preparation room in Bunsen's laboratory at Heidelberg in 1855, but it was linked to the main lecture hall by a serving hatch, which could be covered by a blackboard during lectures.[29]

Certainly by the 1860s such preparation rooms were commonplace. Both of the laboratories designed by Hofmann had preparation rooms linked to the main lecture hall, and this was also the case at Leipzig and the Central College in London. Roscoe's new laboratory at Owens College in Manchester, opened in 1873, had a lecture theatre in a different building from the laboratories, which were linked by a bridge, but there was a preparation laboratory behind the lecture theatre.[30] The laboratories erected at South Kensington in 1871 did not have a linked preparation room, but the laboratories constructed in 1885 and 1907 did.[31] The laboratories at the Eidgenossische Technische Hochschule, Zurich, opened in 1886, also had a preparation room next to the lecture hall.[32] This kind of arrangement appears to have been less common in the USA in the 1880s. There were linked preparation laboratories at Yale and Cornell (with the professor's laboratory being linked in the same way), but they were absent at Lehigh University and at Massachusetts Institute of Technology (MIT).[33]

Significantly, these preparation rooms were in general only linked to the main lecture theatre. The preparation rooms for the secondary lecture halls, if they existed at all, were usually located across a corridor. This shows the increasing importance of the main auditorium in university chemistry departments. From the eighteenth to the mid-nineteenth century, lecture halls were reasonably large but not overwhelming so. This was still the case at Heidelberg in the 1850s. The auditorium is about the same size as the main laboratories and is to one side rather than in the centre of the building. In fact, the relationship between the auditorium and the laboratories is much the same as in Uppsala a century earlier. With the creation of the 'chemical palaces' of the 1860s (see chapter Six), the main auditorium became much larger and was sometimes in the centre of the building, as in Berlin or Zurich, or in one corner, as at Leipzig or Yale. The growing size of the lecture hall reflected increasing student numbers, but also the increasing importance, even self-importance, of the leading chemistry professors. This was certainly true of Liebig, who created a huge auditorium in order to give lectures to adoring crowds, following in the footsteps of Davy and Faraday at the RI.

Preparation rooms can still be found even in modern laboratories, for example at Rostock University in Germany, but as the lecture-demonstration has declined, partly because of health and safety concerns, so the preparation room has largely faded away in Britain. To give one example, part of the 1907 laboratories at Imperial College (formerly the Royal College of Science) has survived, and its lecture theatre was refurbished in 2009 to create a modern auditorium, but the former preparation laboratory was sealed off from the lecture theatre and converted into an office in 2006.[34]

Demonstrating Volumes

In the summer of 1876, despite a ferocious heatwave in mid-August, when it reached 92°F in London,[35] nearly quarter of a million people flocked to visit an exhibition in South Kensington. This was not an art exhibition or even a successor to the 1862 International Exhibition held on the same site, but a collection of old and new scientific instruments and apparatus – the 'Special Loan Collection of Scientific Apparatus' – gathered from the four corners of the Earth.[36] Some of the exhibits were clearly historical, for example 'Otto von Guericke's original Air Pump' and Galileo's telescope, but many of them were simply contemporary examples of apparatus in general use. One section of the exhibition was called 'Educational Appliances'.[37] The subsection 'Apparatus for Teaching Chemistry' was dominated by a series of demonstration apparatus in glass designed by Hofmann. This is not entirely surprising as Hofmann was the chairman of the German organizing committee. In 2010 someone asked on the Yahoo! Answers website, 'What is a Hofmann apparatus?'[38] The person responding assumed it was the Hofmann electrolysis apparatus, also misleadingly called the Hofmann voltameter. Certainly, if you look on the Internet, it appears that the Hofmann electrolysis apparatus is the only one that is at all well known today, although perhaps less so than 40 years ago, when it was a common demonstration apparatus in schools. It was the first item listed in the Hofmann demonstration apparatus group in the Special Loans Exhibition catalogue, which indicates that it was considered to be the most important in 1876. It has remained popular because it achieves the important and visually interesting electrolysis relatively quickly and hence within the timespan of a school lesson. It is also useful for teachers because it shows that water is made up of one volume of oxygen and two volumes of hydrogen. This is in fact the point behind most of these apparatus – namely reactions that combine or form gases and their relative proportions – but they cover much more ground than just the decomposition of water. As the Hofmann electrolysis apparatus shows, gases are good for lecture-demonstrations because the relatively small amounts of gas (by weight) have easily visible volumes at room temperature. Furthermore, due to Avogrado's law, equal volumes of gases contain an equal number of molecules; hence the reaction of gases will involve simple ratios of volumes, which makes for a good visual demonstration.

As Hofmann was a leading university professor and an organic chemist, it might seem strange that his name was associated with apparatus that demonstrated relatively simple reactions in inorganic chemistry. In 1865, however, just as he was on the point of returning to Germany, he published *Introduction to Modern Chemistry, Experimental and Theoretic: Embodying Twelve Lectures Delivered in the Royal College of Chemistry, London* as a valedictory record of his teaching in London.[39] This volume illustrated several of the demonstration apparatus in the Loan Exhibition, and its use and that of similar

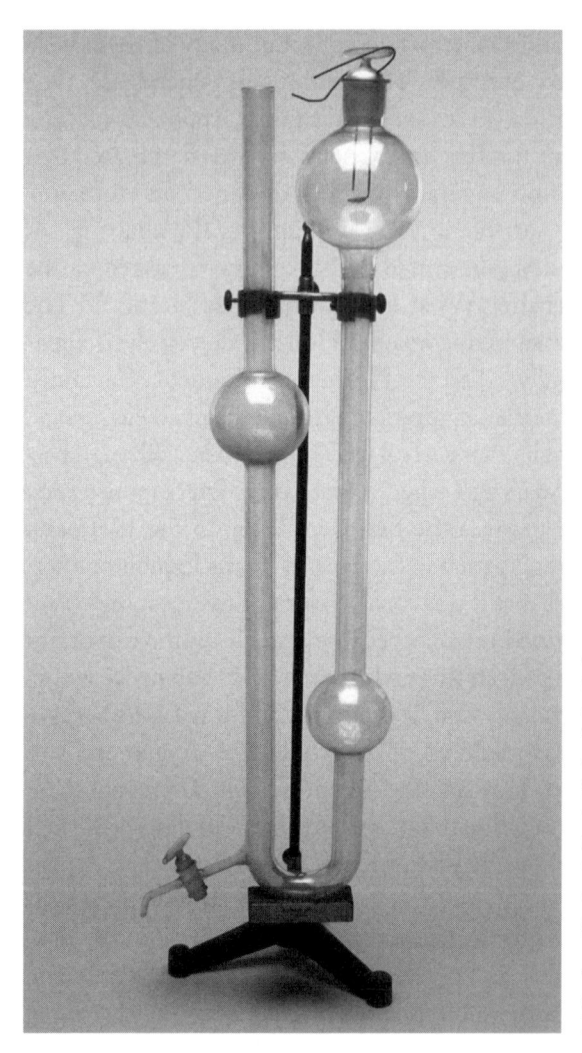

42 Hofmann's apparatus to demonstrate that the volumes of oxygen that enter into the composition of carbon and sulphur dioxide are equal to the respective volumes of these compound gases (inv. no. 1876-0220), which was displayed in the Special Loan Exhibition, 1876.

textbooks by school and university teachers created a demand for off-the-shelf apparatus to demonstrate the reactions discussed by Hofmann. Clearly, when he was looking for examples to show in the London exhibition, Hofmann turned to local apparatus manufacturers – namely C. F. Giessler & Son, Warmbrunn, Quilitz & Co. and Julius Schober. However, other manufacturers made similar apparatus, for example the E. H. Sargent & Company of Chicago listed twelve examples of Hofmann demonstration apparatus in its 1914 catalogue.[40] Apart from the electrolysis apparatus (the term 'voltameter' is not used in the Loan Exhibition catalogue), the apparatus ranged from 'Apparatus for showing that ammonia consists of three volumes of hydrogen and one volume of nitrogen' (Warmbrunn, Quilitz & Co.) and 'Hofmann's Apparatus for demonstrating that the volumes of oxygen which enter into the composition of carbon and sulphur dioxide are equal to the respective gas volumes of these compound gases' (Julius Schober), to 'Hofmann's Apparatus for liquefying sulphurous anhydride' (Giessler & Son). My association with these demonstration apparatus arose in 1994 when my assistant at the Science Museum discovered two of them in a cupboard with a note that they were to be disposed of ('Board of Surveyed' in museum-speak) in December 1931. I decided to keep them as fine examples of late nineteenth-century demonstration apparatus, and now that they have been restored they look beautiful, even if their purpose is not always clear to the casual viewer (illus. 42).

A Moving Boundary

Up to the end of the eighteenth century the boundary between laboratory and lecture space was fluid. Some lecturers, such as Barlet and Bergman, lectured in the laboratory; others, including Wallerius, used a lecture hall either out of choice or because of the university's arrangements. It is hard to say if one arrangement was more 'modern' than the other. In around the beginning of the nineteenth century, attempts were made to have a joint space with a working laboratory alongside the lecture hall. This arrangement did not prosper and was replaced by a laboratory attached to the lecture hall, but completely separate from it – the preparation room. This service laboratory arose at the same time as the laboratory-lecture hall and indeed in the same place, the RI, and became a standard feature of the model established in Germany in the 1860s.

4
Training Chemists
Justus Liebig and Giessen, 1840s

Although the laboratory established at the University of Giessen by Justus Liebig in 1824 has always been well known (illus. 43), not least because it has been preserved as a museum, it was a much cited paper by Jack Morrell in 1972 about 'the chemist-breeders' that brought Giessen to the forefront of the history of chemistry.[1] Morrell held up Giessen under Liebig as an early exemplar of a successful research school, namely one that 'bred' well-trained chemists in contrast to his contemporary, Thomas Thomson, in Glasgow. Sixteen years later Joseph Fruton offered a more detailed examination of Liebig's research group.[2] However, the fittings of the laboratory have never attracted much attention. Even in Liebig's time the famous chemist was the draw, not the laboratory. Thus, the American traveller George Calvert was eager to meet Liebig:

> On the way back to Frankfort, we stopped for the night at Giessen. It would have been a satisfaction to have availed myself of the genial accessibility of German professors, to visit Liebig, one of the stoutest living scientific pioneers, – one of the precocious band that with the sharp edge of thought are hewing for their fellow men paths into untrodden domain, – one of that bold brotherhood of discoverers who, in the holy privacy of the laboratory and the closet, reveal new truths by light struck from the contact of genius with Nature.[3]

Unfortunately Calvert 'arrived late and tired' and missed his opportunity. The attraction is Liebig himself, the 'living scientific pioneer', not his laboratory. Calvert was a philosopher not a chemist, but his

43 The Giessen laboratory, c. 1840.

focus on Liebig as the animating genius of Giessen rather than on the laboratory is typical for this period. Nonetheless, Liebig's Giessen laboratory was influential, especially for the general adoption of the fume cupboard, and as the setting for a new method of training chemists. One of the laboratories influenced by Giessen was the Birkbeck Laboratory at University College London (UCL), which opened six years later. It was the first laboratory to be designed completely for the new teaching drills of organic and inorganic chemistry, and the first to be clearly based around the iconic features of the bench and bottle rack (reagent shelf). The fume cupboard became a key element of the chemical laboratory by the 1860s, and its history is traced in this chapter.

Justus Liebig

Born in Darmstadt in May 1803, Justus Liebig was the son of a dealer in pharmaceutical and chemical products (illus. 44).[4] His father apprenticed him to an apothecary, and Liebig then studied chemistry at Bonn. He followed his teacher Karl Kastner to Erlangen, where he took his doctorate in 1822. Dissatisfied with the quality of the teaching he had received, Liebig went to Paris to study and work under Joseph Gay-Lussac. With the help of Alexander von Humboldt he obtained a professorship at Giessen in 1824, where he remained until he was given the chair in chemistry at Munich in 1852.

44 Oil painting of Freiherr Justus von Liebig by Alexander Craig after a chalk drawing by S. Laurence, c. 1856 (inv. no. 1903-0147).

Liebig began by seeking to systematize organic chemistry using the chemical formulae introduced by Jakob Berzelius. In 1832 Liebig and Friedrich Wöhler proposed the existence of a benzoyl radical that remained stable during chemical reactions. With the parallel work of Robert Bunsen on the cacodyl radical (see page 97), the concept of radicals in chemistry was now firmly established; in the hands of others – including Liebig's student Wilhelm Hofmann – it led to the development of modern organic chemical structure theory. Unfortunately, Liebig's study of the ethyl radical was less successful and led to controversies with Jean-Baptiste Dumas and Berzelius. In order to make his approach work, he needed to improve the quantitative analysis of organic compounds, beginning with the development of his famous potash-bulb for the determination of the amount of carbon present.

Liebig then turned to biochemistry and agricultural chemistry, building on the work of earlier chemists such as Antoine Lavoisier in biochemistry and Humphry Davy in agricultural chemistry. His views on these subjects were less surefooted than in organic chemistry, and perhaps his success in the latter field had increased his confidence in his opinions to an excessive degree. His views on agriculture, specifically on the importance of inorganic salts, were initially warmly received, then rejected, which Liebig found very painful. His views on human nutrition were also ill founded. He placed too much importance on protein in the diet, leading to the promotion of his meat extract (a forerunner of the Oxo cube), which was not as nutritious as he believed. As he got older Liebig became more rigid in his views and more combative, which was unfortunate at a time when his ideas were often incorrect and subject to criticism. As a result he achieved relatively little after his move to Munich, having created the first world-class school of organic chemistry at Giessen.

The Nitrogen Problem

If the eighteenth century had succeeded in improving the analysis of inorganic compounds and minerals, the chemicals derived from plants, animals and marine organisms (the so-called organic compounds) were less tractable. Lavoisier had begun to make some progress with the analysis of natural products before his investigations were cut short by his arrest and execution. The crudeness of organic analysis mattered little up to the end of the eighteenth century, as minerals and other inorganic compounds, such as the new gases, were the main preoccupation of chemists. The situation changed dramatically with the isolation of morphine by the German pharmacist Friedrich Sertürner in 1805. Not only was morphine the first of the alkaloids to be discovered, but it was also the first organic compound found to be basic like the inorganic compound ammonia. Both the alkaloids and the organic bases were to become central to organic chemistry during the nineteenth century. Another thirteen years were to elapse before the French pharmacists Pierre Pelletier and Jean Caventou isolated strychnine and brucine from the *Strychnos nux-vomica*, and it

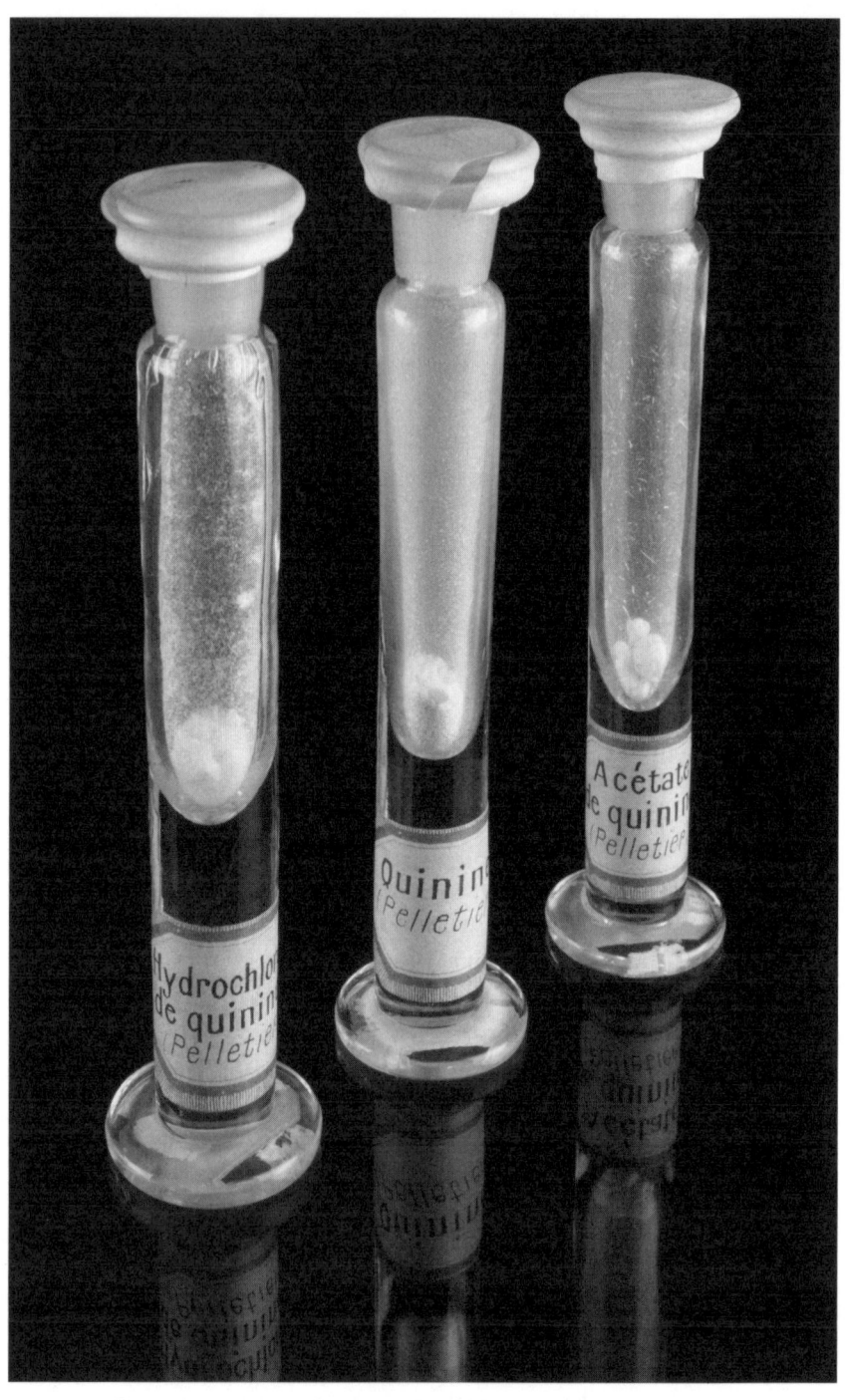

45 Original preparation of quinine salts and base, by Pierre Joseph Pelletier. France, 1810–40. (inv. no. A182555).

was three more years before they extracted quinine from Jesuit's bark, the bark of the cincona tree (illus. 45). At this point, however, the discovery of new organic compounds, both natural products and entirely novel chemicals, took off.[5]

In order to both understand the nature of these chemicals and put them into families of compounds, chemists needed to know their empirical formulae; in other words the relative proportions of carbon, hydrogen, oxygen and other elements. For the development of organic chemistry it was unfortunate that chemists in the 1820s and '30s were most interested in physiologically active and commercially valuable alkaloids. Most alkaloids have complex structures, some of which were not elucidated until the 1930s or even later. Furthermore, they nearly all contain nitrogen, and at the time it was very difficult to calculate the proportion of nitrogen present. Liebig encountered exactly this problem when he tried to carry out the elemental analysis of hippuric acid (from horse urine) in 1829.

Faced with these increasingly difficult analyses, Liebig knew that he would have to perfect the process of determining the empirical formula and develop a reliable way of determining the nitrogen content in order to unlock the chemistry of the alkaloids. He used copper oxide as the oxidant (the reagent that 'burns' the sample), replaced the small, unsatisfactory calcium chloride bulb used by Berzelius to absorb the water with a long thin tube of the drying agent, and most important of all, developed an ingenious apparatus of five bulbs – the *Kaliapparat* (potash apparatus) – to hold the potassium hydroxide solution that absorbed the carbon dioxide. In this way Liebig was able to increase the rate at which the analyses were carried out, although it still took a whole day to complete each one. The reliability of the results was enhanced partly by using gravimetric methods and partly by increasing the amount of the test compound used. Liebig's method was published in his analytical monograph in German in 1837, and two years later in an English version.[6] In 1840 Liebig's glassblower, Carl Ettling, demonstrated the manufacture of the Liebig bulbs to a Scottish colleague during the meeting of the British Association for the Advancement of Science in Glasgow.[7]

The nitrogen content of the compound was obtained by collecting the nitrogen gas evolved (in a second run) free from the other gases formed in the combustion. Dumas had devised a reliable method for doing this in 1831. He forced the gases out of the combustion tube by adding lead carbonate (to generate extra carbon dioxide), and reduced any nitrogen oxides present by passing them over activated copper and collecting the nitrogen over an alkaline solution, thereby absorbing the carbon dioxide. This method was eventually superseded by the sulphuric acid digestion first developed in 1883 by Johan Kjeldahl. His nitrogen determination is still employed today using the special flask he made for this procedure in 1888.

The Former Guardhouse

Liebig realized that the successful development of organic chemistry would also require the training of chemists skilled enough to produce consistently reliable results using his techniques. This was a new kind of training space, one that produced excellent postgraduate chemists rather than proficient undergraduate students. This is the context in which he created a new laboratory in Giessen, but it was mainly notable for the new methods of training it used rather than for its innovative design.[8] Nonetheless, despite its relatively small size it was one of the most advanced laboratories of its time. William Gregory, a leading Scottish chemist, remarked:

> I regret to say, that hardly any university in this country, and but few on the continent, can be said to possess one-half of the necessary accommodation. In the laboratory of Giessen, built and furnished under the superintendence of Professor Liebig, all conveniences above mentioned, and a good many more, have been liberally supplied.[9]

When Liebig arrived in Giessen in 1824, he was unable to use the existing small laboratory due to opposition by the senior professor, Wilhelm Zimmermann. Eventually the university gave him a disused guardhouse erected in 1819. Liebig lived on the first floor, and the ground floor was converted into a modest laboratory. A balance room and a small private laboratory were built as a small extension in 1834. Seven years later it was described by Carl Wilhelm Bergemann as

> a small room where Professor Liebig himself can work. It contains apparatus and arrangements of reagents, balances,

and so forth, further smelting and drying ovens, as well as writing desks.[10]

The most important expansion took place in 1839, when a large laboratory and a smaller laboratory 'for the less advanced students' were constructed in a new wing. While it may have been a larger laboratory, it was certainly not large by later nineteenth-century standards, only measuring about 580 sq ft (54 sq m).[11]

A key source for the interior of the large laboratory is the well-known lithograph made by Wilhelm Trautschold and Hugo von Ritgen (illus. 46, 47). The lithograph was not published until 1842, three years after the laboratory was opened, but certain details (for instance the number of drawers in the benches) suggest that it might be a prospective idealization of the laboratory rather than a sketch from life. Be that as it may, the furnaces used for heating operations can be seen at the back of the picture. The furnaces also produced the draught needed by the fume cupboards in the two back corners of the picture. These were the first glass-fronted fume cupboards, each with a frame that could be raised and lowered, which later became universal in chemical laboratories. The benches along the walls are much more like the laboratory benches that came to dominate chemists' lives for more than a century after 1865. There are bottle racks, rather like bookcases, at the backs of the benches for holding chemicals. Under the benches are small sinks fitted with taps, but draining into conical tubs, which were presumably taken outside and emptied at the end of each day. According to Bergemann the benches

> are equipped with compartments and pigeonholes of varying sizes in which the workers store the objects being used in their investigations, and on the same are shelves of reagents which are in such good supply that usually they can be shared in the experiments of every two students. Under these [benches] are to be found recesses, and in these containers from which water can be drawn and drained through a tube.[12]

Hence the idea of a cupboard under the benches for storing apparatus and chemicals also seems to have originated at Giessen. In the central foreground are four tables, which were mostly used for bulky operations such as distillation. The two tables at the front were made

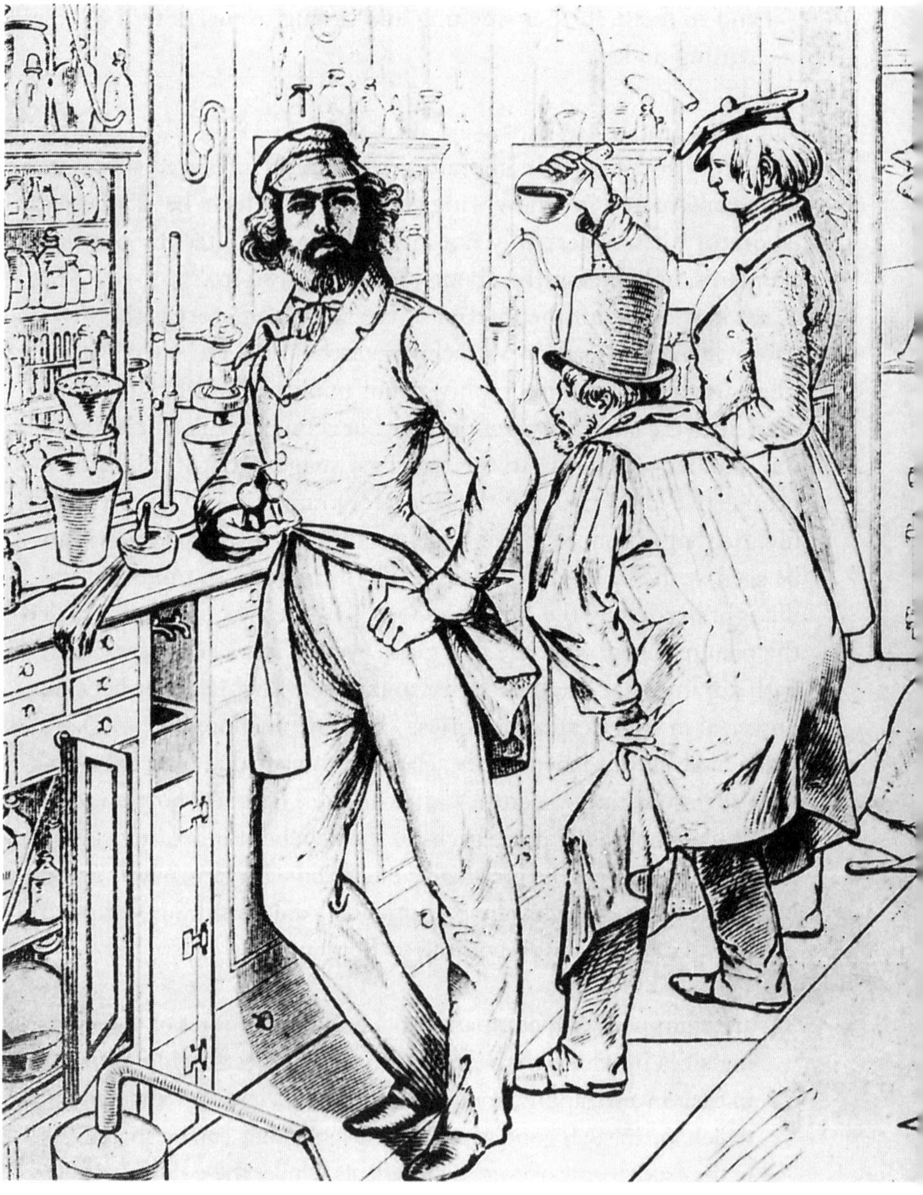

46 The left half of Wilhelm Trautschold's print of the new laboratory at Giessen, 1842.

47 The right half of Trautschold's print of the Giessen laboratory.

lower for the shorter workers (as the architect put it), and perhaps also for the taller apparatus. The architect of the laboratory, Paul Hofmann (the father of Wilhelm Hofmann), described these tables as being 'without fittings', as shown in the lithograph, but they were probably replaced by laboratory benches within a few years. The large, typically German iron stove, which heated the laboratory in the winter, can be seen in the middle of the room. Against the fourth wall (which we cannot see), there was a large drying kiln:

> It is so constructed that a large sandbath can be raised to a high temperature and from this the warmth can be conducted through several flues and pipes into particular rooms, where it can be regulated for several purposes. The whole iron-plated furnace is similarly provided with sliding windows so that the items to be dried are protected from dust and contamination.[13]

Finally, there was an open hatch at the back of the laboratory, which allowed Liebig to communicate with researchers in the laboratory from his office. This orifice, not unlike the serving hatch between the kitchen and the dining room, could have been a 'spy window', a feature discussed in chapter Six, but Liebig seems to have considered it only as a means of communication, not scrutiny, although the line between engagement and surveillance is often a very fine one.

Mitigating the Dangers of Chemistry

Robert Bunsen took up the chair of chemistry at the Höheren Gewerbeschule in Kassel in 1836, and he embarked on the study of the evil-smelling substance Cadet's liquid.[14] The French pharmacist Louis Cadet de Gassicourt made the poisonous brown fuming fluid that bears his name in 1760 by distilling a mixture of potassium acetate and arsenic(III) oxide. Bunsen showed that it is possible to make an oxide and a chloride from the original compound, which contained only arsenic and hydrogen. Berzelius coined the name 'kakodyle' (from the classical Greek for foul smelling, *kakodes*) for the unchanging chemical radical in these reactions. This was subsequently changed to 'cacodyl'. Bunsen's research appeared to present proof of the radical theory of organic chemistry, and he believed that he had prepared free cacodyl as the equivalent of a metallic element, as Berzelius's radical theory had demanded, but he was in fact wrong: cacodyl is actually a dimer of the underlying radical $(CH_3)_2As\text{-}$.[15] In the course of his research Bunsen had to breathe through thin glass tubes to avoid being poisoned by the vile liquid, and did some of the work in the open air. While carrying out a combustion analysis of a compound, he was injured in an explosion and lost his sight in one eye. The risks of carrying out chemical reactions were clearly increasing, and a way had to be found of conducting such research safely. The fume cupboard provided the solution. The standard fume cupboard, although simple in design and using only basic materials, is adequate enough to protect a chemist from the violent energy and thick fumes of the thermite reaction between iron oxide and aluminium powder.

The universal use of the term fume cupboard is relatively new, dating from the 1890s, and even now the term fume hood is still used in the USA.[16] The terms fume closet and draft enclosure were also used earlier. It is, however, the American term fume hood that betrays the origin of the fume cupboard. Anyone who has sat near a domestic wood or coal fire will have noticed the often fierce upwards draught generated by the heat that is needed to remove the smoke from the fire. Those who have visited the kitchen of an old mansion may be familiar with the large cowl above the fireplace with its roasting spit. These impressive fireplace hoods are the forerunners of the modern fume cupboard. From the kitchen fireplace it was only a small step to using them above large chemical furnaces. Whether they were first used to remove the smoke of furnaces, and their ability to remove chemical fumes was only noticed afterwards, is now impossible to say, but the builders of these hoods soon incorporated metal or brick venting pipes to specifically remove the noxious fumes. While we know of examples of such hoods from the seventeenth century, their use continued well into the nineteenth century. They were effective if also rather large, and they only fell out of use when the need for a furnace disappeared from the laboratory. The heat of the furnace could, however, be substituted by gas jets placed under a hood, as in Edmond Frémy's laboratory as late as 1882.

By the mid-eighteenth century a new type of fume hood had appeared, which was not limited to the area above a furnace and usually ran down one side of the laboratory, and in some cases along almost the whole length of the laboratory. This type can be seen in the engraving of the laboratory in the *Encyclopédie* in the 1760s (see illus. 24), in Priestley's laboratory in Northumberland, Pennsylvania, in the 1790s, and in the London Institution in London in 1819 (see illus. 39). Even more innovative was the freestanding, bell-shaped hood in the centre of the laboratory. This can be seen very clearly in the engraving of Lewis's laboratory in 1765 (see illus. 17). This design was used at the University of Padua but does not seem to have been widely adopted.

The Giessen fume cupboards seen on the right at the back of illus. 47 represented a halfway house between the furnace-based fume hoods and the later fume cupboards. Instead of traditional furnaces, they used cast-iron stoves to provide the necessary heat for the draught. Where they were truly groundbreaking was in the provision of glass doors. This is the first known case of a fume cupboard with a glass partition between the chemist and the contents of the fume cupboard. The arrangement was almost exactly reflected in the more convenient central-island fume cupboards at the laboratory of the University of Zurich at Rämistrasse that Carl Löwig moved to in 1842 (illus. 48).[17] It was from Giessen, however, that the glass-fronted fume cupboard spread to Rostock University.[18] The central-island arrangement at the University of Zurich did not catch on in the larger laboratories of the 1860s. In Leipzig they appear to be at the ends (although in illus. 120 they look more like store cupboards than fume cupboards), an arrangement that soon became a standard elsewhere;

48 University of Zurich Laboratory, c. 1850. Carl Löwig is on the far left and Eduard Schweizer is in the middle wearing a fez.

in Berlin they were on the outer walls inset into the windows, a potentially chilly arrangement in a cold Prussian winter.

It has been argued that it was the rise of organic chemistry (and organic synthesis in particular) that spurred the introduction of the fume cupboards into the laboratory.[19] I am not convinced by this argument. Organic chemistry was not perceived as being especially dangerous in the mid-nineteenth century, and the dangers were often not recognized until well into the twentieth century. I have vivid memories of hydrolysing phenyl esters (as an 'unknown') in the open laboratory in 1975, surrounded by pungent white clouds of free phenols, and with leprous white skin from a phenol burn peeling off my hand. Furthermore, inorganic chemistry posed as many dangers, not least in the form of hydrogen sulphide gas, as discussed later (see page 101).

The classic fume cupboard had a window at the front that could be pulled down while the dangerous operation was being carried out. It was vented either by rising hot air from a gas jet or a central furnace, in which case the gases were drawn into a flue. This led to a proliferation of chimneys in modern laboratories; strikingly so at the Lehigh University laboratory building of 1884 (illus. 49). The arrangement need not be so elaborate, however, for a simpler fume cupboard could be created by putting it on an external wall and, when in use, simply closing the front window after opening the external window. There was also a limited move towards providing individual fume extraction. At Bonn, after the original plans were drawn up, it was decided to introduce a number of small 'evaporation niches' (fume hoods) because the students could not be bothered to go to the large fume cupboards. As Hofmann remarked:

> The evolution of vapours ... recurs very often and must, particularly in a laboratory where a great number of operators are at work, ultimately contaminate the air so as to render ordinary ventilation totally insufficient.[20]

They were small fume cupboards (with window panes that could be pulled down), which were placed between the external windows on the outer wall; the gases were removed up a tall earthenware pipe in the wall. The draught was generated by a small gas flame in the funnel at the base of this pipe. Despite the risks of using ether in such a fume cupboard, they remained in use well into the twentieth century, until they were finally ousted by fume cupboards vented by electric fans. At the University of Nottingham in 1885, the students had individual tin hoods that were rather like modern cooker hoods (illus. 50). They must have been both expensive and difficult to maintain, and were not widely adopted. These individual hoods were generally connected with the use of hydrogen sulphide in group analysis. The opposite extreme was a massive fume hood looming over all the benches,

49 Exterior of the chemical laboratory building at Lehigh University, Bethlehem, Pennsylvania, showing the numerous chimneys required for fume cupboards, 1893.

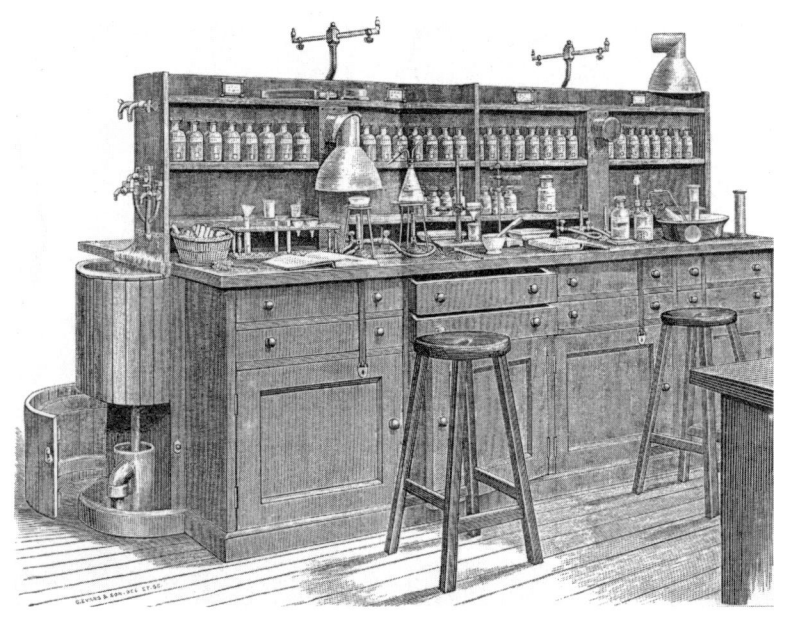

50 Individual fume hoods at the University of Nottingham, 1885.

which can be seen in a picture of a chemical laboratory in the new Teviot Place medical building at Edinburgh in 1889 (illus. 51).[21] It is very likely that the major use of this remarkable hood was the removal of hydrogen sulphide, as a contemporary caricature of the professor of chemistry Alexander Crum Brown shows him generating this compound. This would have also been very expensive and was not generally adopted either.

The use of hydrogen sulphide in elementary inorganic analysis was a major problem in the teaching laboratories, as this foul-smelling gas is at least as poisonous as hydrogen cyanide. When I was at school in the final days of group analysis in the late 1960s, we obtained our hydrogen sulphide from an ancient Kipp's apparatus in the fume cupboard. In a university teaching laboratory however, with many students, providing enough fume cupboards was expensive and time consuming for the students. The elaborate solutions adopted by Nottingham and Edinburgh were too expensive. A cheaper alternative was provided by hydrogen sulphide cabinets on the benches, which contained the hydrogen sulphide gas taps and acted as small fume cupboards.

Overleaf: 51 The chemical laboratory in the Teviot Place medical building, Edinburgh, showing the large fume hood above the bench, 1900.

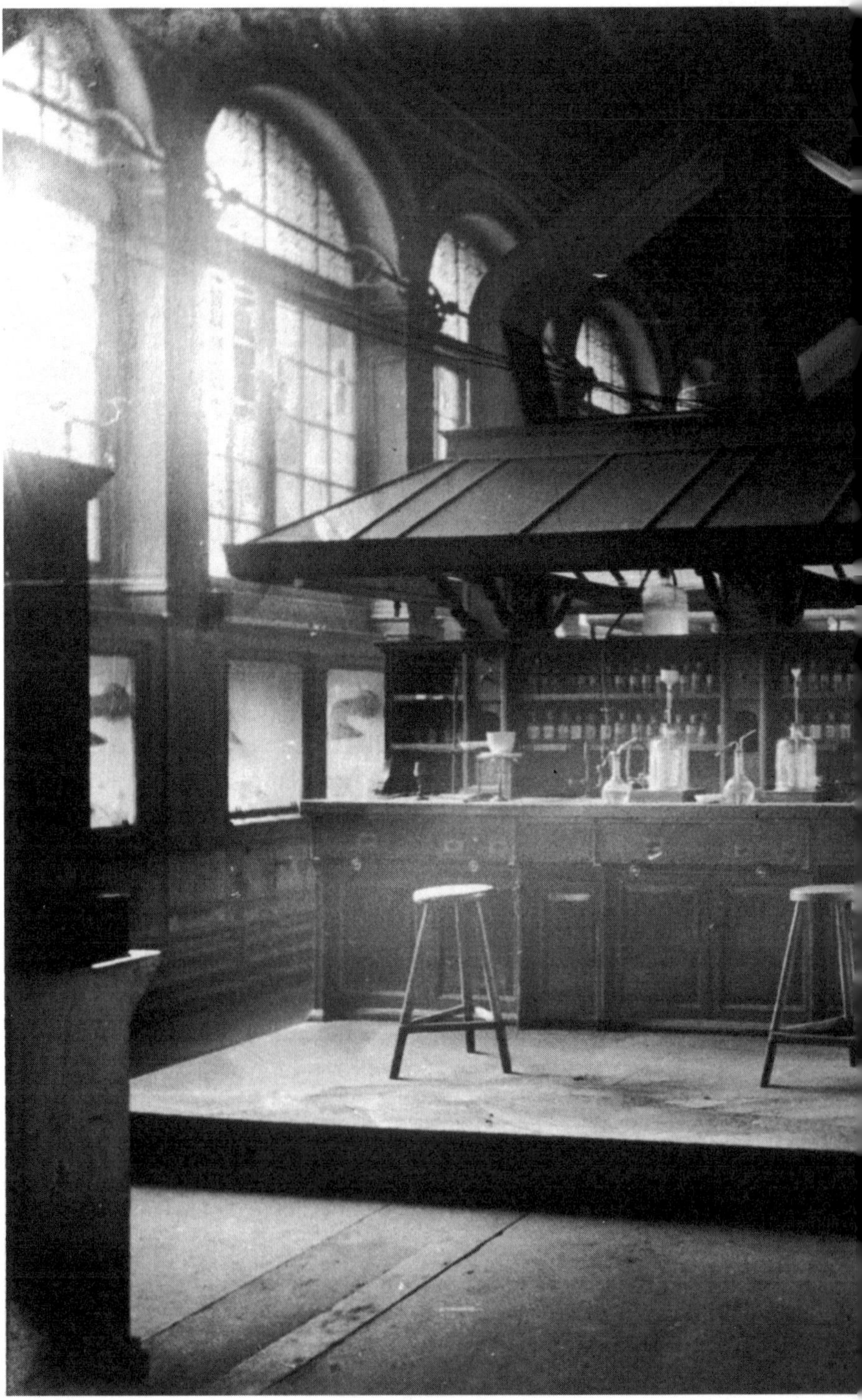

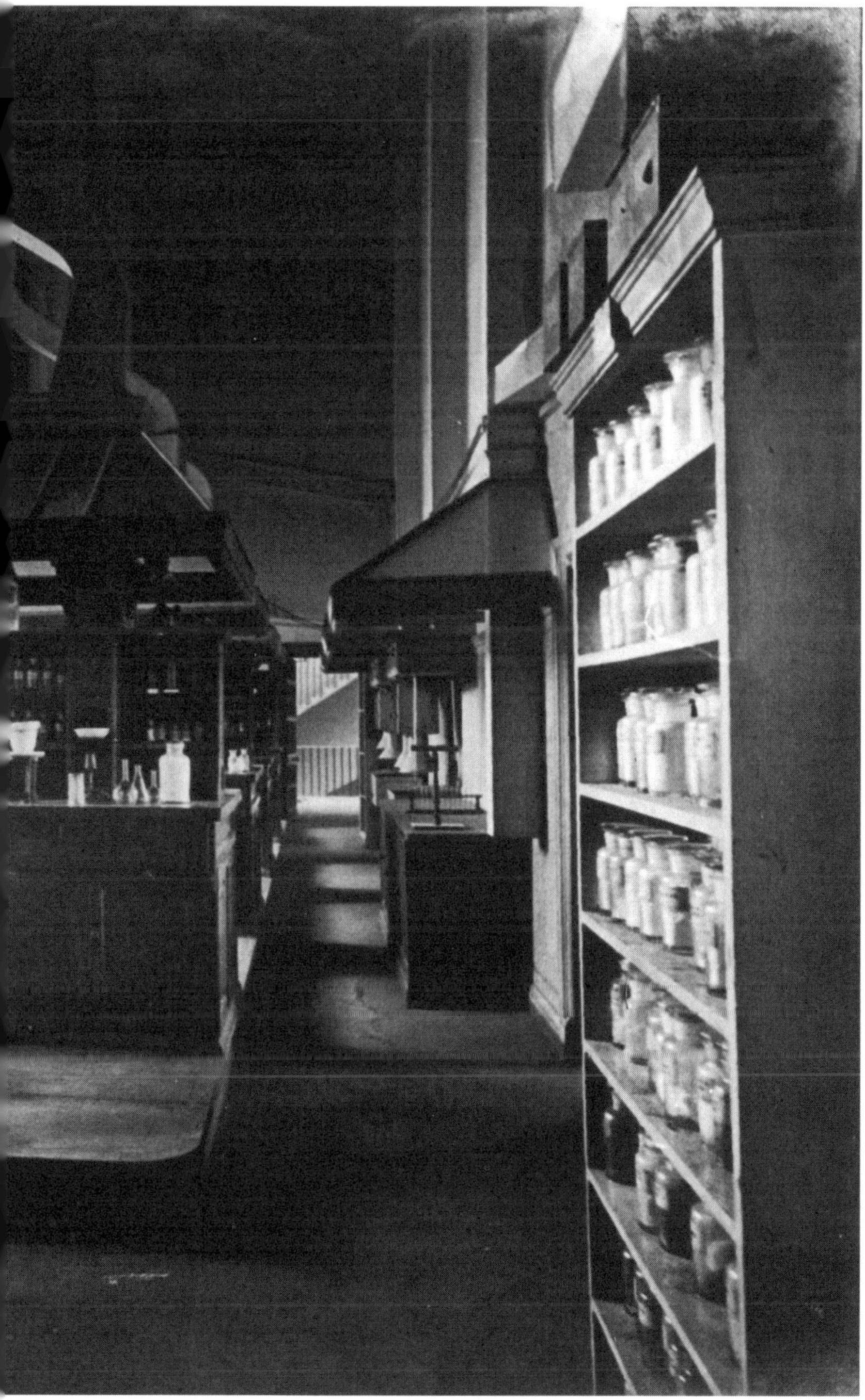

William Tilden described a set-up along these lines in the Imperial College laboratories, which he had designed himself.[22] The flue pipes usually rose from the cabinets, as in Sydney (illus. 52), but at Imperial College in South Kensington, London, they ran downwards into the benches (illus. 53). The hydrogen sulphide at Imperial College was supplied by pipes from a central generator (and gas-holder) in a special room, thus avoiding the hazard of recharging numerous Kipp's apparatus, which would not have fitted into the small boxes anyway. According to Catherine Jackson this centralized system was first introduced by Hermann Kolbe at Leipzig in 1868, and was not limited to laboratories using hydrogen sulphide cabinets.[23] In the laboratories at the University of Wisconsin from around 1916, there were hydrogen sulphide lines with a dedicated fume hood on the side of the laboratory. This system was still dangerous. According to Aaron Ihde, the janitor in charge of the central generator was found dead on the floor one day, killed by the fumes. The students had individual hoods similar to the ones at Nottingham (but square rather than circular), instead of cabinets.[24]

Comfort and safety was also provided by the careful design of the ventilation in the laboratory, so that the air was changed frequently.[25]

52 The main laboratory at Sydney University, showing the large hydrogen sulphide cabinets and the vertical flue pipes, *c.* 1915.

53 Teaching laboratory at Imperial College, showing the hydrogen sulphide cabinets on the benches, c. 1915.

Reference to ventilation is made frequently in the books by William Chandler and Edward Robins.[26] The reason for this concern with ventilation was two-fold. Fume cupboards were not used as a matter of routine as they were considered to be inconvenient. One major problem is that you have to work at arms-length, which is very tiring. Emil Erlenmeyer told Edward Festing that 'it was a great thing to have the sashes of the closets made open so high that a student can work with his head inside the closet'; seemingly forgetting the whole point of the fume cupboards![27] In any event they were relatively few in number – in Bonn, for example, there were only eight for twenty students – and the main working area remained the bench.[28] This meant that many fumes were not removed by the fume cupboard, most notably hydrogen sulphide and solvent fumes. Furthermore there were fumes that could not be removed by the fume cupboard anyway, such as the fumes and smoke from furnaces and stoves, and carbon monoxide from burners. The basic ventilation system employed in most laboratories used the heating system to push the air upwards towards flues in the walls.[29] In the 1880s there was a movement towards mechanical ventilation systems that used fans, helped by the introduction of electricity. However, such systems conflicted with the

54 Two East German students carrying out an outdoor chemistry experiment, Rostock University, 1974.

draught from fume cupboards and were often abandoned for this reason. In the laboratories at the University of Illinois, erected in 1901, the hood (fume cupboard) flues were used to extract the air from the room, almost certainly to avoid this clash:

> In the larger laboratories the air is brought in at the corners of the room, and distributed at various openings along the ceiling. The foul air is forced out through the hood flues. In this way there is an even distribution of the fresh air in all parts of the room and the air is changed six times per hour.[30]

The term 'hood' was clearly unknown in Britain at the time, as Tilden added in a footnote that 'This is understood to refer to chambers for carrying off fumes'. In addition, there were special ventilation conduits with exhaust fans on the benches in the elementary chemistry laboratories, which changed the air eleven times an hour. A standard test for ventilation was to release ammonia and hydrogen chloride fumes into the laboratory and note how quickly the resulting clouds dispersed.[31]

There was another way of conducting dangerous experiments that was particularly favoured at Berlin – this involved carrying out an experiment in the open air and presumably standing well back when the reaction was taking place.[32] At its most basic, an experiment could be carried out in an open space outside the laboratory or on the roof (Bunsen's laboratory had a small verandah).[33] In a legal action brought by BASF against the Manchester dye manufacturer Ivan Levinstein for alleged infringement of BASF's fast red AV patent, an experiment showing the sulphonation of dyes carried out in the judge's chambers created a cloud of sulphur trioxide fumes, forcing everyone to evacuate as quickly as possible.[34] The judge later ordered another experiment involving sulphonation to be carried out by an expert. This time Henry Roscoe conducted the experiment on the roof of his laboratory at Owens College, Manchester. Hofmann, however, insisted on the design of open-air balconies on the higher floors of his laboratory, arranged into Italianate arcades (or loggias, to use Hofmann's own term) for the operation of open-air experiments. This arrangement was immediately copied at the new Normal School of Science in South Kensington, and these arcades can still be seen at the top of the building that is now the Henry Cole wing of the V&A (see illus. 75). There were open-air rooms at Bonn, but according to Festing, 'For a great part of the year they are too cold for work, and the gas flames are too much exposed to burn well'.[35] Similarly at Leipzig the open-air work was done on the roof, but 'judging by the fact that the water supply pipe was burst, apparently by the winter frost, this place does not seem to be much used'.[36] Open-air experiments have now largely died out as they were neither convenient nor safe, but they still took place in East Germany as late as the 1970s (illus. 54).[37]

Group Analysis

The training of undergraduate chemists was also changing in this period. As the number of students taking chemistry increased, there was a growing belief that chemistry was best taught at an undergraduate level by 'drilling' – making students carry out repetitive exercises until they were fluent at carrying out the analyses. Indeed, given the large number of students and the small number of teachers, there was no other option.[38] Fluency is usually associated with learning languages, and the grammar drills of Latin teaching were surely a source of inspiration for this approach. The other origin of the drill approach was of course military training, which had become much more professional during the Napoleonic Wars. The long hours of drilling would have been crushingly boring for the students (and rather dangerous), but enabled them to carry out a range of chemical operations skilfully.

The new set of techniques in qualitative inorganic analysis, later called 'group analysis', in Britain appeared to be ideally suited for this purpose. Group analysis covered most of the common metals and their characteristic reactions. The scheme had the virtue of being cheap; only the simplest apparatus was needed. A rack of test tubes, a filter funnel, a Bunsen burner and a few small beakers and conical flasks served the student throughout the course. As each analysis proceeded, hydrogen sulphide would be boiled off, impregnating the atmosphere of the laboratory with the characteristic smell of rotten eggs. Moreover, the release of ammonia and hydrochloric acids, both group reagents, caused a pall of white mist (ammonium chloride) to linger over the benches. Exposure to these conditions became a rite of passage for chemistry students. Just as medical students are faced with a human dissection early in their training, so it is likely that chemistry teachers viewed the boring nature of the drills and the noxious fumes as a way of weeding out the less committed and less skilful students.

Group analysis is the qualitative analysis of metallic compounds (usually in solution) using a predetermined set of tests arranged by groups of metals.[39] The foundations of this approach had been laid by Torbern Bergman and later chemists such as Martin Klaproth in Berlin and Nicolas Vauquelin in Paris, but their schemes embraced only a few metals. The system that was to occupy generations of chemistry students was proposed by Heinrich Rose in his *Handbuch der analytischen Chemie*.[40] It was extended by Remigius Fresenius in his *Anleitung zur qualitativen chemischen Analyse*.[41] This scheme, which used five groups of tests, was a relatively effective method of quickly determining the composition of inorganic samples, but it really flourished as a method of teaching inorganic chemistry to undergraduates. Both Rose and Fresenius, in the prefaces of their books and in their teaching, urged the student to view each analysis as a piece of research, using the group system merely as a framework within which to work. This attitude was reflected also in Robert Galloway's *A Manual of Qualitative Analysis*,[42] which was packed with practical detail and rich in descriptive inorganic chemistry. However, the situation was about to change decisively. The pressure of teaching large classes of varied ability, especially in England, led to a more mechanical approach. The earlier textbooks were replaced by brief synopses from which all but the most essential detail had been removed. The pioneer of this abbreviated approach was Leibig's successor at Giessen, Heinrich Will. He published an introduction to group analysis in 1846.[43] A brief booklet of tables, in effect flowcharts of group analysis, followed in the same year.[44] Initially 26 tables long, by the eighth edition of 1869 he had reduced it to 13 tables. This was an almost inevitable result of group analysis being used to drill students in universities and increasingly, in a simplified form, in schools.

The Birkbeck Laboratory

What impact did group analysis have on the design of the laboratory? It would be perfectly possible to carry out group analysis on a table, but it would not be very convenient, as the test tubes and other apparatus could easily become jumbled together. Ideally there would be a shelf for the reagents used in group analysis and access to a fume cupboard (assuming the gas-generation apparatus was stored in one). Benches would help to segregate one student from another to discourage joint working, a consideration that would also favour at least modest partitions. A set-up along exactly these lines can be seen in the illustration of the new Birkbeck Laboratory at UCL in 1846 (illus. 55), only a few years after the widespread adoption of group analysis. The construction of the Birkbeck Laboratory was funded by a donation by George Birkbeck and it must not be confused with Birkbeck College (which is the descendant of the London Mechanics Institute, of which Birkbeck was the leading founder). In his *Curiosities of London* John Timbs remarked that the laboratory 'completed from the plan of Prof. Donaldson, in 1845, combines all the recent improvements of our own schools with that of Professor Liebig, at Giessen'.[45] Thomas Donaldson was the first professor of architecture at UCL and the first president of the Royal Institute of British Architects. George Fownes became professor of practical chemistry at UCL in 1846, which effectively put him in charge of

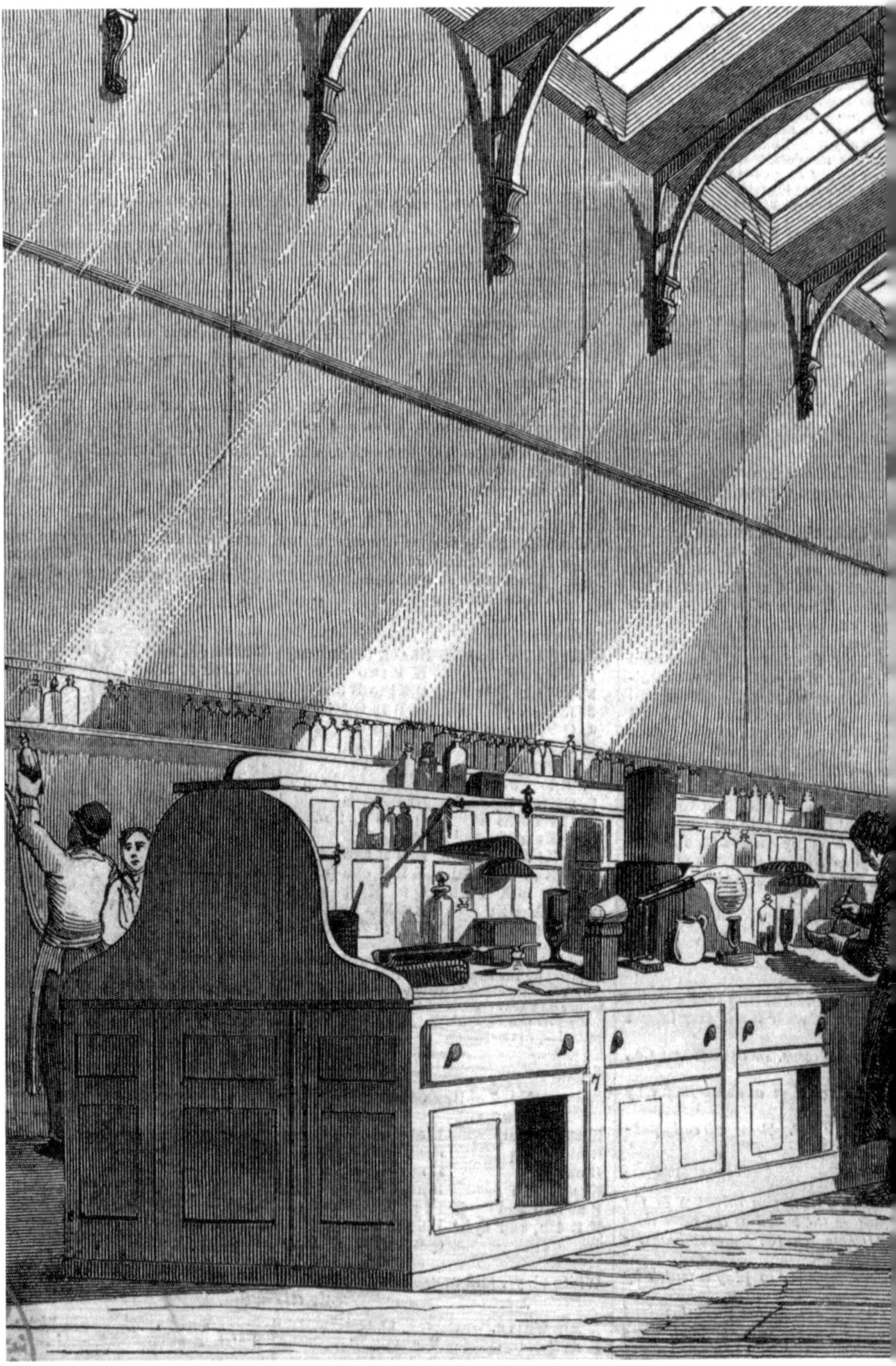

55 Birkbeck Laboratory, University College, London, 1846.

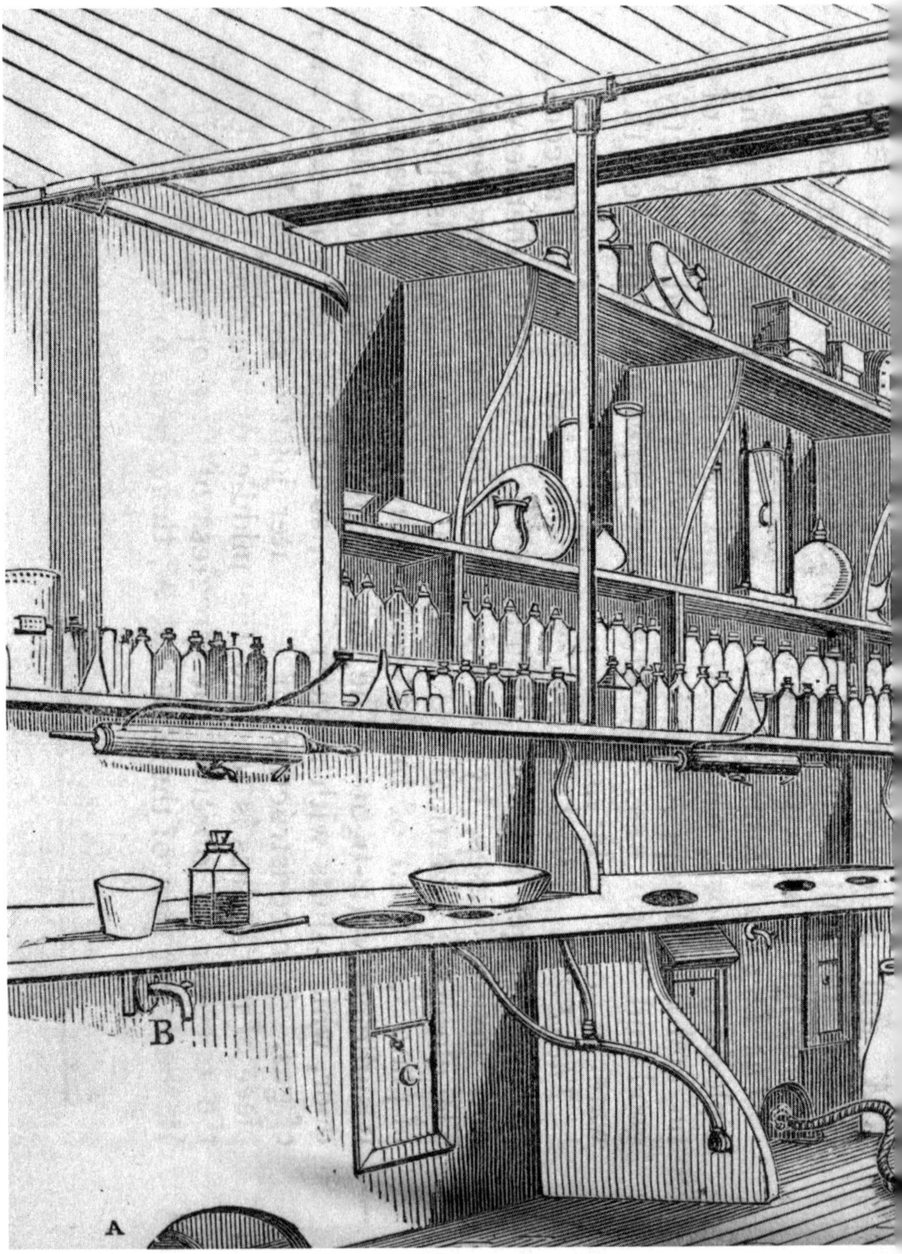
B
C
A

56 Laboratory of the Pharmaceutical Society, 1845.

running the laboratory work. He had spent some time at Giessen in 1839 and had also worked in the laboratory of the Pharmaceutical Society, which in many respects marks a halfway house between Giessen and the Birkbeck Laboratory. The Pharmaceutical Society's laboratory had been opened in October 1844 – just before the Royal College of Chemistry – in Bloomsbury Square, London (illus. 56).[46] With two sets of benches, it had room for eighteen students according to the ground plan, but was designed for eight students. It had a drying closet (which can be seen at the back of illus. 56) and a hydrogen sulphide closet (a forerunner of the fume cupboard). The laboratory had been designed by Theophilus Redwood, the professor of pharmacy, who was another important influence on the Birkbeck Laboratory. Sadly, Fownes was already suffering from tuberculosis and he retired from his chair soon afterwards. He died at the beginning of 1849 at the early age of 33.[47] He was succeeded by the better known chemist Alexander Williamson.

The room in the Birkbeck Laboratory is large enough to accommodate 24 students,[48] and the high ceiling would be useful for carrying away the fumes generated. There are high shelves around the room for large reagent bottles, and a shelf above each bench for smaller bottles. Every bench has drawers and a cupboard at each end for apparatus. The benches are separated by a low partition. There were also fume cupboards based on the ones used at the laboratory of the Pharmaceutical Society. However, the only operation clearly visible in the picture involves a retort cooled by an early Liebig condenser, which implies an organic chemical operation rather than inorganic analysis. An advertisement for the Birkbeck Laboratory in *Chemical News* in September 1865 stated:

> The Course of Instruction in this department is for the assistance of Senior Students in the pursuit of all branches of Chemical Investigation, more especially Organic Research, and for the instruction of less advanced pupils in Elementary Analysis.[49]

It went on to say under the description of the elementary course that:

> The ordinary methods of inorganic analysis are especially dwelt on, and solutions frequently given to the Class for

analysis. All the experiments and analyses are repeated by each Student or by not more than two Students jointly.

So here we have a laboratory design that was capable of combining the latest organic chemical research with the basic drilling in inorganic analysis.

Postgraduate Training of Chemists

Few chemists undertook postgraduate training in the eighteenth century; indeed, few people even took a complete undergraduate degree in chemistry. Many people simply studied some chemistry out of curiosity or a desire to broaden their education, or to learn the part of the chemistry curriculum that would be useful for their careers as either physicians or manufacturers.[50] The reasons for this are quite simple. It cost money to take chemistry courses, whether at a university or a private institution such as Bryan Higgins's School of Practical Chemistry in Soho, London, and there was neither public recognition nor much value in actually having a chemistry degree. After the Napoleonic Wars the situation began to change.[51] The social and commercial value of having a degree increased as part of the process we call 'professionalization'. Certainly, chemists who aimed to teach the subject at universities or technical colleges now needed to have a degree. The situation was patchy, however, and varied from country to country. In Germany a diploma was obtained as the first degree. By contrast, in England it was still difficult to take a chemistry degree. By the mid-nineteenth century, however, it was possible to obtain undergraduate training in chemistry in most European and American universities, even at Oxford.

The research degree of PhD came to Britain after the First World War, although the University of London DSc obtained by written examination or submission of published papers was available from the 1860s. The qualifications of council members of the Society of Chemical Industry, the (Royal) Institute of Chemistry and the Chemical Society over the period between 1877 and 1971 show a definite shift towards having a doctorate of British origin from the early twentieth century.[52] Between 1877 and 1886 roughly one-third of council members had doctorates, of which about two-thirds were German PhDs. From 1887 to 1917, the same proportion held a doctorate; however, 45 per cent held German PhDs, and 35 per cent held British doctorates, while more than 15 per cent held both German and British doctorates. The proportion with a doctorate rose to almost 50 per cent between the wars, and reached two-thirds of the total by the 1960s. While the German PhD remained important for interwar council members (at about a third of the total), practically all the doctorates were British PhDs after the Second World War. It has to be remembered, however, that council members were the elite of British chemistry and would generally have qualified some years before election. The proportion of doctorates among the ordinary membership of the societies was always much lower, but increased fourfold over the whole period. By the 1960s, 45 per cent of the sample of ordinary members of the three chemical organizations had doctorates, and nearly a third were American PhDs, in contrast to council members.

Germany took a different path in this period, establishing the doctorate as the normal degree for chemists, especially those who planned to teach in universities or polytechnics.[53] The doctor of philosophy degree was the middle stage of an academic progression that began with the diploma and ended with the *Habilitation* thesis, only after which could the chemist apply for salaried academic posts. This was completely foreign to the English system, where the doctorate was only firmly established after the Second World War, and even as late as the 1960s chemists were often appointed to university posts before completing their doctorates. Doctorates were still regarded with some suspicion by the public at large as late as the 1970s, and it was not considered to be good form to use the title of doctor if one was the holder of the rather foreign doctorate of philosophy (as opposed to the longer established doctorates of letters or medicine). To be sure, there were English chemists who went to Germany to train under the best professors and thereby gain German doctorates, but they were the exception rather than the rule. They had, however, come to dominate the professoriate of the leading schools of chemistry in England by the late nineteenth century. By contrast, the United States gradually began to introduce the German doctoral system in chemistry after the Civil War.

The teaching of postgraduate students was different from the mass teaching of undergraduates. The postgraduates needed more space and equipment. They also needed their own research projects, which went beyond the research exercises given to undergraduates at some universities. This quickly led to a system whereby a professor did his research through his students rather than through his own efforts. It had a major effect on chemical research. The amount of research a given professor could carry out greatly increased. Instead of being limited by the amount of time he could spend in the laboratory, it was only limited by the size of his research group, which in turn was governed by the laboratory space available and the reputation of the professor. Furthermore, through his students a successful professor would have a strong influence on how chemistry – both teaching and research – was carried out, and hence on laboratory design.

Although Friedrich Stromeyer was not the first research professor of this kind, he had developed a similar system in inorganic analytical chemistry at Göttingen two decades earlier.[54] Liebig was the personification of this system even in his own time. By establishing his reputation through his skilled analyses and his contributions to organic chemistry, Liebig attracted numerous bright students from all over Germany, Britain, the United States, Russia and other countries. He gave them good topics to research that not only set them on the path to successful academic careers, but even dominated their later research, as in the case of Hofmann's lifelong study of aniline compounds. Liebig's Giessen laboratory was thus groundbreaking not for its design or its fitting, advanced as it was, but for the system of training it housed.

Having become the first superstar professor in chemistry, Liebig at the height of his powers accepted an invitation by Bavaria to take up a chair in Munich in 1852. If the Bavarians thought he would create a new and even better version of Giessen in their capital, they were disappointed. Tired of teaching individual students, Liebig wanted to focus on his own research and give lectures. His laboratory in Munich was surprisingly old fashioned. The key feature of his new building was the large and lavishly ornamented lecture theatre shown in illus. 41.

Almost There

By the 1830s the old furnace-centred laboratory that had endured in a stable form for more than 200 years was being swept away, but no single design had yet arisen to replace it. Liebig's new laboratory at Giessen was influential, but it was not itself the design that was to dominate chemistry for the next 150 years. There are various reasons for this: it was only a laboratory, not an entire new building, and it was only suitable for a moderate number of researchers; it was large enough for its time, but not compared with the research teams of the future. Furthermore, it was centred on organic analysis rather than the synthetic organic chemistry that dominated chemistry – and German chemistry in particular – in the century that followed. Nevertheless, certain key elements of the final design were falling into place: fume cupboards, benches, cupboards under the benches and, in the London laboratories that copied Giessen, even bottle racks. These laboratories were, however, only a foreshadowing of the great mansions of organic chemistry that were to follow two decades later. Three key elements remained to be developed, including the increase in scale of chemical laboratories and the general layout of the benches within the laboratory. The third component – the supply of running water and gas to the benches – is the focus of the next chapter.

5
Modern Conveniences
Robert Bunsen and Heidelberg, 1850s

When Robert Bunsen established a new laboratory at Heidelberg in 1855 he introduced mains water and gas.[1] He then invented two key pieces of apparatus that relied on running gas or water: the Bunsen burner and the filter pump. With the hot, stable Bunsen burner flame, he was able to develop spectroscopy, and his pump transformed the filtering of crystals, leading to the dominance of crystalline substances in organic chemistry. Heidelberg was, however, more than just a laboratory with modern conveniences. Thanks to Bunsen's students, especially the foreign ones, Heidelberg became the template for the chemical laboratories that sprang up in the 1860s and '70s.

Utilities Matter

Nothing can be more important in a modern laboratory than mains electricity, running water and gas, yet the crucial role of utilities is almost completely overlooked in the history of laboratories.[2] It is remarkable how late running water and gas entered the chemical laboratory. Even Liebig's laboratory did not differ very much from a garden shed in terms of its utilities, especially gas and water. There was an intermediate stage in the introduction of many utilities: running water from elevated cisterns (as at Marburg), electricity from batteries, and burners that used ethanol or paraffin (kerosene). While there has at least been a muted recognition of the importance of water and gas, the significance of piped steam has been completely overlooked. The introduction of piped steam as a utility allowed chemists to abandon the furnace, thus creating a new freedom in laboratory design. While some attention has been paid (although not much) to the introduction of mains electricity, the use of large electric batteries in chemical laboratories has been largely ignored as a general source of electrical power, except in accounts of specific laboratories.

One might expect laboratories to be early adopters of new technologies, but in the case of utilities they could only adopt these advances if they were locally available. Beginning in London in 1812, gas lighting was already present in several British cities by 1819, and Manchester had set up the first municipal undertaking in 1817.[3] The Prince Regent used gas lighting in the Royal Pavilion in Brighton in 1821. The Continent, however, was slower to adopt gas. In the Netherlands, for example, a few gas plants had been set up by the early 1820s, but the industry only began to grow in the 1840s.[4] Paris was an early adopter of gas, but as the academic laboratories in Paris were not renovated and no new laboratories were built in this period, the opportunity to use gas was not taken up.[5] Petrus Johannes van Kerckhoff in Groningen, Netherlands, renovated the almost-new laboratory of his predecessor in 1851, removing the 'alchemist's chimney'. Two years later, when the local gasworks was opened, he immediately connected his laboratory to the gas mains.[6] In this context Heidelberg as a small German town was neither early nor particularly late. This timing meant, however, that gas and mains water arrived together. Piped water was a result of the growing realization of the dangers of waterborne disease and hence a measure to improve public health. While some kind of water supply had existed since Roman times, modern water supply began in the 1840s and in England was controlled at least in theory by the Waterworks Act of 1847.[7] Amsterdam, for example, was connected to mains water in 1850.[8] In England mains electricity began in a few places in the early 1880s, but the pioneering power stations soon closed down and the industry only really got underway in the early 1890s.[9] To a greater extent than gas, laboratories were quick to take up electricity, partly because they had been using electricity from batteries or dynamos for many years.

From Friary to Laboratory

The chemical laboratories at Heidelberg when Bunsen arrived in October 1852 were located in a former Dominican friary. Henry Enfield Roscoe, at the end of the nineteenth century, recalled his experience of working with Bunsen at Heidelberg five decades earlier:

57 Interior of analytical chemistry laboratory at Technische Hochschule Karlsruhe, c. 1900.

It was roomy enough; the old refectory was the main laboratory, the chapel was divided in two, one half became the lecture-room and the other a storehouse and museum. Soon the number of students increased and further extensions were needed, so the cloisters were enclosed by windows and working benches placed below them. Beneath the stone floor at our feet slept the dead monks [sic, actually friars], and on their tombstones we threw our waste precipitates! There was no gas in Heidelberg in those days; nor any town's water supply. We worked with Berzelius's spirit lamps, made our combustions with charcoal ... and went for water to the pump in the yard. Nevertheless, with all these so-called drawbacks, we were able to work easily and accurately.[10]

As the small state of Baden was keen to attract a leading chemist, Bunsen was offered a new laboratory building if he accepted the chair at Heidelberg. Furthermore, some thought had already been given to the construction of new laboratories. As will be discussed in chapter Eleven, the situation at Heidelberg was remarkably similar in these respects to the situation at Stanford University a century later. Baden's enthusiasm for erecting new laboratories was also part of the economic and cultural competition between the German states (see page

151). The construction of a new chemical institute began in May 1854, and it was opened a year later. The architect was Heinrich Lang, who had just designed the chemistry buildings at Technische Hochschule Karlsruhe, which served as a model (illus. 57). Lang was closely associated with the Technische Hochschule, both as a student and a teacher.[11] He had studied under Heinrich Hübsch, who had himself studied in Rome. Hübsch adopted the Romanesque Rundbogenstil (round arch style), but he also used an understated Neoclassical style, for example in the Trinkhalle in Baden-Baden (built in 1839–42), which has similarities to the Heidelberg laboratory building. The difference between the architects can be seen in the main building of the Technische Hochschule designed by Hübsch in 1833, with its heavy round arches, and the Neoclassical style of the chemistry laboratory building in Heidelberg (illus. 58). Lang remained consistent in his

58 Exterior of the new chemistry laboratory building at Heidelberg with the castle in the background, 1858.

style, and the Fichte and Kant Gymnasia (schools) in Karlsruhe are very similar to the Heidelberg laboratory building. Bunsen drew on his experience of designing the new laboratories at Breslau and made suggestions. As Roscoe noted, the fame of Bunsen drew students from all over the world and 'the new building soon became inconveniently crowded, and many applications [by students] for working benches had to be refused'.[12] The laboratories were more lavish than Liebig's new laboratory in Munich, and Prussia scrambled to catch up (see page 148).

In addition to a preparation room and a balance room, the new building also had special rooms for gas analysis and electrochemical work. There was an element of doubling up – the preparation room

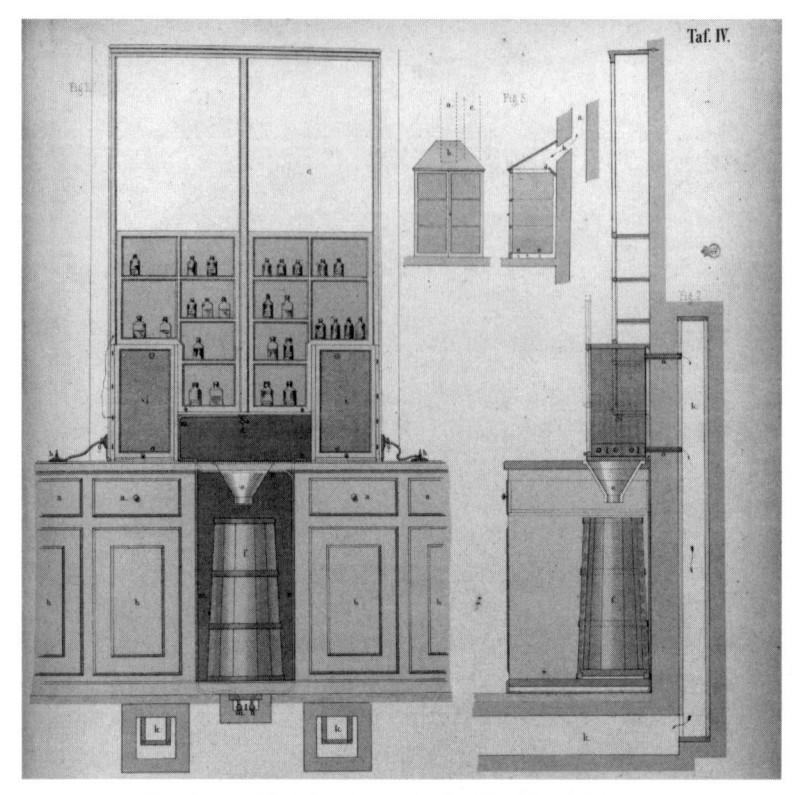

59 Benches and bottle cabinets in the Heidelberg laboratory, showing the waste-disposal system, 1858.

was also the chemical store and the balance room was also the library. There were two large laboratories with 28 and 22 workplaces for teaching; the smaller one was for beginners. The director, Bunsen, had his private laboratory and there was also a small laboratory (not for research as one might assume, but for the carrying out of operations that were inadmissible in the large laboratories). As Christine Nawa has pointed out, this was the aptly named *Stinkzimmer*, a room for smelly and dangerous experiments – possibly a result of Bunsen's early work on cacodyl (see page 97).[13] Alongside the scientific workspaces there was a small and unassuming lecture theatre, and the 'handsome' living quarters of the director alongside the laboratory in the southern wing, reached by climbing marble stairs.[14] Edward Festing later remarked: 'Professor Bunsen is of the opinion that a laboratory should not be too "elegant," and his certainly is not so; but it looks thoroughly workmanlike.'[15] The laboratories had running water and gas, but as in Giessen the sinks drained into conical wooden barrels (illus. 59). The reagents were kept in glass-fronted cupboards – small ones above the bench and larger ones against the walls. The fume cupboards were against the inner walls and large pieces of apparatus were placed on long benches running under the windows.

A Practical Chemist

Chemists are usually classified by nationality and sub-discipline – a German organic chemist, a British physical chemist or a Russian inorganic chemist – yet such pigeonholing is often misleading. Joseph Black was as French as Lavoisier since he was born and partly brought up in Bordeaux. Eduard Mitscherlich was born in Jever, then an enclave of Russia. Bunsen is also difficult to stereotype. Robert Wilhelm Eberhard Bunsen was born in March 1811 in Göttingen in the Kingdom of Hanover, which was ruled by George III of Great Britain and Ireland, albeit then under French occupation as part of its puppet state of Westphalia (illus. 60).[16] Bunsen's father was the head librarian and professor of modern philology at the internationally renowned University of Göttingen and had served in George III's army. Bunsen went to the Gymnasium at Holzminden, then to his father's university in Göttingen to study the sciences, including chemistry under Friedrich Stromeyer. He took his doctorate on the basis of a prize-winning essay on hygrometers, a topic that would today be regarded as belonging to physics.[17]

This disciplinary ambiguity continued throughout Bunsen's career and his research cannot be classified under any consistent disciplinary heading, although he started his career as an organic chemist and is sometimes considered to be an inorganic or physical chemist. As a result of his work on the foul compounds cacodyl oxide and cacodyl chloride, Bunsen thought he had prepared free cacodyl as the equivalent of a metallic element. He was in fact wrong (see page 97), but his research spurred his student Edward Frankland to seek the free ethyl radical, a search that ultimately led to the modern theory of valency.[18] Bunsen also developed iodometry (analytical chemistry), and the Bunsen cell (physical chemistry), which was praised by Jules Verne in *Twenty Thousand Leagues Under the Sea*.[19] Bunsen was particularly concerned with the application of chemistry to industry. In particular he studied the chemistry of the blast furnace – both the charcoal-using furnaces of Germany and, with the help of Lyon

60 Photogravure of Robert Bunsen after a painting by J. Marx, with a framed painting of Heidelberg in the background, 1910.

Playfair, the coal-fuelled furnaces in Britain. He pointed out that the presence of carbon monoxide in the exhaust gas indicated a lack of efficiency and that the gases contained recoverable ammonia. He developed new methods of gas analysis for this purpose, employing a straight-tube eudiometer and balls of solid absorbents hanging on platinum wires.

This was not unusual, as Bunsen was a keen developer of chemical instrumentation. He created the Bunsen 'grease spot' photometer in 1843; the 'constant level' hot-water bath, whereby the water lost by evaporation was continually replenished, in 1850; a bizarre looking thermostat for the measurement of the specific heat of gases

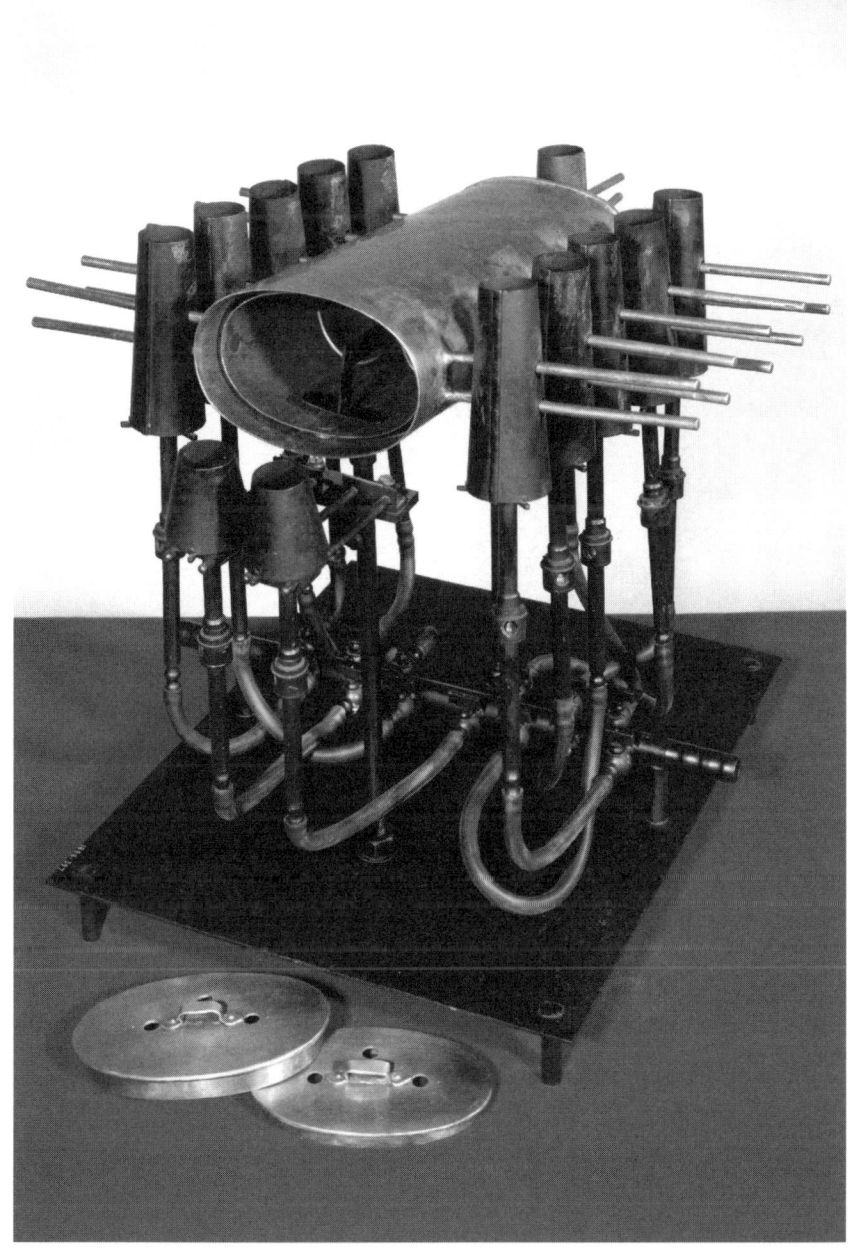

61 Bunsen thermostat supplied to the Science Museum
by the Desaga company (inv. no. 1888-0307).

62 *A Dream of Toasted Cheese*, drawn by Beatrix Potter
for her 'Uncle Harry', 1899.

in 1867 (illus. 61); and the Bunsen ice calorimeter in 1870. Bunsen's technician Peter Desaga, an established scientific instrument maker in the town, manufactured and marketed Bunsen's inventions. All these inventions pale in significance, however, to Bunsen's invention of the Bunsen burner (illus. 62).

In the early nineteenth century there were only two types of laboratory heating device: wick-based lamps fed by alcohol and the laboratory furnace. The Swiss scientist Ami Argand developed an oil lamp in 1784 that mixed oil vapour with air, thereby increasing the temperature of the flame.[20] This hot flame also made it attractive as a heater for chemical furnaces, although it does not seem to have been widely adopted as such.[21] The introduction of illuminating gas obtained from the destructive distillation of coal in around 1820 presented an opportunity for a new type of burner. Michael Faraday sketched an early gas burner in his *Chemical Manipulations*, with a jet of gas mixing with air and burning above an inverted, stemless metal funnel.[22] Bunsen created a reliable gas burner based on these principles in 1855, when Heidelberg was first supplied with coal gas. He worked with Roscoe and Desaga. Roscoe takes up the story:

> Returning from my Easter vacation in London, I brought back with me an Argand burner with copper chimney and wire gauze top, which was the form commonly used in English laboratories at that time for working with a smokeless flame. This arrangement did not please Bunsen ... the flame was flickering, it was too large, and the gas was so much diluted with air that the flame-temperature was greatly depressed. He would make a burner in which the mixture of gas and air would burn at the top of the tube without any gauze whatsoever, giving a steady, small and hot non-luminous flame under conditions such that it not only would burn without striking down when the gas supply was turned on full, but also then the supply was diminished until only a minute flame was left. This was a difficult ... problem to solve, but after many fruitless attempts, and many tedious trials, he succeeded.[23]

In Bunsen's original burner the proportions of air and gas were controlled by closing air-holes in the base by small stoppers. The familiar rotating collar for adjusting the air supply was introduced by the instrument maker John Joseph Griffin a few years later.

Bunsen finally retired at the age of 78 in 1889. When he died in August 1899, *The Times* remarked that

> he was passionately devoted to his laboratory and his students and formed a school of chemistry second only ... to that of Liebig at Giessen. It was in his patient teaching that his genius

showed itself strongest and the pupils from all parts of the world who were fortunate to work under him ... all conceived for him the deepest feelings of respect and affection.[24]

The Analysis of Very Little

When chemists were struggling to analyse inorganic samples such as minerals or the products of industrial processes in the early nineteenth century, they had two major problems: how to quickly find out what elements were present, as this information makes analysis much easier, and how to carry out the analysis with very small samples. Blowpipe analysis offered a partial solution and Bunsen was a keen user of the blowpipe – to the extent that a specific room was created for blowpipe work (and 'other rough work') in his new laboratory.[25] One important aspect of blowpipe analysis is the coloration of the hot flame produced by the blast of the blowpipe. It had been recognized since 1752, when Thomas Melvill noted that common salt in alcohol produced a yellow flame, that metallic salts in a fairly hot flame will produce a colour. If one could relate these colours to specific elements and, even better, if one could work out what elements were present in a hot flame, this would be a powerful analytical technique that would require very little material.[26] Not all metals have flame colours, but a significant number do; mainly the alkali and alkaline earth metals. One major problem with using this flame test to detect metals is that even low levels of contamination with sodium produce a yellow flame and overwhelm the other metals, as Melvill discovered. By 1800 eight flame colours had been related to specific elements, namely boron (flame colour noted in 1732), sodium (1752), copper (1755), potassium (1758), magnesium (1767), and calcium, strontium and barium (all 1793).

William Hyde Wollaston, best known for his work on platinum, was the first to relate the dark lines in the solar spectrum to the bright lines produced by terrestrial material in 1802. It was perhaps no accident that his father, the Revd Francis Wollaston, was an astronomer. Another astronomer, John Herschel, speculated in 1823 that the flame colours were caused by the vaporized salts, but also assumed that all such flames turned yellow in a very hot flame, not realizing that this phenomenon was caused by traces of sodium. Flame tests were only appropriate for a small group of metals, and before the development of the Bunsen burner only comparatively cool flames were readily available. The use of a cobalt-blue glass filter to detect potassium in the presence of traces of sodium arrived on the scene relatively late. Rowlandson Cartmell, working in Bunsen's laboratory, suggested the employment of an indigo solution in a hollow glass wedge, or the rather more portable cobalt glass to filter out the yellow sodium flame in 1858.[27]

The photography pioneer William Henry Fox Talbot was the first to suggest the use of flame colours for qualitative chemical analysis. In 1834 he also became the first scientist to use a prism to study these coloured flames, from which he was able to distinguish lithium and strontium (which have similar flame colours) by their spectra. The American physician David Alter argued in 1854 that an element's spectrum was unique and could be used to identify elements in the Northern Lights or even the stars. Apart from Bunsen and Gustav Kirchhoff, the most important figure in the early history of spectroscopy was William Allen Miller of King's College London. From 1845 onwards he used spark spectra, first described by Charles Wheatstone in 1835, to study the spectra of many elements, including those that did not produce a strongly coloured flame. His work was, however, vitiated by his failure to recognize that many of his results were contaminated with sodium. Later, in 1864, he assisted his neighbour William Huggins in obtaining the spectra of planetary nebulae and stars.[28] By 1856 William Swan was able to show that the yellow spectral 'D' line could be used to detect sodium down to a concentration of less than one part per million, a hitherto unimaginable level of sensitivity in chemical analysis. William Crookes, at that time as much a photographer as a chemist, devised a camera incorporating a prism to capture spectra photographically in 1856.[29]

As a result of the provision of mains gas at Heidelberg, the spectral analysis of metal salts was transformed. The experimental partnership of Kirchhoff and Bunsen began in Breslau in 1851, where Bunsen briefly held the chemical chair before moving to Heidelberg. With Bunsen's support, Kirchhoff was appointed professor of physics at Heidelberg in 1854 (illus. 63). Following the research of Cartmell, Bunsen was analysing materials on the basis of their distinctive colours in burner flames, using various filters of coloured glass to differentiate between flames with similar colours, when Kirchhoff suggested that the spectroscope would be better suited to the purpose than filters. Effectively he was doing no more than echoing Talbot some 25 years earlier, and proposing the use of the new Bunsen burner with its hot, colourless flame of low luminosity rather than the electric sparks employed by Miller. However, Bunsen and Kirchhoff went beyond their predecessors in the carefulness of their preparations, recrystallizing their substances several times before examining their spectra, and in experimenting with different fuels and fuel mixtures for the flame. Kirchhoff was also able to show that the emission spectrum of a given element was simply the reverse of the absorption spectrum, as seen in the dark lines of solar spectrum, first noted almost six decades earlier by Wollaston:

> Kirchhoff has made a most beautiful and most unexpected discovery: he has found out the cause of the dark lines in the solar spectrum, and has been able both to strengthen these lines artificially in the solar spectrum and to cause their appearance in a continuous spectrum of a flame, their position being identical with those of the Fraunhofer's lines. Thus the way is pointed out by which the material composition of the sun and fixed stars can be ascertained with the same degree of certainty as we can ascertain by means of our reagents the presence of SO_3 and Cl. By this method, too, the composition of terrestrial matter can be ascertained, and the component parts distinguished, with as great ease and delicacy as is the case with the matter contained in the sun; thus I have been able to detect lithium in 20 grams of sea-water.[30]

Their original apparatus was very simple.[31] The chemist observed the spectrum of a sample heated in a Bunsen burner flame. The spectrum was produced by restricting the light emitted from the flame to a narrow slit using a collimator tube. This light beam was allowed to pass through a flint glass prism, thereby forming the spectrum, which was then focused in a small telescope. The tubes and the prism were fixed on top of a heavy stand (the box arrangement used in the first experiments was soon dropped in favour of a cast-iron stand; see illus. 64). A third tube illuminated by a candle was soon added to the basic apparatus. This projected a scale onto the image of the spectrum from which the wavelengths could be read off. Since

63 Left to right: Gustav Kirchhoff, Robert Bunsen and Henry Roscoe, 1862.

then, the apparatus has changed very little, except for the introduction of the direct-vision spectroscope, which incorporated the prism into the telescope, thereby allowing direct observation of the spectrum, at the cost of decreased accuracy.

Using highly purified salts as standards, Bunsen and Kirchhoff were able to map out the line spectra for several elements (illus. 65).[32] They then began to examine spa waters and minerals, looking for lines that could not be attributed to known elements. In the spring of 1860 they studied samples from the nearby spa at Bad Dürkheim. At first they could only observe the lines of the known alkali and alkaline earth metals, but when the alkaline earth metals and most of the lithium had been removed, they saw two remarkable blue lines close together for the first time. Bunsen called this new element 'cæsium', after *cæsius*, the Latin for the light blue of the upper sky.[33] Not long afterwards, in early 1861, Bunsen and Kirchhoff examined lepidolite, a mica-like fluorosilicate mineral found in Saxony. After removing the potassium present in the mineral, they observed two violet lines in the spectrum, which they recognized as a new element. Bunsen called it 'rubidium', after the Latin *rubidus* for ruby red, based on the beautiful pair of deep red lines in its spectrum.[34] Subsequent investigations showed that rubidium was actually quite widespread but only in low concentrations. In contrast to cæsium, which was only isolated in 1882, Bunsen prepared rubidium metal in 1863.

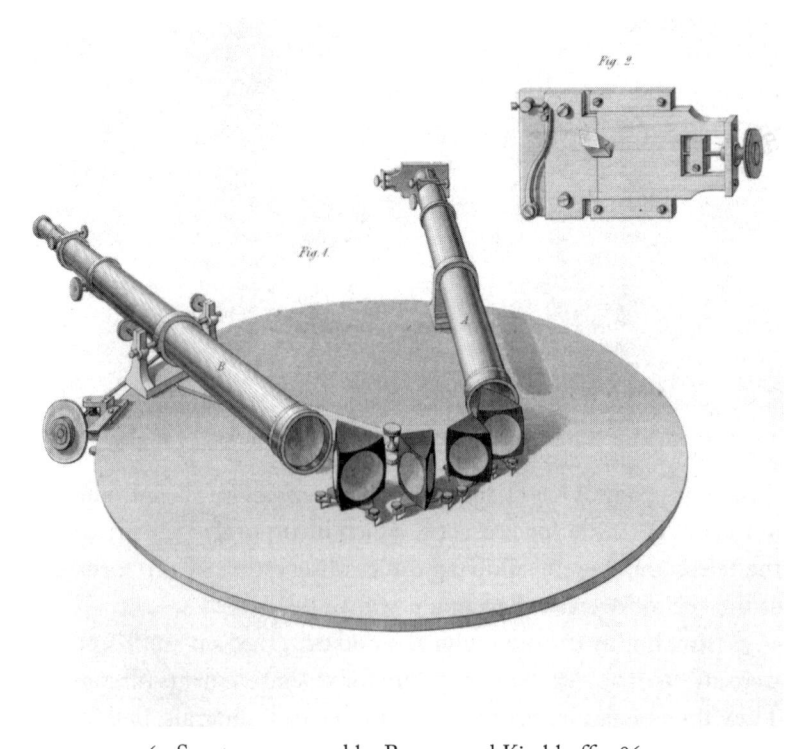

64 Spectroscope used by Bunsen and Kirchhoff, 1861.

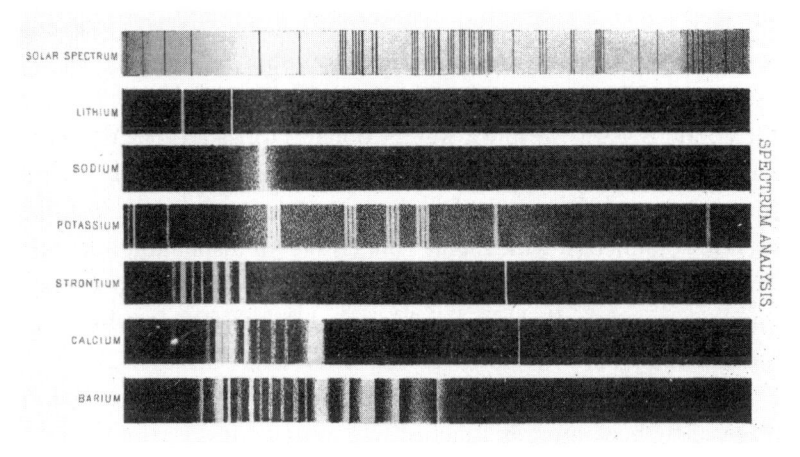

65 Alkali metal and alkaline earth spectra, 1869.

Water Power

While its impact on spectroscopic analysis may have been revolutionary, the Bunsen burner was also valuable in organic chemistry as a source of heat. Running water also had a major impact in this field. We met the alembic and the retort in chapter One. The vapours produced by the retort are cooled in its characteristic long spout, but clearly this is less than completely effective. Sometimes an alembic or a retort with a particularly long spout was cooled by passing it through a barrel filled with water, but it seems that the idea of cooling a condenser tube by running cold water over it was first proposed by Christian Weigel in 1771. He used an outer metal tube to confine the running water, which was sealed with gypsum at the bottom and left open at the top end. Nonetheless he directed the water in the opposite direction to the condensing vapour; the countercurrent principle used in modern condensers. In 1836 Friedrich Mohr was the first to seal the ends of the outer tubes with corks, presumably with the help of the cork-borer he had just invented. Liebig did much to popularize what is now universally known as the Liebig condenser, a term in common use by 1852. Up to this time, however, the water had to be poured from a water tank into a funnel. The water coming out at the other end could be collected and returned to the tank. This was obviously very laborious and required frequent attention. To be able to attach the inlet tubing to a tap and let the water flowing out of the condenser go down into the drains was a major breakthrough (although not in the Heidelberg laboratory, which still lacked a plumbed-in drainage system). Hard glass was probably first used for the outer sleeve in the early 1860s, as it appears in the catalogue of the apparatus supplier Griffin in 1866.[35] Until this time the condenser was simply a water-cooled retort adapter. Although flasks and tubes had previously been used as an ad hoc replacement for retorts, round-bottomed flasks with side-arms intended to be used in conjunction with Liebig condensers for distillation were first listed in the 1877 edition of Griffin's catalogue.[36] The all-glass sealed condenser now called the 'Liebig condenser' was introduced in around 1885.

The humble filter pump (or water aspirator) is simple to use. A piece of rubber tubing is attached to a tap, and at the other end the tubing is attached to a small, T-shaped piece of metal (which is the pump), and the tubing from the middle outlet produces the vacuum. It is generally held that Bunsen described this pump for the first time in a paper on the washing of precipitates in 1868, but was he the inventor? The attribution of the pump to Bunsen led to an exchange of correspondence in 1873 between two Hanover-born chemists: Bunsen and Hermann Sprengel, who was working in London.[37] Sprengel had introduced a more powerful version of the water pump in 1865; it used drops of mercury in a capillary tube, but he pointed out that he had used water for his pump as far back as 1860. Indeed, he had started with water, but abandoned it for various reasons (including the need to dispose of the large volume of water produced) and used mercury instead. He had wondered if a waterfall might make a good water pump (this evokes a splendid image of a group of water-sodden chemists gamely carrying out a filtration behind a large waterfall!). However, his concerns about the force of the water and the need for disposal were already overcome by the introduction of pressurized mains water and drains, a breakthrough that was soon seized upon by the inventive Bunsen.

The introduction of the filter pump into the laboratory allowed chemists to create a moderately strong vacuum using cheap and safe equipment as long as they had access to pressurized running water. This vacuum was particularly useful in organic chemistry. It enabled chemists to reduce the pressure during the distillation of liquids (with a Liebig condenser of course) and thus lower the distillation temperature. The temperature reduction allowed chemists to distil liquids that would have decomposed (or otherwise undergone undesirable reactions) at the normal boiling point; the biggest breakthrough in the purification of liquid mixtures until the introduction of liquid-liquid chromatography in the 1940s.

The filter pump also enabled the purification of solids by speeding up the filtration of crystals. The importance of recrystallization in organic chemistry has been largely overlooked by historians. Chemical compounds, both organic and inorganic, are often purified by taking the original crystalline product of a reaction, redissolving it in warm, fresh solvent, then recrystallizing it from the solution. The recrystallization is often accelerated by adding a second 'non-solvent' liquid or a seed crystal (or even, according to popular chemical mythology, dandruff from the chemist's beard), before cooling the solution to precipitate the product. Because this recrystallization process was one of the few reliable methods of purifying organic compounds in the nineteenth century, chemists preferred to produce compounds that were easy to crystallize. Consequently, they developed special reagents that converted liquid compounds or compounds with poorly marked melting points into easily crystallized products for identification purposes. In sugar chemistry, for example, one thinks of the use of phenylhydrazine by Emil Fischer in 1875 and dinitrophenylhydrazine by Oscar Brady in 1926. If the crystalline derivate could be readily converted back to the original non-crystalline compounds, such as the acid salts of amines or the bisulphite addition product of aldehydes, this was also a simple way of purifying the parent compounds.

The crystals are filtered off from the supernatant solvent using a Büchner funnel; invariably used with a glass flask with a side-arm that is in turn attached to a water pump.[38] The funnel was invented by Ernst Büchner, who is often confused with the more famous German biochemist Eduard Buchner. The basic set-up was first described by the Swiss chemist Jules Piccard in 1865 using a Woulfe bottle and a paper filter funnel. In 1888 Robert Hirsch placed a porcelain (or iron or glass) perforated plate – based on the plate developed by Otto Witt two years earlier – in the standard funnel (illus. 66). Büchner then changed the shape of the funnel to produce the wide-bottomed funnel we now generally use. The filter pump transformed a slow process that was rarely complete into a rapid one which produced dry crystals. In this way, the ability to produce well-formed pure crystals became the hallmark of a good organic chemist.

THE MATTER FACTORY

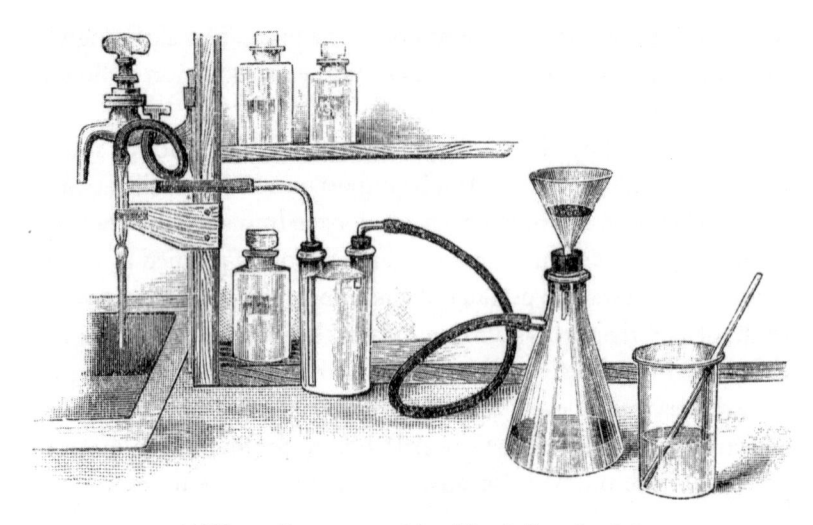

66 Water-filter pump with a Hirsch funnel, 1898.

A professor could quickly check the work of his students and assistants by asking to see the crystals.[39]

Steam Pipes

As we have seen, the chemical laboratory was dominated by the furnace from the days of the alchemists up to the end of the eighteenth century. The furnace heated the laboratory, provided heat for distillation, reactions and drying and, inter alia, produced distilled water if required. Yet the classical laboratory of the mid-nineteenth century did not have a furnace at all. Indeed, one could argue that this type of laboratory, with its long, relatively narrow rooms and multiple benches, would not have been possible as long as the furnace was in use; yet this revolution is never mentioned in the historical literature. In effect, the furnace was replaced by piped steam that heated the laboratories, provided a source of heat for operations and was also used to heat stills, autoclaves and drying ovens. Its big advantages were the lack of a naked flame, a consistent temperature and no danger of overheating. If piped steam was available, there was no need for an individual steam generator, which could dry out. When passed between two layers of metal, piped steam was used for funnels, calorimeters and viscometers for viscous liquids or solids. Piped steam was also used for driving centrifuges. Very neatly, the condensed steam was used as distilled water, an important commodity in the laboratory even today. Piped steam could, however, also be dangerous, as an accident at MIT as late as 2008 demonstrated.[40]

The use of steam in the laboratory independently of the furnace (small-scale use of steam from the furnace had long been in use) appears to have been pioneered at the laboratory of the Society of Apothecaries in London under William Brande in 1815.[41] This demand for steam seems to have arisen from the importance of distillation in the laboratory; there was a still house with six stills, and there was also a need to steam distil plant material. It was certainly not created as a result of wanting to avoid using the furnace, as the laboratory was also supplied with numerous furnaces for sublimation, fusion and calcination. There was, however, a concern with fire prevention, and this too may have promoted the use of steam as a way of reducing the use of naked flames. To some extent this introduction of steam may have paralleled the growing use of piped steam in the kitchen. In her cook's dictionary of 1823, Mary Eaton recommends the use of steam for similar reasons, namely no ashes, lack of fire and ability to use gentle heat. She particularly recommended its use for cooking vegetables, and attributes the development of the apparatus to 'an ingenious economist of Derby'.[42]

While the use of steam in this way seems an obvious innovation, especially if a boiler was installed for heating, the first academic laboratory where the supply of steam to the laboratory is clearly shown was at Berlin, designed in 1864. Even Hofmann's other new laboratory at Bonn was not supplied with steam, nor was it apparently on tap in Heidelberg. Wilhelm Hofmann implies at first that steam was only available for specific drying ovens.[43] Later in his report he says that 'Especial attention was bestowed on the laying out of a simple and readily accessible system of mains for the convenient supply of water steam gas and sulphuretted hydrogen', which suggests a more general access to piped steam.[44] In Leipzig, steam was supplied to laboratories and specific rooms but not to the bench. Piped steam had become routine in academic laboratories by the 1880s, but was eventually displaced by electricity.

Introduction of Mains Electricity

The RI was a pioneer in the use of utilities – it generated its own gas as early as 1816. Humphry Davy had obtained a powerful battery for the RI in 1808, so it is not surprising that it was one of the first laboratories to have mains electricity, in 1892.[45] The use of electricity produced by electrochemical cells was in common use by the 1850s. For example, Bunsen's laboratories had DC electricity from a cupboard housing several galvanic cells. Ingeniously, the gas and water pipes were isolated electrically and attached to the two poles of the galvanic cells.[46] An electric current was produced by attaching wires to the water and gas pipes at the bench. Hofmann remarked a decade later that the electric light (presumably an arc lamp) was 'now rapidly becoming an indispensable appliance of the lecture table'.[47] The first laboratories to use mains electricity as a utility generated their own electricity on site rather than drawing their supply from an external power station. Part of the reason for this rapid adoption of electricity was the need to use it for heating and lighting. The new laboratories of the 1860s were particularly problematic when it came to heat and light; being tall, long and narrow, they were not well designed to use a central stove or fire. Furthermore, these large spaces were difficult to illuminate with the relatively feeble gas light, although gas lighting remained in use at the chemical laboratories at Cambridge University until after the Second World War. Nonetheless, it is noteworthy that none of the laboratories surveyed by William Henry Chandler, which were all erected in the 1880s, had mains electricity installed – not even the laboratories at Cornell, which opened in 1890. The Eidgenossische Technische Hochschule in Zurich had two electric dynamos in the electrochemical room, but this is the only reference to electricity in Chandler's book.[48]

A good example of the introduction of generated electricity, and possibly the first British laboratory to introduce it, was Finsbury Technical College, which opened in 1883.[49] It was essentially an evening college of science founded by the Guilds of the City of London, which were under political pressure to invest their funds in technical education. The founding professors, the physicist William Ayrton and the chemist Henry Armstrong, were both familiar with leading modern laboratories (the physics laboratory of the Royal Engineering College in Tokyo and Kolbe's chemical laboratory in

Leipzig respectively). The architect Edward Robins was a friend of Ayrton and Armstrong, and was closely involved with the reform of technical education. Probably at Ayrton's insistence, the new laboratory was supplied with a gas-powered dynamo in the basement and Edisonian electric lighting in the laboratories. The laboratories were heated by a hot-air system that was also supposed to draw air from the fume cupboards up a master flue, although it did not work very well. Armstrong installed an electric fan to increase the draught, thus showing the usefulness of having mains electricity installed. The use of electricity at Finsbury influenced the new Central Institution at South Kensington, which was designed by Alfred Waterhouse and opened in 1884. However, externally generated electricity was still unavailable. Siemens's pioneering power station at Godalming in Surrey, opened in 1881, closed down in 1884, and Edison's Holborn Viaduct power station in London, which opened soon after Godalming, was shut down in 1886.[50] It was not until the early 1890s that the general provision of mains electricity became established (as noted, the RI was linked up to the mains electricity supply in 1892).

Electricity had become sufficiently established in laboratories by the 1920s that contemporary manuals of laboratory design devoted considerable space to the provision of mains electricity. Alan Munby remarked that:

> A generous provision is indispensable for modern laboratory work, electric motors being required for refrigerating plant, centrifuges, mixing, and stirring, timing pendulums, kymographs and many other purposes. Special lighting again for projectors and microscope work may involve preciable [sic] current consumption.[51]

So the 'lamp for the lecture table' in the guise of a projector remained a major consumer of electricity in the laboratory in the interwar period. One rather doubts if this is the case for modern laptop projectors! In fact, the main consumers of electricity in modern laboratories are the fume cupboards (see chapter Twelve).

Giving Instruments a Name

The Bunsen burner, along with the Liebig condenser, is probably the best-known chemical apparatus with a name attached to it. Yet the burner was invented by Bunsen (with the help of Roscoe and probably Desaga), whereas Liebig had little to do with the development of the Liebig condenser and never used the hard-glass apparatus that goes by that name today. So how do chemical apparatus and instruments end up with names, and why? Chemical apparatus are not unique in being named in this way: medical instruments and telescopes have long borne the names of their inventors, such as the Newtonian, Cassegrain and Gregorian. Furthermore such naming was relatively late, and started to become common in the 1840s and '50s, around the time of the invention of the Bunsen burner. Seemingly earlier apparatus such as the Woulfe bottle and the Berzelius beaker were named years after their eponymous inventors first produced them. In *Chemical Manipulation*, Michael Faraday uses names mainly for blowpipes, although he does mention Woulfe's and Nooth's apparatus for gas generation.[52] Named instruments began to appear in apparatus catalogues, notably John Joseph Griffin's *Chemical Handicraft*, which was effectively a catalogue.[53] If Griffin's catalogue of 1866 is compared with the catalogue of William Russell Bland and Charles Albert Long twelve years earlier, the difference is striking.[54] Whereas Bland and Long's catalogue uses names conservatively along the lines of Faraday, Griffin's is replete with named apparatus. The named instruments in Bland and Long, as in the similar American catalogue of Edward Kent,[55] are largely confined to arsenic-testing apparatus, lamps for heating purposes and blowpipes.

The introduction of such catalogues on a large scale implies mass production, as apparatus was becoming standardized, more numerous and more diverse. Instead of your local supplier – like Bland and Long or Kent – selling you an (unnamed) beaker or a flask, there were numerous pieces of glassware available. The apparatus sellers had to find a way of distinguishing between these different types. Of course the differences could be shown in the pictures, but the cost of engraving and lack of space prevented every item from being shown in pictorial form. And how would they be indicated in the index? The sellers could have used a code of some kind, such as

ABX-24 for one beaker and ACF-19 for another, but such codes would have been unique to one firm and would have been confusing, as a plethora of codes would have soon sprung up. Sensibly the apparatus sellers realized that there was a virtue in attaching names to the different pieces of apparatus – they roll off the tongue better and stick in the memory.

It might be thought that naming a piece of apparatus had the aim of making it appear that the chemist in question somehow endorsed the item or the seller. However, it appears that this was not generally the case, with the notable exception of Peter Desaga, who doubtless did benefit from his association with Bunsen. It seems to have been accepted on all sides that giving a name to a piece of apparatus was a purely pragmatic process without any such endorsement or any kind of historical priority. If it was not by a direct connection with the putative inventor or user, how did the apparatus sellers get these combinations of names and apparatus? In many cases the chemist in question had introduced the apparatus in a textbook, for example Hofmann and his demonstration apparatus (see page 83). When the textbook was used, a demand would be created for the apparatus shown in it, and manufacturers would soon put them on sale with the appropriate names attached. In other cases it is likely that the name had attached itself to an apparatus through popular use in laboratories – if a chemist wanted to get a piece of apparatus from the storeroom, he needed a name for it. In some cases, especially in the late nineteenth and early twentieth centuries, manufacturers and sellers evidently went back to the original publication to find out the name of the putative inventor (or first user). Soon some pieces of apparatus had a whole raft of names attached to them, each one signifying some minor difference of detail in its construction. A good example is melting point tubes, which were named after Anschütz, Roth, Streatfeild and Thiele (see chapter Eight). This list also shows how many names attached to apparatus were very obscure in contrast to Liebig or Bunsen.

Such are the advantages of naming apparatus that it is perhaps surprising that the process has declined in recent decades. Part of the reason for this is the growing standardization of apparatus – clearly manufacturers are no longer willing to make several variants of the same thing or to carry the stock required. Technology has also played a role. As new types of instrument are introduced, usually by large companies, they are given either numbers such as the Varian A60 NMR machine (see page 308), or poetical but essentially meaningless names invented by a marketing department. Computers have also played a role in making the use of codes feasible. All apparatus can now be illustrated online, and codes such as AFC-19-238 are now easy to use. Of course the Bunsen burner and Liebig condenser have survived, as has the Erlenmeyer flask in America (in Britain it is simply called a 'conical flask').

Breaking Free

As discussed in this chapter, utilities really do matter. In the hands of the inventive genius Robert Bunsen, the arrival of mains water and mains gas in the Heidelberg laboratory wrought important changes in chemistry. Bunsen and Kirchhoff developed atomic spectroscopy using the hot flame of the new Bunsen burner, thereby transforming analytical chemistry and astronomy. At the same time Bunsen also invented a filter pump that utilized the pressure in mains water, thus enabling organic chemists to produce high-quality crystals quickly and cleanly, turning synthetic organic chemistry (and qualitative organic analysis) into a science based on crystals.

From the perspective of the history of the laboratory, however, these new utilities (and here we must include piped steam) enabled laboratories to break free of the furnace – even Giessen needed furnaces to power the fume cupboards. The way was now free for the laboratory architect to plumb water, gas and even electricity (from a central battery) to each bench place. The fume cupboards could be operated using a small gas flame. It was now possible to heat large laboratories with piped steam and introduce gas lighting. With the addition of plumbed-in waste disposal instead of the barrels used in Giessen and Heidelberg, the scene was set for the arrival of the classical chemical laboratory.

Compelling reasons were needed by states in order for them to provide the money to build these scientific behemoths. These states, initially mainly German states, had to be convinced that chemistry was central to agricultural and industrial progress; that in supporting chemistry they would gain an economic edge over rival states. This process was already underway, as Bavaria had enticed Justus Liebig to Munich and Bunsen had been persuaded to move to

Baden. By the early 1860s Prussia sought to overtake its rivals in chemistry, and at the same time states such as Saxony, defeated on the battlefield, redoubled their efforts to compete in this field. The spark that set off this laboratory revolution was the desire of Wilhelm Hofmann to return to Germany after the death of his patron Prince Albert at the end of 1861, coinciding with two leading chemistry chairs in Germany becoming vacant.

6

The Chemical Palace
Wilhelm Hofmann and Berlin, 1860s

'This is a mansion of a noble kind – a palace', exclaimed the flamboyant editor of *Chemical News* William Crookes in his magazine in January 1869.[1] Crookes, an independent scientist with a small laboratory in his north London home,[2] might seem an unlikely champion of the shift in scale of academic laboratories that took place, principally in Germany, in the late 1860s, but his enthusiasm for the new laboratory building in Berlin was unbounded. He declared that 'the idea is destroyed that a cellar or any hole may do for a laboratory'. Before this time, the laboratory was remarkable only for being a dedicated space for scientific work and teaching. Justus Liebig's laboratory in Giessen and Robert Bunsen's laboratory building in Heidelberg may have been groundbreaking in other respects, but they were fairly small buildings.

A sequence of events between 1863 and 1867 led to the construction of three excellent large laboratory buildings in Germany. In 1863 Karl-Gustav-Christoph Bischof, professor of chemistry at the University of Bonn, retired and his chair had already been offered by the Prussian government to Wilhelm Hofmann. In addition to an excellent salary, Hofmann also demanded the most expensive and up-to-date laboratory building in Germany.[3] The university architect August Dieckhoff was appointed to design the new laboratory and he worked well with Hofmann. There is little doubt, however, that the moving force behind the grand design was Hofmann himself, the son of the architect who designed the extension of Liebig's laboratory in Giessen. In the autumn of 1863 he travelled across Germany and Switzerland, visiting his alma mater, Giessen, Lang's laboratory buildings at Karlsruhe and Heidelberg, Liebig's new laboratory in Munich, Göttingen, Zurich and finally the 'splendid institution just completed' at Greifswald, a small town on the Baltic that was, significantly, in the same state as faraway Bonn (illus. 67).[4]

Dieckhoff designed the laboratory building as a Neoclassical mansion not dissimilar to the villa in Bonn that he designed for the businessman Albrecht Troost in 1861 (which later became the official residence of the President of the Federal Republic).[5] It is unsurprising that Dieckhoff produced a Neoclassical palace (illus. 68), as he had studied architecture in Berlin under the great Neoclassical architect Karl Friedrich Schinkel.[6] As the contemporary laboratory building at Leipzig shows, however, the Neoclassical style was considered appropriate for chemical laboratories, on the one hand linking them to palaces and grand museums, and on the other to the intellectual lineage of ancient Greece and the more recent Enlightenment. This connection between Neoclassicism and science is also shown by the classical frontage of the RI in London, which was installed in 1837. While the design was formally made by the architect Lewis Vulliamy, Michael Faraday played an important role and insisted on Corinthian columns, possibly inspired by a visit to Rome he had made with Humphry Davy in 1814.[7]

67 Teaching laboratory ('Der grosser Arbeitsaal') at the University of Greifswald, 1864.

Hofmann's situation was then transformed by the death of Eilhard Mitscherlich, professor of chemistry at the University of Berlin, in August 1863, which was quickly followed by the demise of the other professor of chemistry at Berlin, Heinrich Rose, almost exactly five months later. Berlin became irresistible to Hofmann. Not only did the university have a good reputation, but by this time it was clear that it was becoming the most important city in Germany, at least in political terms. Hofmann negotiated another lavish laboratory building as part of the deal for taking the Berlin chair, and extracted a promise from the Prussian government that the Bonn laboratory building would still be built. In the meantime, Hofmann kept his option on the Bonn chair in case he did not like Berlin (and perhaps also to ensure that the government kept its promise). Eventually, in 1867, August Kekulé (1829–1896) was appointed to the chair at Bonn, as its splendid new laboratory building neared completion.

Although Hofmann's new chemistry laboratory building in Berlin was erected on a constricted site just off the famous Unter den Linden, it was on a completely different scale from its predecessors. Once again it was designed by the university architect, in this case Friedrich Albert Cremer, who was already famous for the new School of Anatomy. Although Cremer was also a student of the Berlin Bauakademie, he was a generation younger than Dieckhoff and he adopted the Rundbogenstil (round arch style) and the related Romanesque style of August Soller and his fellow student

68 Frontage of the Chemical Institute in Bonn, *c.* 1903.

August Orth.[8] One cannot help thinking that the Rundbogenstil, based on early Christian basilicas, worked better for churches than for laboratories – Cremer later restored Limburg Cathedral. Furthermore, the cramped site of the Berlin laboratory did not lend itself to grand gestures, so he focused on the ornamentation of the facade, which Hofmann described, probably tactfully, as 'rich, but simple and tasteful'.[9] One feature of the facade was medallions of famous chemists, the selection of which gave Hofmann considerable trouble.[10] It was decided that contemporary chemists would be given busts in the entrance hall (understandably as later chemists could be easily added to the collection) and historical chemists would be added to the facade. Given Hofmann's internationalist views, it is not surprising that foreign chemists (mainly French and British) greatly outnumbered the German chemists. Both the placing of busts in laboratories and putting names, medallions or busts of famous chemists on the walls were widely copied, notably in Aston Webb's buildings for the Royal College of Science at South Kensington four decades later.[11]

By now, of course, Hofmann was experienced at designing laboratories. Crookes declaimed:

> Let us take Jermyn Street School of Mines, and multiply it by from two to three, and probably the total space occupied by the new laboratory will be found. And yet the laboratory in Jermyn Street is only a small room in a corner.[12]

Yet even before Crookes had published his article, an even larger laboratory was opened in the Saxon city of Leipzig.

These new laboratory buildings represented the culmination of two centuries of development of the chemical laboratory. They would have been recognizable to any chemist of the twentieth century in a way that Lavoisier's or even Liebig's Giessen laboratory would not have been. This similarity was on two levels. The buildings combined teaching and research to create an institution in which many chemists would spend their lives, leaving only to eat, drink and sleep (and some chemists such as Robert Burns Woodward argued even that was only barely acceptable). At the level of the individual laboratory, they contained almost everything that a twentieth-century chemist associated with the term 'chemical laboratory': fume cupboards, rows

69 Chemical laboratory, Eidgenössische Technische Hochschule, Zurich, 1905.

of laboratory benches each supplied with gas and water taps, tiers of bottles on the benches and washbasins at the ends of the benches (illus. 69). On the other hand, these buildings were so large that they contained many specialized rooms such as darkrooms for spectroscopes and polarimeters. A particular feature of the new laboratories was the provision for dangerous experiments ranging from open-air balconies, '*Verbrennungsnische*' (draught closets) and '*Stinkzimmers*' common in the nineteenth century, to the fume cupboards that are found in every laboratory today.[13]

One novel feature of the new laboratory buildings was plumbed-in waste disposal. As in the case of gas and water, this aspect of the laboratory was dependent on the wider infrastructure of its location. It was only after city-wide sewage systems were introduced in the 1850s and '60s to prevent diseases such as cholera and the pollution of waterways that it became possible to remove waste water from the laboratory continuously, rather than by emptying barrels every day. Through forced ventilation, fume cupboards and plumbing, all laboratory waste was now being swept away into the atmosphere and the sewer. Add the improved heating and lighting, and laboratories were becoming much more comfortable places to work in.

Why Germany?

The most important reason for the building of the new chemical palaces in Germany was the political and economic situation in the region. As the rise of Prussia and eventually German reunification (effectively from 1866 onwards) made open warfare impossible, rivalry between the states was carried out by other means. The smaller German states were also concerned about being able to compete economically with Prussia, with its coalfields in Silesia and Westphalia. Faced with a rapidly growing population, the states were additionally anxious to find work for their populations, rather than risk their internal migration to Prussia or external emigration to the United States. Furthermore, the ruling class was worried about revolts provoked by hunger as a result of the population growth, and wanted to promote higher yields in agriculture as a precaution. Germany as a whole was worried about military and economic competition from Britain and France, and increasingly the United States. Science was seen as the solution to all these problems and chemistry had a clear utility value, not only for what chemistry itself could produce, but also as a service science for important industries such as agriculture, coal distillation, steel manufacture and railways.[14]

Before the revolution of 1848, Saxony, Bavaria and Hanover had been important members of the German Federation; they were not as large or important as Prussia or Austria, but as long as a balance existed between the two largest states, the medium sized states could retain their independence and international standing. After 1848 the balance began to swing in favour of Prussia, especially after the appointment of Otto von Bismarck as minister-president in 1862. After the war of 1866, Hanover lost its independent existence and the other states effectively became satraps of Prussia, albeit with nominally sovereign heads. All the other German states had initially reacted to the rise of Prussia by seeking economic development by any means, including scientific research. Bavaria, for example, attracted Liebig to Munich in 1852, and the chemical company BASF to Ludwigshafen in 1865. After the accession of Ludwig II, however, the powerless monarch found compensation in building elaborate palaces, thereby nearly bankrupting the state. Saxony, by contrast, continued down the road of supporting science and industry, of which Hermann Kolbe's laboratory was a leading example. Rather than erecting quasi-medieval palaces, the Saxony of King Johann constructed a series of scientific palaces at the University of Leipzig.[15] Adolphe Wurtz remarked in a letter to the French Minister of Public Instruction that Saxony 'seems intent on conquering, in scientific and intellectual authority what it has just lost in political autonomy'.[16] At the same time, of course, the rising power of Prussia was constructing similar edifices in Bonn and in Berlin, which was soon to become the Imperial capital.[17] While interstate rivalry had long been a factor in promoting universities and science in particular in the German states, the great laboratories of the 1860s marked a major change of scale.

Thanks to the existence of the German states, German science escaped the central planning of countries such as France and to some extent even Britain. This provided room for experimentation, and the existence of other excellent laboratories nearby – not only in Germany but also in Switzerland – enabled professors seeking to erect new laboratory buildings to find out what was effective in these buildings and choose the best design elements from them.[18] Furthermore, German chemistry, in contrast to French chemistry, was still outward looking. German universities welcomed foreign students and German chemists went abroad at least up to the 1860s.[19] Germany also benefited from a generation of enthusiastic and well-trained chemists who had learned their craft at the pioneering laboratories of Liebig or Bunsen. Other countries, including Britain, the United States and France, also had students from Liebig and Bunsen, but inevitably not in such numbers.

Because of the existence of rival universities in competing states there was fierce competition for professors in Germany in much the same way that American universities now compete for elite football players. To attract the best professors the various states had to offer the finest new laboratory buildings and the funds to run them. Due to the ability of German students to move freely between universities the professors had to work hard to attract – and crucially keep – the best students by offering the best facilities. The market in foreign students exerted a similar pressure. All these pressures, arising from competition, inexorably led to the construction of the 'chemical palaces' in the 1860s in Germany. It is obvious that this situation had a lot to do with the state of affairs in Germany and relatively little with the current position of chemistry. The synthetic dye industry, which was so important by the 1880s, was too small in the 1860s to play a major role, but the potential represented by the synthetic dyes produced by Hofmann and his fellow organic chemists made chemistry more attractive to the competing German states. Once established, the German organic chemistry institutes were supported by the rising dye industry from the 1870s, as their profits grew and the industry began to consolidate around a number of firms that were able to make a success from the manufacture of synthetic alizarin.[20]

An Ambitious Chemist

Chemistry changed dramatically between the 1840s and '70s. The leadership of the field moved from France to Germany and laboratory buildings became much larger. Organic chemistry became the predominant sector of academic chemistry and was increasingly concerned with organic synthesis and aromatic chemistry. Chemical structures became a crucial tool for the organic chemist.[21] Aromatic chemistry (the study of compounds containing a benzene ring) was associated with the rise of the synthetic dye industry, in which Germany also became the world leader in the 1870s. Remarkably, all these changes were to some degree associated with one person: (August) Wilhelm Hofmann.

Hofmann was born in Giessen in 1818, only seven years after Bunsen but effectively a generation younger.[22] As stated earlier (see page 96), his father was the architect of the extension to Liebig's laboratory in 1839 and Hofmann entered this laboratory in the very same year, despite having enrolled at Giessen to study law and languages. A former student of Liebig, Ernst Sell, owned a tar distillery and sent samples to his old laboratory for analysis. When Liebig gave these samples to Hofmann, he unwittingly changed his student's life and created a new industry. As a student of Liebig, Hofmann was interested in organic compounds containing nitrogen. Alkaloids were biologically active, expensive and contained nitrogen. Other compounds containing nitrogen might be alkaloids, be as active as alkaloids or at least might be capable of being turned into alkaloids. After extracting a kilogramme of chemically basic material from Sell's coal tar, Hofmann found a nitrogen-containing basic compound that had been isolated before from other sources (such as indigo), but only in small quantities. He now focused on the reactions of this aniline (as it was soon called) and received his PhD for his studies in 1841. In order to understand the relationship between the different aniline compounds, Hofmann now turned to the new type theory of chemical structure. This theory postulated that organic compounds were variants of a simple compound, most notably water and ammonia. Ethers were the water type with two alkyl radicals (or groups) replacing the hydrogens of water, and the (theoretical) replacement of one or more hydrogens in ammonia produced amines. The type theory also allowed the development of simple structural formulae with curly brackets joining the alkyl radicals (and/or hydrogen) with the oxygen or nitrogen atom.[23]

At this point Hofmann found a junior position at Bonn University, but his life was about to take another unexpected turn. The story goes that during a Beethoven festival in Bonn, Prince Albert decided to look up his old rooms at the university and found them occupied by Hofmann.[24] According to most authors, Prince Albert offered the post of director of the new College of Chemistry in London to Hofmann on the recommendation of Liebig.[25] Either way, Hofmann left Bonn for an uncertain future in London's Mayfair; indeed, it was so uncertain that he only took the post after being offered two years unpaid leave from his university. While in London Hofmann

70 *Carte de visite* photograph of Wilhelm Hofmann, possibly taken just before he left for Germany, *c.* 1865.

continued to explore the chemistry of aniline and the other coal-tar compounds, notably benzene. His students included Frederick Abel, who invented cordite; Charles Blachford Mansfield, who pioneered the commercial distillation of benzene (but was burnt to death as a result);[26] William Henry Perkin, who founded the synthetic dye industry, and William Crookes, who discovered thallium and set up *Chemical News*.

Prince Albert died in December 1861 and it was becoming clear that changes – both political and chemical – were taking place in Germany. Hofmann then decided to return to his old university and, as already mentioned, was offered the chair and a new laboratory (illus. 70). In a farewell lecture at the RI, with the Prince of Wales in the audience, he displayed his new molecular models made with table croquet balls.[27] Most of the colours he used from his table croquet set remain the colours used by molecular modellers. Having moved to Berlin rather than Bonn, Hofmann developed his position there by engaging in both politics and industry. He ensured the pre-eminence of Berlin as a centre of academic chemistry and became an advisor to the Imperial government on chemical matters, for example the patent law of 1877. He was ennobled as August Wilhelm von Hofmann by the Emperor Friedrich – the husband of Vicky, the daughter of Queen Victoria and Prince Albert – during his short reign in 1888. Hofmann worked closely with the Berlin dye makers AG für Anilinfabrikation (Agfa) founded by Carl Martius, who had known Hofmann in England, and one of Hofmann's students, Paul Mendelssohn Bartholdy (son of the composer), to make new dyes and to discover the composition of dyes made by rival firms such as BASF. Having served as president of the Chemical Society in London in 1861–3, Hofmann sought to build up the German chemical community, and he was the first president of the German Chemical Society when it was founded in 1867. He became its president for the fourteenth time in 1890, not long before he died in May 1892.

Aisles and Benches

The most striking feature of the new laboratories, which is rarely commented on, is their layout. For the most part earlier chemical laboratories were organized either along the edges or around the

71 Bench and bottle racks at Leipzig University, 1868.

centre; or indeed both, as shown in a striking manner by the Zurich laboratory of the 1840s with its central-island fume cupboard (see illus. 48). Even Bunsen's laboratory had the benches in the centre of the room. By contrast, the Berlin and Leipzig laboratories of the 1860s had an aisle down the middle with doors at each end, and no benches or other functional features around the walls except fume cupboards. The symmetry of the laboratory changed from being centrosymmetric to twofold reflection symmetry. In fact these laboratories had a layout similar to that of hospital wards. Just as the beds in a ward were placed close together but far enough apart to allow staff to move in and out, so the benches were placed relatively close together, but with a gap wide enough to allow students or researchers to work at both benches. Above all, the classical laboratory and the classical hospital ward had a wide central aisle that allowed the consultant or professor, accompanied by his retinue of assistants, to progress from bed to bed or bench to bench.[28] Of course, the wide aisle was also convenient for the movement of people or equipment, whether it be a medicine trolley, or a trolley of apparatus or Winchester (Boston round) bottles. This new arrangement enabled professors to keep a much closer eye on their students – the professor could easily move from bench to bench, but with their vision blocked by bottle racks the students would be caught unawares. Almost a century later the famous organic chemist Derek Barton perfected this system by wearing crepe-soled shoes that allowed him to creep up behind students unheard, then hiss 'show me the crystals!'[29]

However, the most iconic feature of the 'classical' laboratory is the bench and bottle rack (illus. 71). From first entering the school laboratory to the day they retired, late nineteenth- and twentieth-century chemists were likely to spend time working at the bench, using reagents from bottles lined up on a shelf at eye level in front of them. The bench clearly evolved from the long tables and counters of eighteenth-century laboratories, but the laboratory bench in the laboratories of the 1860s and '70s was a relatively new invention. As described earlier (see page 114), the first laboratory benches with a central divider and cupboards with doors under the bench rather than table legs arose at the Pharmaceutical Society in London in 1844, and the Birkbeck Laboratory of UCL in 1846. While these benches were similar to the classical bench in almost all respects, they do not look like the classical bench as portrayed in an engraving of the Leipzig bench (see illus. 71). Certain features of the classical bench are still missing in that period, specifically the two-tier bottle racks and the washbasins.

Clearly, the tops of these benches had to be impervious to liquids and resistant to acids, alkalis and the heat from the burners.[30] Most laboratory furniture was made from deal, a wood obtained from pine trees and the material used for old-style school desks. Berzelius's table was made from deal, but at least today it is coated in a light blue paint.[31] Deal was often used as a cheap option for the bench top as well, but Edward Robins claimed that 'the economy is more apparent than real'.[32] Rather than mahogany, as one might have thought, teak was considered to be the best wood, for example by Henry Roscoe. Rangoon teak was also recommended by Alan Munby.[33] Despite Roscoe's preference for teak, Owens College used oak, presumably due to cost. Nottingham University College was unusual in using American walnut for all its laboratory furniture including the bench top. The wood was usually sealed by ironing in paraffin wax, as natural oils and waxes were attacked by acids and alkalis.

The formerly asbestos mat (called 'asbestos cardboard' by Robins), ubiquitous in laboratories up to the 1970s, was introduced to prevent the wax being melted by heat, rather than to protect the wood itself. Robins favoured lead sheets on top of the bench, as at Finsbury College and BASF, but many chemists objected on the grounds that the lead would buckle ('blow') under a burner and that glass objects would break on a lead surface. Any blowing could be prevented by using an asbestos mat, and Henry Armstrong dismissed the danger to glass as imaginary. Robins stated that slate was not suitable for academic laboratories, but it was used in the United States as it was cheap.[34] However, it was easily broken by the heat reflected by the evaporating dishes. Soapstone benches were popular in the eastern USA; most recently under the trade name Alberene.[35] Tiled bench tops were apparently popular in France.[36] It is stated that birch or maple strips were used for laboratory bench tops in the 1950s, at least in America.[37] They were treated by alternating coats of potassium chlorate/copper sulphate and aniline oil/hydrochloric acid, before being blackened with potassium dichromate solution. It has been claimed that black benches were not popular in Britain and Continental Europe, but the same recipe is given in a book published in Britain (written by an Australian), although William Ferguson remarked that the black bench had become a matter of debate.[38] After the Second World War compressed asbestos cement compositions (under trade names such as Colorlith or Coloceran) became popular because they were cheap.[39] Glass bench tops had existed in the nineteenth century, but since the early 1960s a heat-resistant glass ceramic material (Pyroceram) has been used for benches.[40] Modern laboratory benches are usually made from an epoxy or phenolic resin composite material.[41] The floors of fume cupboards were more often protected by stone, tiles or lead than were the open benches.

Bottle Racks

So when did the now-familiar bottle racks come into use? Effectively there are two types of bottle rack. One lines the walls of the laboratory; the other is the bench-top shelf used by the individual student or chemist. Wall racks and rudimentary bench racks can be seen in most pictures of laboratories in the 1840s. The Scottish chemist Andrew Ure, in 1821, remarked in the entry on the laboratory in his dictionary:

> To the walls of the laboratory ought to be fastened shelves of different breadths and heights; or these shelves may be suspended by hooks. These shelves are to contain glass vessels and the products of operations and ought to be in as great a number as is possible. In a laboratory where many experiments are made there cannot be too many shelves.[42]

In Faraday's laboratory at the RI in the 1840s the wall racks were very prominent and looked similar to bookcases, from which they were surely derived (see illus. 30). Faraday did not use a bench for experimental work, so the question of bench racks did not arise. By contrast, in Liebig's new laboratory in Giessen in the same period, there were laboratory benches along the wall – although worktables were also used – so bench racks effectively displaced the wall racks. This was also the case at the Pharmaceutical Society in London, although the lower shelf seems to have been used for apparatus rather than reagents (see illus. 56). In the illustration of the Birkbeck Laboratory of 1846 (see illus. 55), both types of rack can be clearly seen in use for the first time, with very high shelves on the walls for the large bottles and individual shelves above the benches for the students' own bottles (although this is far from clear in the rather flat depiction of the bench in the engraving). In Bunsen's laboratory at Heidelberg, chemicals were apparently stored in glass-fronted cupboards (rather like Victorian china cabinets) on the bench rather than in racks.

It is only in photographs of the new chemical palaces – at Berlin and Leipzig – that we first encounter the classical two-tiered bottle racks above the bench and the absence of wall racks in the main laboratory, the large bottles either being stored in cupboards under the bench or in dedicated store rooms. To some extent this is surprising, as bottle racks make more sense in undergraduate teaching laboratories where each student is given a uniform set of reagents, usually of a standard concentration. This is particularly true for qualitative inorganic analysis. I recall such standard sets when I was a schoolboy and an undergraduate in the 1970s. Graduate-level organic chemistry does not require such standardized sets, and in any event the volume of liquids required as reagents and solvents was usually greater than the 250 ml bottles generally found in bottle racks, at least before micro-scale preparations became fashionable in the 1960s. Needless to say, some branches of chemistry, notably physical chemistry, do not need bottle racks at all and just use plain benches (or tables). Probably for this reason, some laboratories, especially American ones, have metal racks that may have been removable.

In the picture of the Leipzig bench (see illus. 71) there is a pipe from the washbasin to the floor. This raises the question of the disposal of waste water and other liquids in the laboratory. In his dictionary Ure also mentions the sink:

> The most convenient place for a stone or leaden cistern to contain water is a corner of the laboratory and under it a sink ought to be placed with a pipe by which the water poured into it may discharge itself.[43]

Where did this pipe discharge itself? Ure does not say, but as noted the laboratories at Giessen and Heidelberg used wooden barrels. Further light is shed on the matter by the American chemist Campbell Morfit in a long discussion about the laboratory sink:

> The waste pipe which must be constructed so as to admit of the free discharge of the waste water is led by a gradual descent into a drain which conveys its charge into cess pools or tanks lined with bricks and sunk into the ground. As the emanations of foul air from these pools are noxious they should be placed some distance from the building and kept well covered. If the situation be favorable the drains should empty themselves into a gutter or some running stream which in conducting away the foul matter would relieve the air of the apartments of its noxious effluvia.[44]

By the 1860s, however, public sewers were becoming common and it is very likely that the waste water in Bonn, Berlin and Leipzig flowed into the drains. Hofmann refused to shed light on the matter until such internal arrangements were 'completed and rigorously tested', but it is clear from Festing's description that waste did in fact flow into the drains.[45] The earliest such arrangement mentioned by the architect Edward Robins in his handbook on technical schools was at the new laboratories at Munich constructed by Adolf Baeyer in 1877. The organic chemistry laboratories there used a covered gutter between the bench and the floor:

> The floors for the sake of warmth, are made of wood, with an asphalte border round, three feet wide, on which stand the wooden sinks and digestoriums or draught-closets [fume cupboards]. In the middle of this border is an asphalte gutter-channel; this gutter is trapped at the two opposite ends, and communicates with the main drain. This channel carries away the wastes from sinks and draught-chambers, and also any floor-droppings, being covered only with perforated boards. This arrangement is said to work well in practice, and allows of carrying on large operations in the immediate neighbourhood of the work-tables; the central avenue being the general thoroughfare.

This system may have worked well, but it was probably liable to liberating volatile vapours and possibly mercury vapour into the atmosphere of the laboratory. It is perhaps not surprising that W. H. Perkin, Jnr, a student of Baeyer who copied this arrangement in his laboratories at Manchester and Oxford,[46] probably died of mercury poisoning. Given that smoking was not usually banned in laboratories in this period, ether vapour rising from the gutter would pose a particular risk of fire or explosion.

Specialized Rooms

Another notable feature of the laboratory buildings in Berlin, Bonn and Leipzig was the large number of specialized rooms they contained. Apart from the library, these were generally created for the carrying out of operations that were not suitable for the standard laboratory. Some of these operations were hazardous, such as the making up of chemicals for lectures or experimental use, and the generation of hydrogen sulphide for inorganic analysis in the *Stinkzimmer* – or an even more specialized hydrogen sulphide room or bench-top cabinet.[47] Another dangerous activity requiring its own room was the combustion analysis described in chapter Four, which

72 Combustion analysis room, Eidgenossische Technische Hochschule, Zurich, 1905.

involved heating the powerful oxidant potassium perchlorate with organic compounds – a potentially explosive mixture (illus. 72).

The use of explosive mixtures and so-called bomb ovens, and the heating of sealed tubes, led to the development in the mid-nineteenth century of the ballistic cabinet, or *Schießschrank*, to protect chemists against the risk of injury from an explosion.[48] Unlike the fume cupboard, there was no standard design for these cabinets, but they typically featured strong doors and sides, and were made of metal, strong wire mesh or thick wood, with metal plates to divide them into separate compartments. Sometimes, as at Bonn, a slit covered with mica would be added so that the experiments could be viewed safely. The rooms in which the ballistic cabinets were located were themselves strengthened and equipped with heavy doors.

There was also usually a room for the gas analysis as pioneered by Bunsen. Other rooms involved the use of delicate and expensive equipment, notably the chemical balance in the balance room. This period saw a great transformation in the chemical balance.[49] Since the 1780s the balance's sensitivity had been improved by making its arms very long (the crossbeam could be over a metre long) and also very thin, to keep the weight down. This created a balance that was very susceptible to draughts, and also very slow to settle down after the beam had been raised to make the measurement. Thus balances had to be protected from open windows and vibrations. From the late 1860s onwards the short-beam balance, first developed in 1866 by the German engineer Paul Bunge, was commercialized by firms such as Sartorius, and soon replaced the inconvenient long-beam balance. Bunge ironically made no major breakthrough in balance design. He simply decided that conventional wisdom about beam length, first laid down by Leonhard Euler in 1738, was incorrect, and made a beam that was light simply by being short. He then strengthened it with triangular struts, drawing on his experience of designing bridges. The short-beam balance made the balance room more, not less important. Although the original short-beam balances, using expensive newfangled materials such as aluminium, were not cheap, they were less expensive than the older balances and more of them could be accommodated within a given space. Hence there were more balances, and the balance was used when previously a simple scale would have sufficed.

The darkroom was a new feature of the laboratory in the 1860s, and it arose for two completely different reasons. One, as we have seen, was relevant to spectroscopy. To see the spectrum of the substance under investigation a darkroom was clearly needed. The other new instrument was the polarimeter, which measures the change in the polarization of light as it passes through a solution. To do so, it uses a strong light source such as a sodium flame (replaced when electricity became available by a sodium lamp). The light from the flame is polarized using a prism made from Iceland spar (a Nicol prism), and it passes through the solution tube before hitting another Nicol prism. If you use water in the tube and the two prisms are at right angles to each other, no light will emerge from the polarimeter; this is the null point. If the water is then replaced by an optically active solution that bends the plane of polarization, the second prism will have to be turned to either the left or right to get back to the null point. After the discovery of optical activity by Jean-Baptiste Biot in 1815, it was discovered that some organic compounds were optically active.[50] The sugars were the most common example of such compounds. The polarimeter can be used to measure the angle of

73 Schmidt & Haensch polarimeter, c. 1900 (inv. no. 1963-0278).

rotation of a beam of light passing through a solution of an optically active compound, hence its concentration if the angle of rotation of a standard solution is known. Ludwig Wilhelmy in Berlin used an early polarimeter to measure the rate of inversion of glucose (the acid-catalysed loss of optical activity) in 1850.

Early polarimeters were relatively crude, but Eilhard Mitscherlich introduced the set-up with two Nicol prisms in 1844. The classical saccharimeter (polarimeter), with its prominent brass circle, was introduced by Leon Laurent of Paris in 1874, followed by his 'large model' of 1888. By the 1890s the Berlin instrument firm of Franz Schmidt and Hermann Haensch (founded in 1864) was producing beautiful quartz wedge polarimeters of its own design (illus. 73).[51] Even after the tetrahedral carbon atom was proposed in 1874 by the Dutch chemist Jacobus van't Hoff (and independently by the French chemist Achille Le Bel), the main use of the polarimeter remained the measurement of sugar concentrations – the saccharimeter.[52] Even in the academic chemical laboratory it was employed chiefly by chemists studying the structure of sugars. Much progress was made on the chemistry of sugars in the late nineteenth century, culminating in Emil Fischer's masterly study in 1894 of the aldohexose family of sugars (the most prominent member of which is D-glucose). In this he not only correctly identified the sixteen stereoisomers of the family, but also gave them their correct absolute configuration – despite only having a 50 per cent chance of getting it right, as there was no method then available for discovering this experimentally.[53]

It is obvious that access to these specialized rooms had to be controlled – otherwise unexpected light would enter the darkroom, or the vibrations of people walking through the balance room would upset delicate weighing operations. This was a problem in the Berlin laboratory. Due to its confined site in the middle of Berlin, several rooms had to serve as makeshift corridors and, remarkably, the balance room was one of them.[54] To make matters even worse, the two adjoining laboratories were used for fusions and ignitions, and Robins witnessed a fire that ran along the floor into the balance room. Robins argued that this made Berlin an unsatisfactory model for other laboratory buildings: 'it is now regarded as by no means an example to be imitated even by Berliners'.[55] Despite this, Berlin was seen as a model to be emulated (see chapter Seven). A distinction has to be made between the overall scheme, which was admired, and specific details of the design created by the cramped site, which could be avoided elsewhere.

The Director's Residence and the Spy Window

In his biography of Kolbe, Alan Rocke describes the accommodation in the new laboratory building at Leipzig:

> Fully a seventh of this space was taken up by an opulent director's residence along the front wing of the building, open to light and air on three sides, and consisting of fourteen large rooms plus a cellar and attic, with a small garden outside. A hundred feet of second-floor corridor separated Kolbe's airy private study in his residence from his well-equipped private laboratory.[56]

Hofmann described the accommodation at Bonn in similar terms as

> a magnificent residence for the director consisting of a suite of rooms which as regards number and size could be very seldom met with in a private house.[57]

Later in his report he remarked that 'the director's spacious residence is as richly ornamented, and will in all respects be worthy of the institution to which it belongs'. Hofmann could not resist noting that it enjoyed an 'imposing entrance hall, illuminated by a glass cupola above, and the splendid ballroom', which amply satisfied 'the social requirements of a chemical professor of the second half of the nineteenth century'.[58] Curiously, he is much more reticent about his own new living quarters overlooking the Dorotheenstrasse in Berlin! There had been a tradition in Germany for the chemistry professor to live above or at least near the laboratory building. This had two advantages. The professor was close by if the laboratory was on fire, as Kolbe pointed out.[59] As Jean-Baptiste Dumas was acutely aware, 'living over the shop' also removed the need to commute to work.[60] By the 1860s the professor's accommodation had become an important aspect of the laboratory's design. It was palatial in both its scale and fittings. Hofmann ensured that his new laboratory was placed next to the Academy of Science and close to the royal palaces and the Treasury. The chemistry professor had become a scientific prince or general. His ennoblement, as in the case of Liebig, was a natural consequence of this shift in status.

This was at least partly a consequence of the growth of chemical research groups. Until the 1850s, chemistry professors had relatively small groups of graduate students and co-workers, most of whom moved to another university or a non-academic career after a few years, thus spreading their professor's influence, but keeping the size of his own group small. Dumas had about twenty chemists in his group, and Liebig perhaps rather more by the late 1840s. With the rise of organic chemistry, which was more labour intensive, and the funding for academic chemistry increasing, leading professors took on more students and many of their co-workers remained in their laboratories for their entire careers. The group thus became a large and relatively stable community, with technicians and cleaners at the bottom, graduate students and junior demonstrators (*Privatdozents*) not much higher, then the general teaching staff and the superintendent of the laboratory, and at its apex the leading professors.

One of the most striking features of some late nineteenth-century academic laboratories was the so-called 'spy window' in the director's office, which enabled him to look out over the teaching laboratory.[61] This is reminiscent of Jeremy Bentham's 'panopticon', insofar as the students would have presumably believed that they were being observed even when the professor was actually absent. The first major laboratory to have anything like this spy window was Giessen, where there was a hatch between the professor's office and the main laboratory. Although this has been called a spy window, it neither overlooked the laboratory (it was on the same level), nor opened out onto a general teaching space. The genuine spy window originated at Kolbe's new laboratory in Marburg in 1863. It was described as 'a great glass wall'. Roscoe adopted this feature for his laboratories at Owens College ten years later, and he considered it to be important enough to mention in his evidence to the Devonshire Commission.[62] The spy window was then faithfully copied at the new university laboratory in Newcastle, but not at Leeds. The spy window is sometimes considered to be a 'Germanic' feature of the laboratory, but it does not appear to have existed in Kolbe's laboratories in Leipzig; nor was the spy window a feature of the Hofmann-designed laboratories at Bonn and Berlin.[63]

According to Yoshiyuki Kikuchi the importance of the spy window lies in the light it sheds on the system for the supervision of students in the laboratories.[64] In the 'Marburg system', the responsibility for the supervision of all students lay with the senior professor; hence the window was needed to enable a busy professor to keep an eye on his students' experimental work. By contrast, in the 'UCL system' this responsibility lay with the junior staff, the demonstrators and lecturers, and it was their workplaces that were next to the student laboratories. This was the arrangement at the Birkbeck Laboratory at UCL, and it was copied at the Imperial University in Tokyo (see page 185). Furthermore, the oversight by these demonstrators was on the laboratory floor rather than from a higher window.

A Model Established

In the 1860s two important German organic chemists, Hermann Kolbe and Wilhelm Hofmann, brought together several innovations in laboratory design that had been introduced in the previous three decades, and used them to build monumental laboratory buildings that became models for other laboratory designs. These designs are still familiar to us today because they have been copied (and modified) so many times. That is not to say that Berlin or Leipzig were the only models; Heidelberg in particular was also copied. However, Hofmann was so influential both as an individual and as a symbol of German supremacy in organic chemistry that Berlin became the archetype for chemical laboratory buildings everywhere.

One of the most important aspects of the chemical palaces was the creation of specialized rooms. Of course, there was nothing new in the concept of such rooms – Bunsen had created rooms for gas analysis and electrochemical work. However, their number grew and they became standard features of most laboratory buildings rather than rooms put aside for the professor's own specialisms. On one level this development may seem trivial, as it is obvious that special rooms are needed for spectroscopes or balances, but this increased specialization of laboratory space ironically made these laboratory buildings more flexible. When new types of chemical practice arose, they could be accommodated in such specialized rooms, thus keeping the basic model intact (see chapter Eleven). The scale of these laboratory buildings was also very influential. As Crookes remarked, no cellar or hole would in future suffice as a laboratory. After the 1860s, any state or university that wanted to attract the best chemists and gain prestige from its laboratories would have to construct large laboratory buildings rather than simply fit out a room, as Union College in the United States had done in 1857 (see chapter Seven). At least in this sense, chemistry had indeed become 'big science'. How this 'big science' was transferred abroad is the subject of the next chapter.

7

Laboratory Transfer

Henry Roscoe and Manchester, 1870s

The classical laboratory design explored in the last chapter had become well established by the early twentieth century, because it was the first truly generic laboratory design, copied in numerous countries including Britain, the USA, Japan and eventually France. This was a break with earlier practice. Most laboratories were still designed from scratch in the 1850s, and while they might incorporate features of earlier laboratories, the architects avoided producing facsimiles. Giessen was the most famous laboratory of the 1840s, and it was copied by several universities including the University of Zurich.[1] Such laboratories were not clones of Giessen's laboratory, however, and Liebig's laboratory was not widely imitated outside the German-speaking world.

The situation changed in the 1860s. The German laboratory buildings were now seen as the model to emulate. There are two key aspects of the diffusion of the classical laboratory model in the late nineteenth century. The laboratory had become standardized, and its components – central aisle, fume cupboards, benches, washbasins, bottle racks – were easy to copy. As more and more laboratories were fitted out in this way, the components became standard items in laboratory catalogues and information about them was readily available. There were in fact five modes of transmission. Books, generally short or padded out with reports of chemical research, about most of the major laboratories and some of the more minor ones such as Christiania (now Oslo) were published from the time of Giessen onwards.[2] In addition to these published works, chemists and civil servants from different nations would make official visits to the leading laboratories, mostly in Germany and Switzerland, and produce reports on their findings. Professors who had been students in one of the leading laboratories or, as in the case of Henry Roscoe, worked with the professor in charge of such a laboratory, would contact their former *doktorvater* (supervisor) or visit the laboratory – and of course a professor (or architect) planning a new laboratory could visit laboratories without having been there as a student. No barriers were put in their way, as it was accepted practice for the builders of new laboratory buildings to visit the best examples of existing ones before commencing their work. Finally, different laboratory buildings were built by the same architect, who was able to develop a standard model as he constructed each new laboratory building. In Germany (and to some extent America) there was a university architect, or perhaps, as in the case of Heidelberg, a state architect. In Britain, by contrast, universities were free to choose their architects and some firms began to specialize in laboratory design. Two of the leading architects in this field were Alfred Waterhouse of Manchester and from 1865, London, and Edward Cookworthy Robins of London, who was associated with the Dyers Company.[3]

The Country to Emulate

Between the 1840s and '70s, Germany was transformed from a largely rural patchwork of independent states, into a united industrialized empire which had humiliated France on the battlefield. Even more worryingly for its competitors, it had also routed France and Britain in the new synthetic dye industry and showed signs of forging ahead in the electrical field. Moreover, it was not just a matter of industrial progress, German chemists were also dominating academic organic chemistry, a stronghold of the French up to the 1860s. Perhaps a little simplistically, a connection was often made between German advances in organic chemistry and synthetic dyes. Just as the smaller German states had concluded a decade earlier that the only way they could compete with Prussia was to develop their scientific facilities and especially their academic laboratories, Germany's international competitors now came to the same conclusion. To give one example, the British government set up the Royal Commission on Scientific Instruction and the Advancement of Science, chaired by the 7th Duke of Devonshire (and hence known as the Devonshire Commission) to investigate the position of British scientific education and how it could be improved. The commission issued several volumes of evidence and recommendations up to 1875. The chemists appearing before the commission and its own members, notably Thomas Huxley, believed that better laboratories were needed. This commission was followed by the Royal Commission on Technical Instruction Abroad between 1881 and 1884, with an even more specific remit to consider what other countries were doing, above all Germany.[4] As described later (see page 176), even before these commissions got underway the Department of Science and Art planned to copy the new German laboratory buildings in South Kensington.

Hofmann as Prime Mover

What is most notable in this process is the role played by Wilhelm Hofmann.[5] Rather than trying to preserve a technological advantage for Germany by restricting information about his laboratories, from the very outset he published all the relevant facts in his report on the Bonn and Berlin laboratory buildings. The information was immediately put to use in the construction of the South Kensington laboratory building. Even more importantly, Hofmann welcomed visitors to Berlin for more than two decades and provided them with further information for the design of their own laboratory buildings.

Why was Hofmann so willing to assist others? He was generous and sociable, happy to help his fellow chemists. Perhaps his willingness to help architects stemmed partly from his father being an architect. His assistance to Paul Nénot, the architect of the new Sorbonne laboratory, was a personal favour to his lifelong friend Adolphe Wurtz. Having worked in London and with worldwide contacts, he was a true internationalist. Hofmann was, however, also ambitious for himself and for organic chemistry. He was doubtless proud of the attention that his laboratory attracted, which was probably linked in his memory with the visitors to Liebig and Giessen. Even more importantly, the more laboratory buildings that were constructed according to his design, the more firmly entrenched was his approach to organic chemistry as it became embedded in the practice of chemistry. Just as Lavoisier had ensured the triumph of his combustion theory by changing the language of chemistry, Hofmann immortalized his practice of organic chemistry by creating a standard design of laboratory and laboratory building. Regardless of his motivation, his advocacy for the type of laboratory he had designed in the mid-1860s ensured that it became a template for chemistry laboratories across the globe for decades to come. Even the main laboratories in the new laboratory buildings at Stanford University in California in the 1960s had more than a passing resemblance to the laboratories constructed in Berlin's Dorotheenstrasse almost a century earlier.

A Political Chemist

Henry Enfield Roscoe was born in London in 1833, although he was the scion of a well-established dissenting Liverpool family (illus. 74).[6] His family was also literary – both his father and grandfather were writers and lawyers. The economist William Stanley Jevons, who also studied chemistry at UCL, was his cousin, and the writer Beatrix Potter was his niece. His family fell on hard times after his father died in 1837, but he was able to take chemistry as a dissenter at UCL, and he entered the pioneering Birkbeck laboratory under Thomas Graham and Alexander Williamson (see chapter Four). He then went to Heidelberg to study under Robert Bunsen and specialized in analysis. After spending only six months in Heidelberg, he took his PhD in March 1854.[7] He returned to UCL to become Williamson's assistant following Graham's retirement, but continued to work with Bunsen during the summer vacations. At first it seemed that he would pursue a career as a chemical consultant, but in 1857, after Edward Frankland resigned his chair at the newly founded Owens College in Manchester, Roscoe applied for the position. The college was in a state of crisis and for a while it looked as if it might even close down. Roscoe related the situation with his usual humour:

> The institution was at that time nearly in a state of collapse, and this fact had impressed itself even on the professors. I was standing one evening, preparing myself for my lecture by smoking a cigar at the back gate of the building, when a tramp accosted me and asked me if this was the Manchester Night Asylum. I replied that it was not, but that if he would call again in six months' time, he might find lodgings there![8]

Roscoe set about with characteristic energy to persuade local industrialists of the value of a chemical education. By the time he gave evidence to the Devonshire Commission in 1872, his classes were becoming overcrowded and a new laboratory building was needed. Two years later his assistant, the Bunsen-educated German chemist Carl Schorlemmer, became the first professor of organic chemistry in Britain. By building up the chemistry department, Roscoe effectively saved Owens College single-handedly. Appropriately given his literary family, Roscoe developed a reputation as a textbook author, publishing *Lessons in Elementary Chemistry* (1866), which was even translated into Icelandic,[9] and the multi-volume *A Treatise on Chemistry* with Schorlemmer (1876 onwards). His monograph on *Spectrum Analysis* (1869) was also influential. Owens College became part of Victoria University in 1880 and Roscoe retired in 1887.

Having transformed Owens College into one of the leading centres of higher education and chemical research in Britain, Roscoe became involved in the wider development of science in Britain, serving on the Royal Commission on Noxious Vapours between 1876 and 1878, the Royal Commission on Technical Instruction Abroad (Samuelson Commission) between 1881 and 1884, and the Royal Commission on Secondary Education in England (Bryce Commission) between 1894 and 1896.[10] For this public service he was knighted in 1884. He was active in the Royal Society, of which he served as vice-president in 1881–2 and 1888–90, and was president of the Chemical Society in 1880–82. Furthermore, he was the moving force behind the establishment of the Society of Chemical Industry in 1881 and became its first president. He was also president of the British Association for the Advancement of Science when it met in Manchester in 1887. The son of an MP and active in local affairs, Roscoe became the MP for middle-class Manchester South in 1885, when it was carved out of the former multi-member Manchester constituency. Beatrix Potter confided in her secret journal that 'Uncle Harry's position was most extraordinary, the only Liberal member for Manchester, Liverpool, Bolton, Preston and Stockport'.[11] He used his position in Parliament to campaign for the improvement of scientific and technical education, which resulted in the passage of the Technical Instruction Act of 1889. Roscoe defeated Conservative opponents in three elections, but in 1895 the Unionist coalition put up a Liberal Unionist. He was

74 Sir Henry Enfield Roscoe, 1906.

narrowly defeated by John Campbell, Marquess of Lorne, who ironically had to give up his seat in 1900 when he succeeded to the Dukedom of Argyll.

Roscoe then became vice-chancellor of London University between 1896 and 1902, and wrote his autobiography, which is certainly the most humorous example of a genre that is usually dull. He had always been interested in the history of chemistry – he gave a lecture on the history of the chemical elements in 1875 – and published a biography of John Dalton in 1895. With his political connections, Roscoe was instrumental in setting up the Science Museum as an independent institution in 1909 with Norman Lockyer and Robert Morant, perhaps his most enduring achievement.[12] He was made a Privy Councillor in the same year. Roscoe died at his home in Leatherhead, Surrey, in 1915.

South Kensington and Manchester

Britain was uniquely placed to copy the new laboratories. Wilhelm Hofmann had worked in London for two decades and was well known to the Department of Science and Art (DSA). The secretary of the department, Henry Cole, asked Hofmann to write a report on the laboratory buildings at Bonn and Berlin.[13] Hofmann produced an excellent report, and Edward Festing of the South Kensington Museum, which was then part of the DSA, also visited several German laboratory buildings (and Zurich) in 1870 to gather further information.[14] Hofmann's move to Germany and his subsequent report spurred Cole to bring forward his plan to move the laboratories of the Government College of Mines (an unhappy marriage of the Royal College of Chemistry and the original Government School of Mines) from Oxford Street and Jermyn Street to South Kensington.[15] Berlin was not the only model for the new institution – the Museum of Economic Geology in Jermyn Street, with its glass-roofed lecture hall, was another influence.[16]

As luck would have it, the ideal building had just become available. The government had authorized construction of a school of naval architecture and marine engineering, under the aegis of the Admiralty and the DSA, which was relocated to Greenwich, freeing up the South Kensington building. The original design had been drawn up by the departmental architect Francis Fowke, but he died in December 1865. His plan was then put aside by Cole and a new design based on the laboratory building at Berlin was elaborated by Cole, Richard Redgrave and Fowke's successor as departmental architect, Major-General Henry Scott (illus. 75). The architecture has a similarity to the Royal Albert Hall, which had just been designed by Fowke and Scott, and the new buildings of the South Kensington Museum (this part of the museum later became the Victoria and Albert Museum).

75 Normal School of Science, South Kensington, London, 1871.
This is now the Henry Cole Wing of the Victoria and Albert Museum.

The polychrome brickwork and the general style owe much to the ideas of the exiled German architect Gottfried Semper, and there is more than a passing similarity to Semper's Dresden Opera House built in the same period. The style can be described as a mixture of the Rundbogenstil in the ground-floor arches, with a Neoclassical top floor. It is intended as a grand statement both for science itself and in particular for the Department of Science and Art in the middle of its 'Albertopolis'. The most striking similarity between Berlin and this design lay in the open air loggias, which were included at Cole's insistence.

The plan for a new institution called the National Training College for Science – which aimed to combine science and naval architecture – was developed by Thomas Huxley, Cole and the inspector for science at the DSA, Captain John Donnelly. The Treasury was unhappy with the idea as it had no official authorization and the building work was briefly suspended. The problem was solved when the oversight of construction was transferred to the Office of

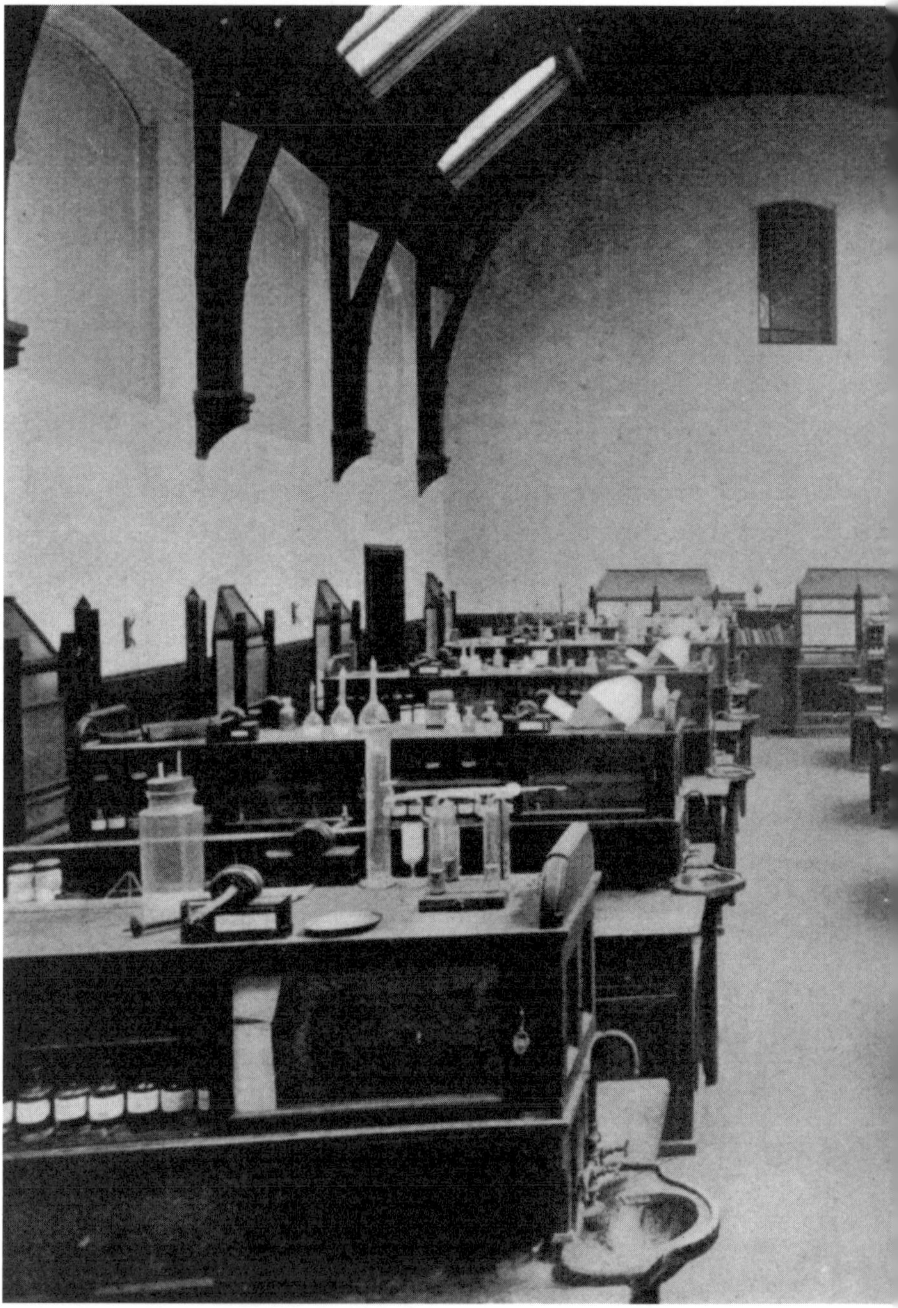

76 The quantitative analysis laboratory, Owens College, Manchester, 1873.

Works in 1870. The laboratory was now supported by the Devonshire Commission, which began its work in May 1870. Huxley was a member of the commission and Norman Lockyer, another supporter of the scheme, was its secretary. Huxley and Frankland pointed out that they needed more space to teach science. As a result of this external pressure the chemistry, physics and biology departments moved to South Kensington at the end of 1872, but the mining engineers and geologists refused to leave Jermyn Street. The chemistry laboratories were on the floor where the open-air loggias were located, below the biologists on the top floor.[17] The new laboratory building was renamed the Normal School of Science in 1881 after the French term for teacher training colleges. It soon became very crowded and new buildings were built on the other side of Exhibition Road, and these opened in 1906. They were designed by Aston Webb with advice from the professor of chemistry Thomas Thorpe (see chapter Ten) and the physicist Arthur Rücker.

The other major chemistry laboratory building constructed in Britain in the period was erected by Roscoe at Owens College, Manchester, after it moved to Oxford Road in 1873. As Roscoe told the Devonshire Commission before the move to Oxford Road, his existing laboratory was 'altogether inadequate. The present building was built to accommodate 35 students and now I have 78 working in that place'.[18] After visiting the German laboratories in the long vacation of 1869, he designed a new laboratory building with the well-established architect Alfred Waterhouse, who according to Roscoe, 'at once understood the special requirements of a laboratory'.[19] As Roscoe had earlier compared Owens College to the local dosshouse, it would have doubtlessly amused him that one of Waterhouse's earlier projects in Manchester was Strangeways Gaol. This laboratory building had room for 100 students and at a pinch could hold 200. The lecture theatre was designed to hold 300 to 400 students. As a former colleague of Bunsen, it is not surprising that the laboratory reflects his master's simplicity:

> There are two long rooms, 30 feet high, lighted both at the side and at the top, which form the chief working laboratories. At the side we have a number of rooms for the various operations: assistants' rooms, library, balance rooms, organic analysis rooms, rooms for spectroscopic work and electrolytic

work; also rooms in the basement which are available for special purposes, as also furnace rooms and large store rooms for apparatus and materials. A class-room for smaller classes is also attached, and a room for a mineralogical cabinet. On the second floor, above this portion, is the laboratory and private rooms of the professor, and that is so arranged that he can see what is going on down below. The main building is to be of stone, but I have insisted upon having my building in brick, because I prefer rather to spend the money on the internal arrangements.[20]

Roscoe's new laboratory building (illus. 76) looked similar to that at Heidelberg; for example, they shared two long laboratories side by side. The furnishings at Owens were far less elaborate than in Leipzig – there were no washbasins, for example, and the bottle racks were very similar to those in a photograph of the Heidelberg laboratory taken in around 1900. Although there were specialized rooms, as Roscoe mentioned in his evidence the emphasis is clearly on teaching rather than on advanced research.

From its bridgehead at Owens, the Heidelberg model spread to other universities in Britain, British schools and even to Germany. Roscoe later recalled that the laboratories he had

designed for Owens College were copied far and wide not only in this country but abroad, and especially at Munich, where [Adolf] Baeyer (Liebig's successor) based his new building on the plans which I sent to him.[21]

In Britain a similar laboratory was designed at Newcastle by Roscoe's student Peter Philips Bedson and the local architect Robert J. Johnson in 1894.[22] Meanwhile in 1878, Waterhouse and the chemist James Campbell Brown designed the 'extraordinary tiered' chemical laboratory building at Liverpool. At Leeds Waterhouse designed the laboratories of Yorkshire College (now Leeds University) with Roscoe's former assistant Thomas Thorpe; they were formally opened in 1885 (illus. 77).[23] It is not clear to what extent these laboratories were based on German designs, but there were similarities with Leipzig (see illus. 112 and 119 in chapter Ten). Waterhouse designed the City and Guilds College in South Kensington in 1881–5, and despite the

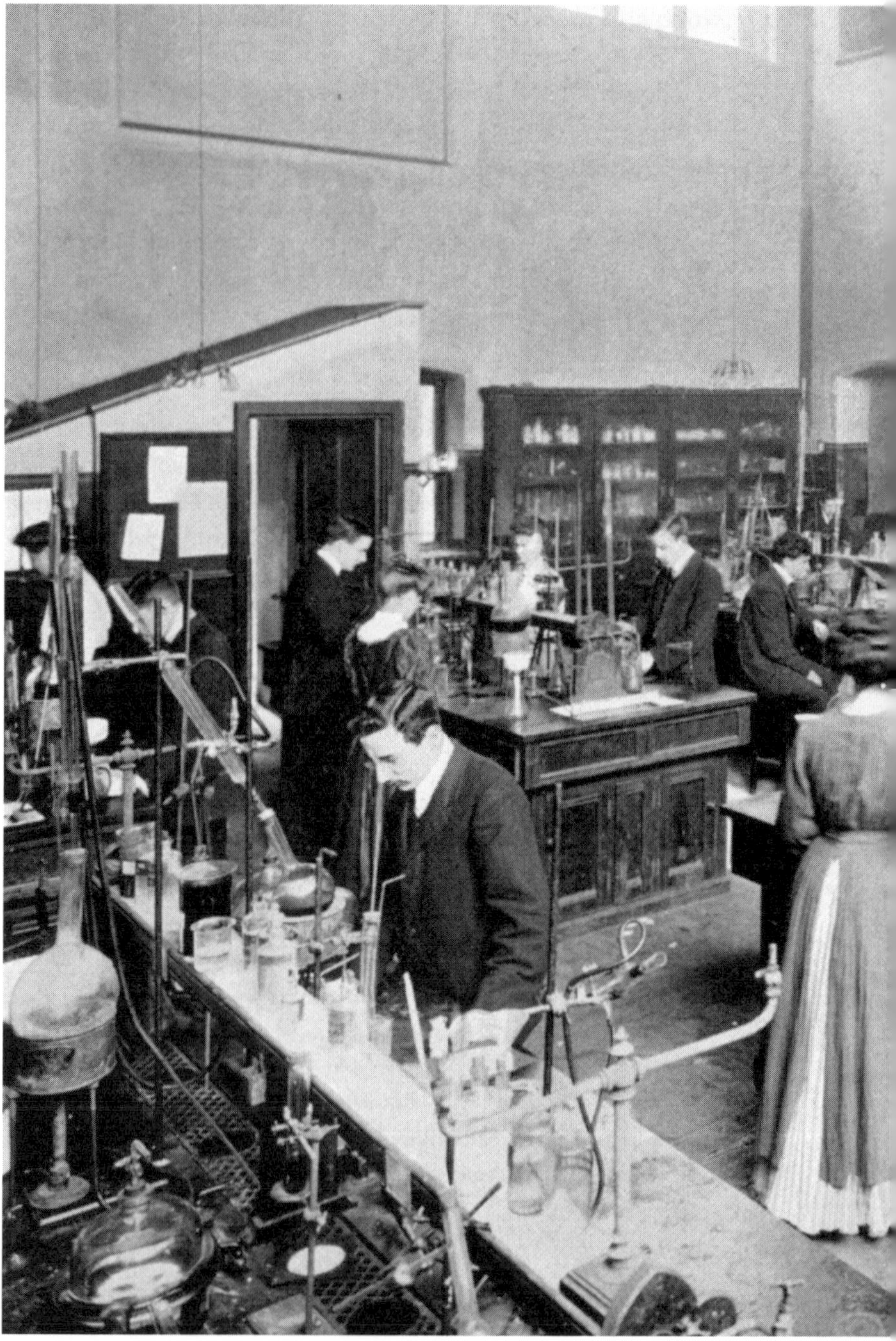

77 Organic research laboratory at Yorkshire College, Leeds, 1908.
A standard laboratory for the period, but note the significant
number of women working in it.

involvement of Roscoe, the main influence appears to have been Mason's College, Birmingham, designed by Jethro Cossins.[24] The Manchester-Heidelberg model returned to Manchester in the shape of the Schorlemmer Laboratory of 1895, which was based by the tireless Waterhouse and William Henry Perkin, Jnr on Baeyer's laboratory in Munich; this in turn, as Roscoe remarked above, was based on the Owens College laboratory of 1873.[25]

Berlin or UCL?

Japan was also linked to the German template (illus. 78). The chemistry laboratory building in Tokyo University's College of Science was designed by the architect Yamaguchi Hanroku with the advice of the pharmaceutical chemist Nagai Nagayoshi. It was completed in 1888 and at the insistence of Nagayoshi brought all the chemistry-based departments under one roof. The close links between the Imperial University and Berlin have been well described by Yoshiyuki Kikuchi:

> A comparison of Yamaguchi's plans of the chemical laboratory at Tokyo and those of Hofmann's chemical laboratory in Berlin shows the extent to which Nagai's design followed the Berlin example. Apart from the unmistakable similarity in shapes, several key elements of Nagai's academic advice are based on his experience in Berlin. For example, the plans included special rooms for a variety of operations, such as balance rooms, spectroscopic (photometric) and blowpipe analysis rooms on the first floor, and gas analysis and combustion analysis rooms on the second floor. There was also to be a chemical museum with showrooms for prepared chemicals and scientific instruments. The design also included preparatory laboratories for demonstrations next door to the central lecture hall and lecture rooms.[26]

Nagai's idea of a central chemistry department was overturned in the reorganization of the university into the Imperial University in March 1886. The link between chemistry and medicine was broken, and the other sciences (and mathematics) now moved into the as yet uncompleted laboratory building, which became the College of

78 Laboratory of Applied Chemistry (completed c. 1896), College of Engineering, Imperial University, Tokyo, 1900.

Science. Nagai himself lost his position, and the construction of the building was taken over by Edward Divers and Sakurai Joji, whose experience lay in England rather than Germany. One important difference was the moving of the professors' offices and laboratories from the area adjoining the student laboratories, where as planned by Nagai they would have the supervisory role discussed in chapter Six, to an area next to the library and classrooms. According to Kikuchi, this resembled the layout of Williamson's new laboratory at UCL, and reflected a view that senior professors delivered lectures and junior professors supervised the students.

Across the Pond

Initially it looked as if the American universities would be avid copiers of German designs. The Giessen-educated Eben Horsford attempted to create a similar laboratory at Lawrence Scientific School based at Harvard, but his attempt was abortive.[27] Charles Joy had trained in Paris, Berlin and Göttingen, where he had taken his PhD under Friedrich Wöhler. On his return to America he was elected professor of chemistry at Union College in Schenectady, New York state. One of his first tasks was to set up a chemical laboratory in Philosophical Hall, which had been erected in 1852 (illus. 79).[28] From the *Annual Report of the Regents of the University of New York* for 1858, we learn that

> Professor Joy was sent to Europe by the trustees expressly to procure apparatus and plans for the laboratory for which his previous residence of four years in Berlin and Gottingen as the pupil of Rose and Woehler gave him peculiar facilities.[29]

The link between the German laboratories and Union College could not have been stronger:

> The laboratory is constructed after plans furnished by Professor Lang the architect of the famous laboratories of Heidelberg and Carlsruhe with such modifications as were suggested by Professors Bunsen and Weltzien. It is furnished with the best modern appliances for acquiring a thorough knowledge of chemistry and the applications of this science to agriculture and the arts.[30]

Even the fittings and apparatus were inspired by Germany and sometimes even obtained from German firms:

> The students have the use of superior balances from Giessen and Berlin, of a Schieck microscope, of an Eckling air pump and of graduated instruments for volumetric analysis. The sand-bath arrangements and the furnaces were copied from Liebig's new laboratory in Munich and enable the operator to work without danger from deleterious gases or acid fumes. A Beindorf apparatus containing the usual steam drying chests, steam digesters and distilling apparatus with numerous improvements was manufactured by Murrle in Pforzheim expressly for the laboratory.[31]

79 Philosophical Hall, Union College, Schenectady, New York, the home of Lang's American laboratory, c. 1900.

80 Quantitative analysis laboratory, Lehigh University, Bethlehem, Pennsylvania, 1893.

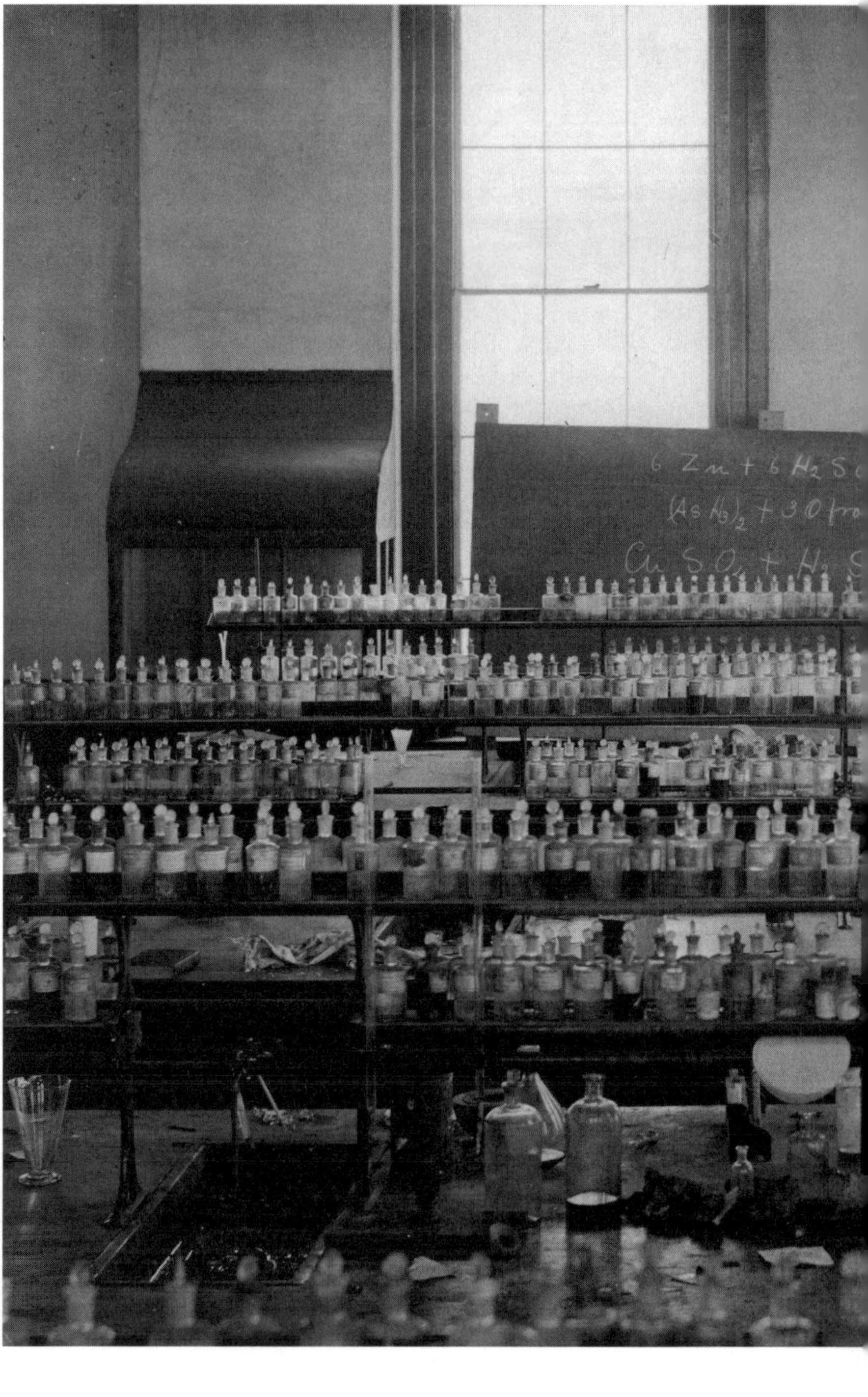

81 Picture of Culver Laboratory, Dartmouth College,
Hanover, New Hampshire, 1871.

82 Laboratory for water and gas analysis, Massachusetts Institute of Technology, Cambridge, Massachusetts, 1893.

This exercise was not cheap. Joy was sent to Europe with $3,000, but he exceeded his budget and the fittings alone cost $7,000 or at least $180,000 in 2010 terms.[32] When it opened in 1857, this was probably the first well-designed academic laboratory in America and it had an importance far beyond Schenectady. Joy had left Union College for Columbia University in 1857 and was replaced by his young colleague Charles F. Chandler, who himself left for Columbia in 1864; there he became one of the most influential chemists in America and later built the Havemeyer Hall laboratory building. However, the Union College laboratory was simply a new laboratory in an existing building, and it does not appear to have been widely emulated in America.

According to William H. Chandler's *Construction of Chemical Laboratories*, there is no indication that any of the four American laboratory buildings described in this volume had any German influence. It may just be that none of the contributors felt that it was relevant to the description of the laboratories or – as this was a report to the Paris Exposition of 1889 – American national feeling (or perhaps tact towards the French as hosts of the exposition) may have made any reference to Germany undesirable. However, more recent studies of the chemical laboratories at Illinois, Columbia and Dartmouth have not mentioned any German influence.[33] We can therefore only seek to trace German influences in the appearance and layouts of the laboratories.

The analytical laboratory in the Kent Laboratory at Yale has a 'German' appearance with a central aisle and wooden bottle racks, similar to that of the Government Chemist's Laboratory in London (see illus. 114).[34] The laboratories at William Chandler's Lehigh also have the central aisle (illus. 80).[35] In other respects the American laboratories do not look as 'German' as their British counterparts. Characteristically, the American laboratories, for example Culver Hall at Dartmouth (illus. 81) and those at Massachusetts Institute of Technology, had long benches that were either centred or shifted towards the wall, thus creating an aisle down one side with the fume cupboards on the other side (illus. 82).[36] It appears that some American laboratories did not have bottle racks (MIT), or had tall metal frames to hold the bottles (Lehigh and Dartmouth), which I suspect may have been removable. MIT kept chemicals in glazed cabinets on the wall that were very similar to the cabinets in Heidelberg. As Kikuchi has described for Tokyo's Imperial University, there were similarities with the German laboratories in terms of the floor plan, which incorporated lecture halls, a library, specialized rooms and a variety of laboratories. Yale, Lehigh, Cornell and Columbia all had chemical museums (see chapter Eight). There were, however, also differences in the layout. None of the laboratories had residences for the director or open spaces for dangerous work. Given the significant move of the professors' offices in Tokyo, it is interesting to find out where they were placed in American laboratories. At Yale and Lehigh they were near the lecture hall, not the laboratory. At MIT and Cornell they were close to the organic chemistry laboratory, but not placed in a way that would make direct oversight of the laboratory possible.

The 'recitation room' was a feature of many American laboratory buildings in this period, including Lehigh, MIT and Dartmouth. This room – common to all academic disciplines including the humanities – was effectively a classroom for a group of students. A professor would give a lecture on a particular topic and engage with

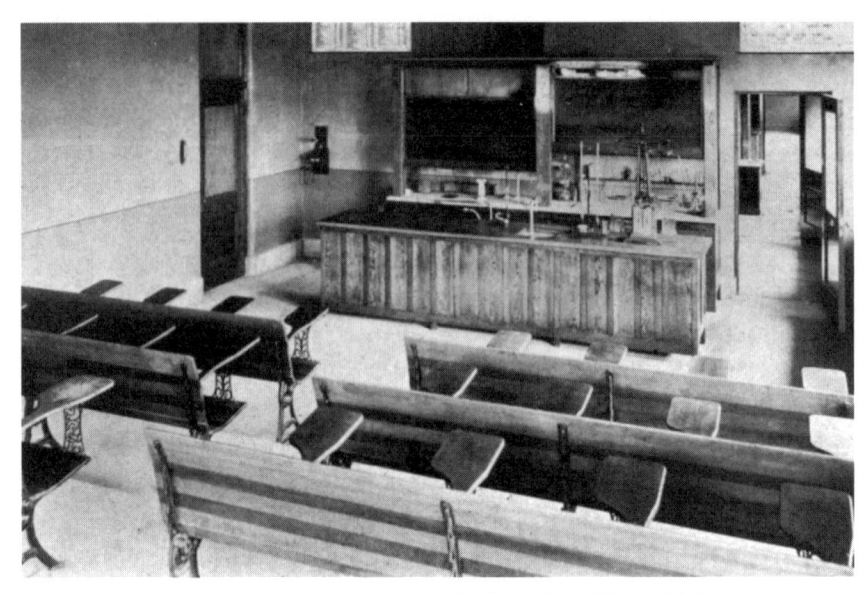

83 Chemistry recitation room at the Iowa State Normal School (now the University of Northern Iowa), Cedar Falls, 1908.

the students in discussion ('recitation') in the course of the lecture. In chemistry, as the picture of the recitation room at the Iowa State Normal School shows (illus. 83), there would usually be a demonstration bench to show experiments and tiered seats, so that the students could see the demonstration. After hearing the lecture and seeing the demonstration, the students would return to the laboratory to carry out their experimental coursework based on the lecture.[37] Recitation rooms did not exist in European universities because their teaching systems were different, and their place was taken by the small lecture theatre.

The German Model Arrives in France

Superficially the situation in France was similar to that in Britain.[38] Just as Roscoe and Festing toured Germany, the French chemist Adolphe Wurtz visited Bonn in December 1867, and went to Berlin and Leipzig the following summer. Just as Huxley campaigned for the Normal School in London, Louis Pasteur campaigned for a new laboratory at the Sorbonne in Paris and even gained the ear of the emperor Napoleon III. Just as Hofmann assisted Cole in London, he visited Paris twice in 1867 and may have acted as a consultant to the French government. However, that is where the resemblances end. To be sure, the Imperial government set up the Ecole Pratique des Hautes Etudes in July 1868, but this was just a reorganization of existing institutions. Two years later the Franco-Prussian War broke out and France descended into a civil war followed by several years of political instability. After years of pleading for a new laboratory, Wurtz was given a new research laboratory at the end of 1877, but it was simply the conversion of an existing building, not the palaces of Berlin or South Kensington. The breakthrough came with the appointment of the Republican Jules Ferry as minister of public instruction in 1879. The renovation of the Sorbonne was finally approved two years later. Wurtz revisited Germany and Austria-Hungary. This new building was the first major commission of Paul Nénot and it remained his most famous. In contrast to other architects mentioned in this volume, Nénot was fairly eclectic in his style. The Beaux-Arts style of the New Sorbonne, with its rich ornamentation based on seventeenth-century French architecture, contrasts dramatically with the neoclassical Palace of the Nations in Geneva that Nénot designed at the end of his career (with other architects). Even the New Sorbonne was a mixture of styles, with a medieval-style Italianate tower that has an observatory perched rather incongruously on top of it. It has a *fin de siècle* grandeur in common with the Ritz Hotel in London (also in the Beaux-Arts style), the Imperial Institute in South Kensington and the headquarters of the Post Office Savings Bank in Hammersmith (see chapter Ten). Nénot consulted Wurtz and other leading French chemists before travelling to Berlin, where Hofmann spent several days with the architect going over the details of the plans. When the new Sorbonne laboratory buildings were completed in 1894, the German model had finally reached France.

The internal arrangement of the laboratories was very different from that of the German laboratories of the 1860s. The laboratories are smaller (even the teaching laboratories), and are dominated by a large central bench rather than rows of benches (illus. 84). There are bottle racks but they are not prominent; there are fume cupboards but they are not numerous. In appearance, the Sorbonne laboratories are similar to three laboratories that were opened in Britain during 1896: the Davy-Faraday Laboratory of the RI in London, the Wellcome research laboratories also in London (see illus. 103)

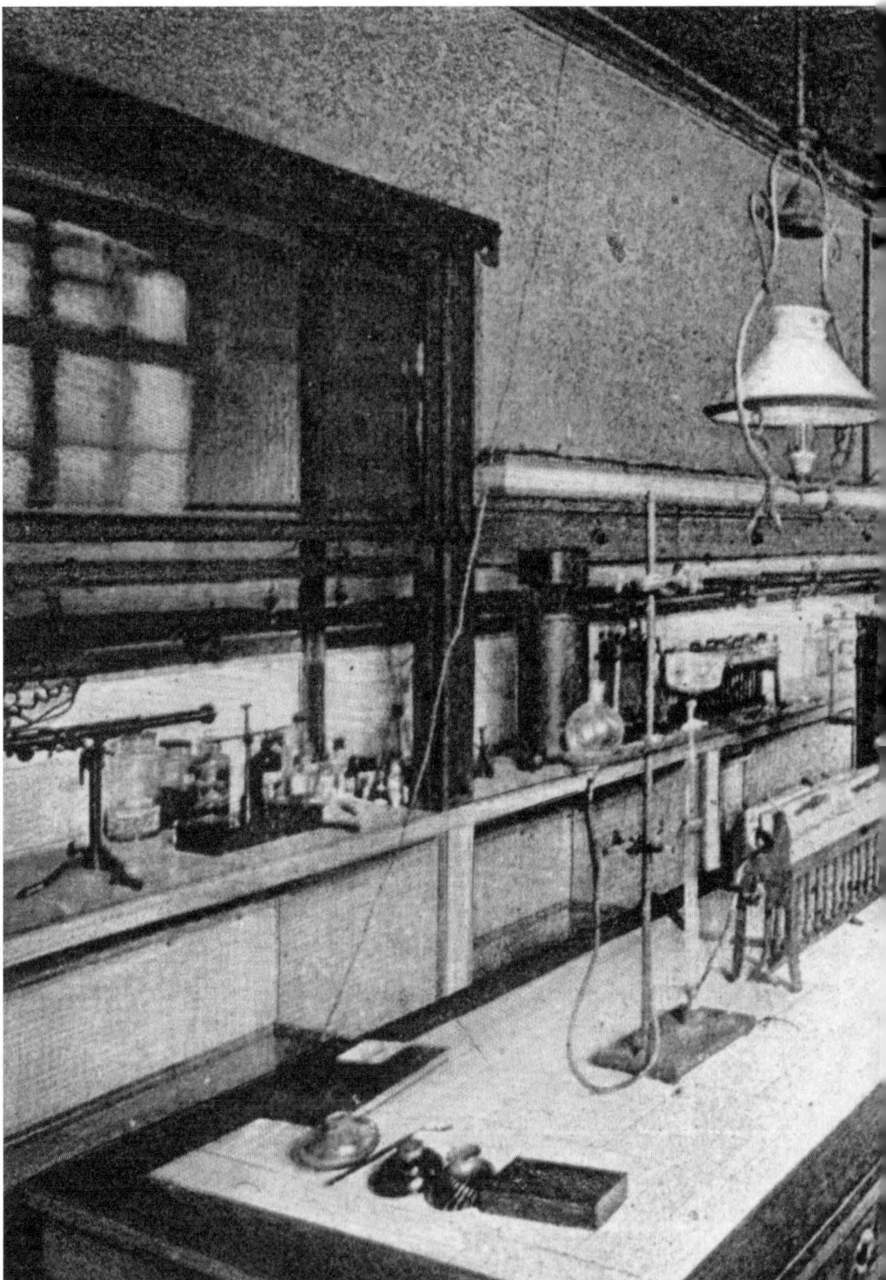

84 Organic chemistry laboratory, New Sorbonne, *c.* 1895.

and the laboratories of the Royal College of Physicians of Edinburgh (RCPE).[39] In contrast to the Sorbonne, however, these laboratories were intended for small groups of researchers rather than students, and they were fitted into existing buildings rather than being purpose-built laboratory buildings.

A Model Shared

Two developments came together to make the diffusion of the new German-style laboratory – what I call the classical laboratory – widespread. One was the standardization of the laboratory and its fittings, which made it straightforward to copy and easy to obtain the necessary fittings from the laboratory suppliers that were springing up in the late nineteenth century. The other was the apprehension of Germany's growing scientific and industrial prowess by other countries, and the need they felt to match it. Clearly Germany was getting it right as far as science was concerned, and its laboratories were seen as a key element of its success. Thia imitation may not have happened and certainly may not have happened as quickly if it had not been for the openness of the Germans, and Hofmann in particular, to foreign enquirers. This openness ensured that the rest of the world largely adopted the German version of the laboratory, rather than trying to create an alternative version.

The German model was not, however, taken up everywhere or immediately. In France delays in the reform of science teaching (rather than any reluctance to use German practice) prevented the adoption of this model for two decades after 1870. In Japan, as a result of Kikuchi's research, it can be seen that there was an initial preference for the German model both in the laboratory design and in the method of supervision, but with a change of personnel it switched to a way of operating that was based on UCL. Even so, the actual laboratory was still clearly based on German designs; as indeed was the case at UCL, although its laboratory stemmed from Giessen not Berlin. After an initial burst of enthusiasm for German laboratories, the Americans seem to have developed their own designs. There were differences in the layouts of the laboratories and the style of the bottle racks, but even these laboratories clearly owed much to the German model. One aspect of the Berlin and Bonn museums that quickly became popular in the USA was the chemical museum; so much so that it became more common in American laboratory buildings than German (or British) ones. This interesting but rarely studied aspect of the laboratory building is the subject of the next chapter.

8
Chemical Museums
Charles Chandler and New York, 1890s

Not every feature of the nineteenth-century laboratory building survived to become part of the standard chemical laboratory building of the late twentieth century. The chemical museum was a major component of any self-respecting late nineteenth-century chemical laboratory building, but hardly any such museums survive today and they have largely been forgotten. With the exception of the Chandler Chemical Museum at Columbia, which is usually treated as being practically unique, they do not feature in the literature of the history of chemistry. Christoph Meinel has even argued that collections of chemicals did not exist, at least in the same way that there were (and are) collections of minerals or butterflies.[1]

What Were Chemical Museums?

The nature of the chemical museum must first be clarified. It was not a science museum in the sense of a museum of the *history* of chemistry we know today – for example the museum at the Chemical Heritage Foundation in Philadelphia and the chemistry galleries of the Deutsches Museum in Munich. Indeed, a few chemistry departments now have historical museums of this type, including the Universities of Cincinnati and Göttingen. Rather, the chemical museum in the nineteenth century was an ahistorical collection of specimens, apparatus currently in use, and models (or illustrations) of molecules and factories. Especially in the late nineteenth century, there was often a strong focus on the applications of chemistry and the products of applied chemistry. Sometimes such collections were not called chemical museums but collections of chemical preparations (*Präparatensammlung*). They were, however, usually in their own room, which was often adjacent to (or underneath) the lecture theatre, suggesting that the difference was largely one of terminology. The term 'museum' was used widely in the eighteenth and nineteenth centuries, and it was entirely reasonable for such collections to be called museums.[2] Many of them had features that are associated with museums even today, including a room dedicated to a display, and display cases to house the specimens.

Not every chemical museum had all these aspects, but broadly speaking they all conformed to this basic model. Indeed, the chemistry section of the science galleries in the Science Museum in South Kensington were like this up to the mid-1920s, and the section on applied chemistry evolved into the industrial chemistry gallery in the new building on Exhibition Road in 1925.[3] One can immediately see that there is a similarity here with geological museums; after all minerals are, in a sense, 'chemicals'. Some laboratory buildings, such as those at Bonn University, put the mineralogical museum alongside the chemical museum; others, for example those at Berlin University, combined them; yet others, as at Lehigh University, put the mineralogical museum on a different floor.

It has to be emphasized that there was nothing unusual about academic museums in this period. Particularly in America, universities would have a range of departmental museums. Apart from the usual zoological, geological and archaeological museums, there might be museums devoted to medicine, pharmacy, pathology, anthropology and other sciences such as physics. The profusion of such museums in this period is well illustrated by the Teviot Place medical building at the University of Edinburgh (completed in 1888), which housed both medicine and chemistry, and contained no less than seven museums, the largest of which was the anatomical museum.[4] The physical museums, which were derived from the physical cabinets of the eighteenth century, tended to be very much about apparatus and their historical use.[5] Of course, this is partly because today's current exhibit is tomorrow's historical exhibit, a process that has also occurred in mainstream science museums.

The Function of Chemical Museums

What were the purposes of the chemical museums? They were not intended for the public, although they were sometimes put on display in public areas after their original purpose had vanished, presumably with the intention of impressing visitors. However, eminent guests were sometimes shown the museum as a noteworthy part of the department. For example, the chemical museum in the College of Engineering at the Imperial University, Tokyo, was shown to the Emperor Meiji in 1896, the year it was founded (illus. 85).[6] The chemical museums do not seem to have been created for taxonomical purposes, for the classification of chemical species, what John Pickstone calls 'museological knowledge'.[7] This may have been true of the earlier museums in the eighteenth century, especially in the case of Torbern Bergman with his background in natural history and geology, but there is nothing about the later chemical museums to suggest that they were taxonomical; their collections were not suited for this purpose or generally arranged in this way. They were mainly used in conjunction with lectures for teaching purposes, as William Crookes makes clear in his desiderata for a chemical museum:

> But the specimens in a chemical museum be so ordered and labelled as to illustrate the types, the notation and the nomenclature of the science; while models of crystals and apparatus, together with diagrams and drawings, will serve to illustrate crystallography, chemical processes and the arts dependant thereon, as well as to familiarize the student with chemical symbols, formulæ and reactions.[8]

The museums were perhaps also intended – especially when they were chemical collections or *Präparatensammlung* – to contain a set of reference samples for use by experienced chemists, like specimens in natural history museums or the 'Black Museum' at Scotland Yard founded in 1875.[9] In at least one early case, a chemical compound isolated elsewhere was deposited as a matter of record in a university's chemical museum:

> In the year 1839 during the examination of the compounds formed by carbazotic acid and the vegetable bases, in order to determine the value of that acid as a discriminative test, the writer ventured to place berberine amongst the alkaloids; and pursuing the subject in the following year, the muriate, acetate, sulphate, and various other salts were formed; the

85 Museum of Applied Chemistry (completed c. 1896), College of Engineering, Imperial University, Tokyo, 1900.

modes of formation communicated to Prof Buchner, with specimens which were deposited in the Chemical Museum of the University of Munich, and the basic properties of berberine considered as fully determined.[10]

Fourteen years later, Hermann Kämmerer referred to a 'simple iodine chloride' that had been stored in the *Präparatensammlung* at Heidelberg for approximately six years.[11]

By the late nineteenth century, however, the majority of the museums were museums of applied chemistry at least in practice and sometimes even in title. It is interesting to ponder why this was the case. Applied chemistry was at its peak in the late nineteenth century, partly as a justification for academic chemistry being the best training for chemical manufacturers and their skilled workers.[12] Yet applied chemistry was difficult to teach in the absence of factory experience and chemical museums offered a way around this problem. Furthermore, while students could be expected to make the most important laboratory chemicals (or at least witness a demonstration of their synthesis), industrial chemicals were more problematic. Keen to put their names in front of prospective employees and to make connections with leading academic chemists, manufacturers were quick to offer collections of their products and the intermediates involved. These donations by chemical companies determined that the museums would be museums of applied chemistry rather than pure chemistry. Indeed, some museums appear to have been set up to house these donations. Precisely because future industrial chemists were taught chemistry in a university setting, chemical museums became almost inevitable.

The Origins of the Chemical Museum

The chemical museum's origins lie in its links with mineralogy in the eighteenth century, as minerals were usually kept in collections and put on display, for example by Sir Hans Sloane.[13] Since this association was particularly close in Sweden and Scotland, we might expect to find early examples of chemical museums in these countries. As noted in chapter Three, Johan Gottschalk Wallerius in Uppsala separated the auditorium from the laboratory in his chemistry building. It might have been expected that Torbern Bergman would have adopted Wallerius's laboratory without any major changes when he became the chemistry professor in 1767, but the building had been gutted by a town fire in the previous year.[14] Only the shell of the building remained, along with fireproof pieces of apparatus such as Hessian crucibles. This gave Bergman a free hand in terms of the internal walls, and he took this opportunity to design a completely new layout. In creating this layout, Bergman had to find room for the Svab mineral collection that had been donated to the university in 1751, but was still waiting to be put on display. He could have put the mineral collection on the first floor he had constructed which, as Wallerius was quick to point out, would have protected it from the risk of damp and flooding from the neighbouring river. However, he chose to use this space as his living accommodation. Given his poor health Bergman may have been more concerned about the effect of the damp on himself than on the mineral specimens.

Crucially, Bergman got rid of the auditorium and replaced it with the new mineralogical museum with two side rooms for other specimens and industrial models respectively. The presence of industrial models and samples of intermediate products from mining and industry shows that Bergman's museum was closer to the contents of the late nineteenth-century museums than simply a mineralogical collection, albeit one with a mineralogical bias. In his autobiography Bergman wrote:

At the other end of the gallery [of minerals] is a room for models, not mechanicals but of apparatus which is used for manipulation of metals. My purpose here was also to get the ovens and tools in models of those crafts which more or less depend on chemistry. Such crafts as the manufacture of glass, porcelain, glazed earthenware, pottery vessels, brick, tobacco pipes, all kind of salts, oils, gunpowder, lampblack, all fire-resistant dyes, but of which hitherto no collection has been made.[15]

He had obtained the idea of adding models and products from seeing the collection of industrial models and samples (for example of textiles) that the professor of economics, Anders Berch, had assembled in the nearby Theatrum œconomicum mechanicum.[16] Public lectures were held in the museum, but most of the chemistry teaching now took place in the laboratory, for which the students had to pay fees. However, the new museum, opened in 1769, can be interpreted as an auditorium in which minerals and models were displayed, making it a more crowded lecture space. Clearly the minerals were important in a period when Sweden was heavily dependent on the income from its mines. Bergman systemized the analysis of minerals and for the first time in a university course taught the use of the blowpipe. Thus, it could be argued, Bergman had created a 'chemical museum' at least partly for teaching purposes. The mineral specimens would have been useful for lectures in chemistry, which in the Swedish (and Scottish) manner of the time were largely concerned with metals, minerals and mineral waters.[17] To some extent this alteration had been imposed on Bergman by the donation of the collection to the university, but it also reflected his view of chemistry as a combination of natural history (the minerals) and mathematics (the crystal structures).

Bergman's display of minerals and models was not completely new. Probably from the late seventeenth century at least, chemical lecturers would have had collections of apparatus, chemicals, materia medica and minerals to show during their lectures – the exact compositions of their collections reflecting their syllabuses. The objects in these teaching collections were very likely stored in similar cabinets or cupboards (*Kunstschränke*), as was the case with the cabinet of John Francis Vigani at Queens' College, Cambridge.[18] As the collections grew, the arrangement for their storage and display became more permanent. Trinity College Dublin (TCD) erected an anatomical theatre and laboratory for the teaching of medicine and chemistry in August 1711.[19] This included a museum on the upper floor, which was probably an anatomical museum. In 1729 William Maple supplied a painted floorcloth for the 'little museum in the laboratory', which implies that there was also a smaller museum associated with chemistry. Three years later Maple also supplied glass frames for Mrs Rawdon's insects and flint glass jars for some reptiles, which suggests that it might have been a more general museum, not unlike the (Old) Ashmolean at Oxford, which had been founded in 1683.[20] It is not clear, however, whether this general museum was the upstairs one or the little museum in the chemical laboratory.

Chemical displays began to appear in Paris at the Jardin du Roi and Académie Royale des Sciences in the late seventeenth century, and at the School of Mines in the eighteenth century. At the time of his death in 1738, Herman Boerhaave at the University of Leiden had a collection of 422 chemical preparations he had prepared himself (along with natural history specimens and apparatus) which he may have used in his lectures.[21] In Florence in the late eighteenth century, Felice Fontana directed the Imperial Regio Museo di Fisica e Storia Naturale. There were exhibits of chemicals and minerals in cases in the chemical rooms, some of which are still on display today.

A Civic Chemist

Charles Frederick Chandler was born in his grandfather's house in Lawrence, Massachusetts, in 1836, into a family that settled in the state in 1637, only seventeen years after the *Mayflower* landed.[22] His father was a dry-goods merchant in New Bedford, Massachusetts, then an important whaling port. Chandler became interested in chemistry at school, studying the minerals around Lancaster and attending lectures given by the famous geologist Louis Agassiz at the New Bedford Lyceum. He then took chemistry under Eben Horford at the Lawrence Scientific School attached to Harvard as he lacked the Latin and Greek required to enter Harvard. While he was there he met Charles Joy, professor of chemistry at Union College, Schenectady. This meeting was to shape the rest of Chandler's life. Joy spoke about his student days in Göttingen and encouraged Chandler to study there. The young student, still only seventeen, sailed on a whaling boat to Europe. He studied under Friedrich Wöhler at Göttingen and, again with the help of Joy, under Heinrich Rose in Berlin, before taking his PhD in mineral chemistry in 1856.

By this time Professor Joy had created a new laboratory designed by Heinrich Lang, the architect of Bunsen's laboratory in Heidelberg (see chapter Seven). Chandler ran this laboratory as the janitor-assistant, and when Joy left soon afterwards for Columbia College, New York, Chandler took his place, becoming the Nott Professor of Chemistry in 1861. Union College had an excellent collection of minerals, which Chandler analysed, and this brought him into contact with Thomas Egleston, who with other prominent New Yorkers was in the process of setting up the Columbia School of Mines. With the support of Professor Joy, they invited Chandler to become professor of geology and analytical and applied chemistry in 1864. The new school grew rapidly and new laboratories, designed by Chandler, were established in a former factory. Chandler was now appointed dean of the School of Mines, a post he held until 1897 (illus. 86). He remained a professor of chemistry until 1911, when he was 74. His major achievement at Columbia was the building of Havemayer Hall for the School of Mines between 1896 and 1898. This splendid building also housed the Chandler Museum, which Chandler had established almost as soon as he arrived at Columbia in order to 'show my boys' what he was talking about in his lectures.[23] Chandler also taught chemistry at the New York College of Pharmacy and the College of Physicians and Surgeons, both of which eventually became part of Columbia University – the College of Pharmacy at Chandler's suggestion.

Moreover, Chandler's activities extended beyond the confines of the three colleges. He was frequently called upon as an expert witness and acted as a chemist or consultant for several companies, including New York Gas Company and Standard Oil. He was one of the pioneers of consulting for industry, an activity that did not make him popular with some of his academic colleagues. He thus helped to establish the strong links between academia and industry that are a notable feature of American chemistry. If all this was not sufficient, he also became a leading campaigner for the improvement of public

Overleaf: 86 Charles Frederick Chandler, centre, with students at the Columbia School of Mines, New York City, 1866.

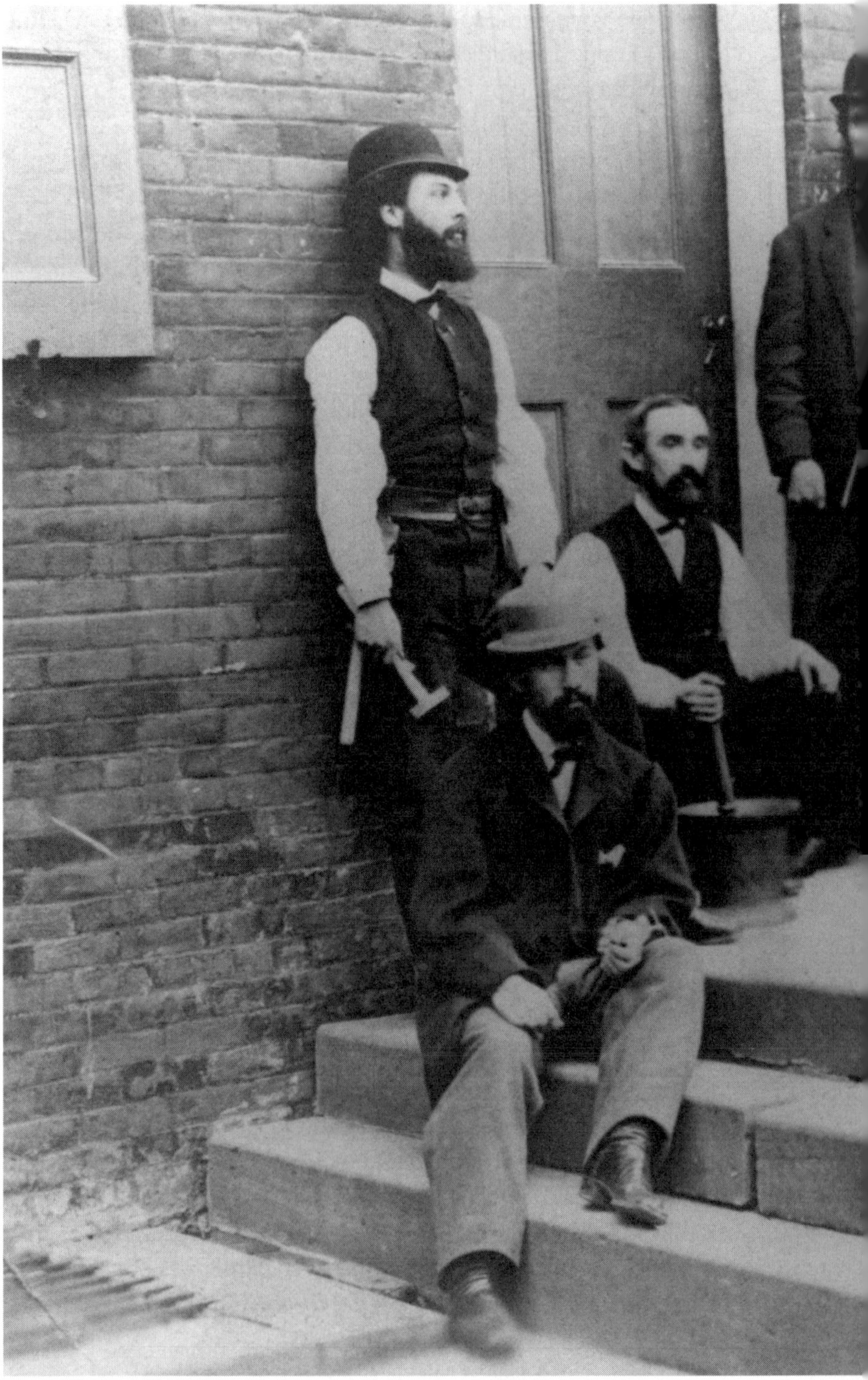

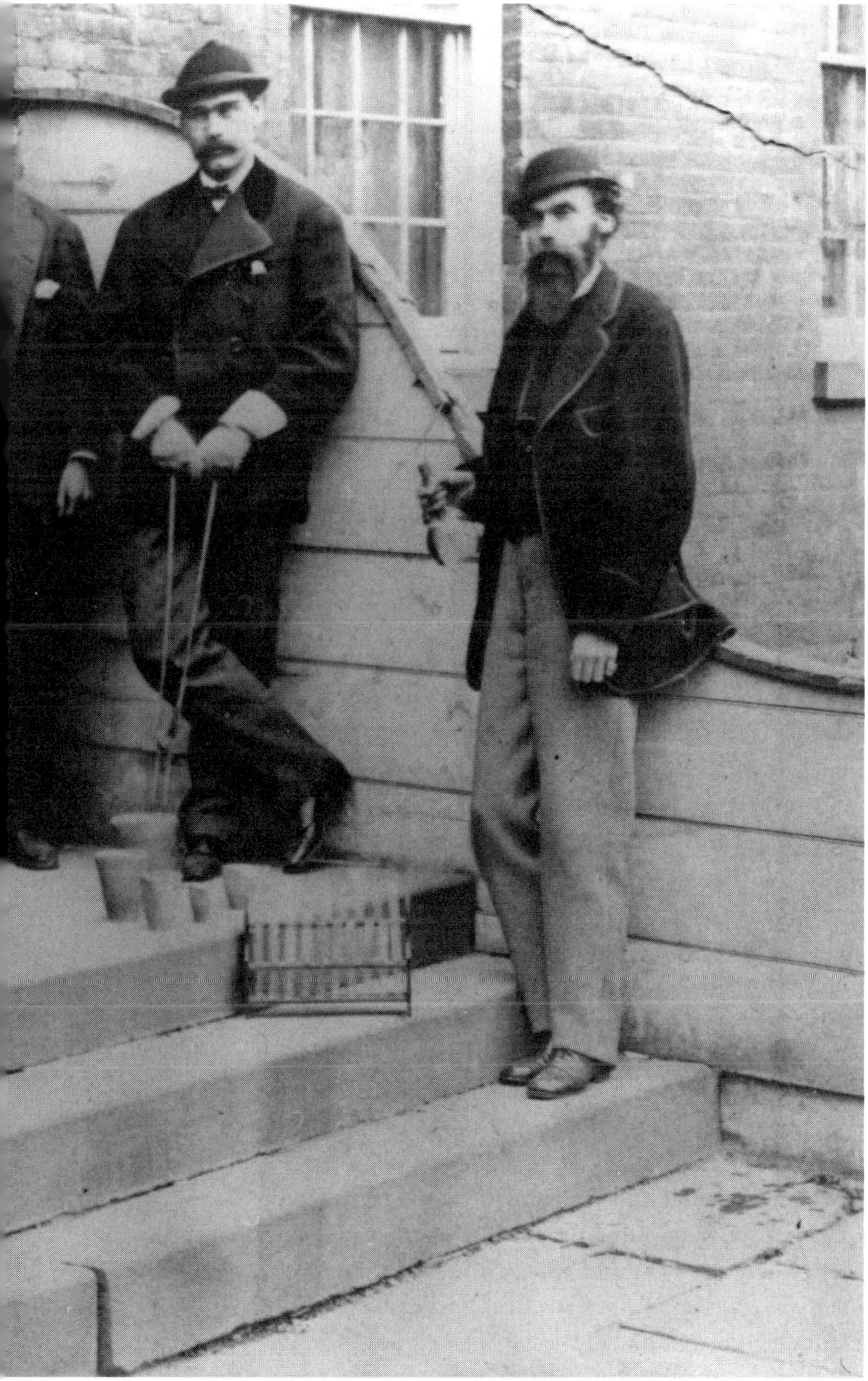

health. Chandler was invited to become the chemist to the New York Metropolitan Board of Health in 1867, and he was appointed president of the board in 1873. He stepped down from this position in 1883. His work in this field was in some ways similar to Thomas Thorpe's work at the Government Chemist's Laboratory in London and Charles Girard's work at the Municipal Laboratory in Paris (see chapter Ten), except of course that Thorpe and Girard held full-time positions, whereas this was only one of Chandler's many activities.

Chandler's activities at the Board of Health were a combination of testing and control. A good example of such work related to the prohibiting of giving watered-down milk to children, which of course necessitated the testing of the milk (see page 276). While alcoholic liquors and the public water supply were well regulated in New York, cosmetics were a major problem because of the presence of lead compounds. Chandler was particularly anxious to prevent kerosene (paraffin) accidents. He had been one of the first promoters of shale oil (the forerunner of modern petroleum) in preference to the expensive whale oil or the dangerously flammable coal-based benzene, having encountered Scottish shale oil in Berlin. Petroleum-based kerosene was fairly safe, but some retailers (ironically working in his father's trade) were mixing it with the cheaper benzene. Chandler stamped out this practice and reduced the death rate from lamp explosions more than three-fold. Chandler's public health activities were not restricted to issues related to chemistry. He clamped down on street markets, regulated the slaughter houses, improved the city's plumbing and raised the standards of the construction of mass housing. He also improved the control of infectious diseases, notably smallpox, and abated New York's air pollution and smells.

Not satisfied with building up a major university or reforming the public health of America's largest city, Chandler also organized America's chemists. In the face of considerable opposition, which included Egleston and his old teacher Horsford, Chandler, his brother William Chandler and William Nichols established the American Chemical Society (ACS) in 1876.[24] Chandler was president of the society between 1881 and 1889. At first the ACS was largely a New York society (just as the Chemical Society was mainly a London-Oxbridge society for many years), but by the 1890s it had become a nationwide organization. Perhaps partly because of this, Chandler set up the Chemists' Club in 1898 to enable chemists to meet together in a social setting and to have access to an excellent library. It certainly brought Chandler in contact with industrialists such as Leo Baekeland and Charles F. Squibb. In some respects it was a chemical version of the Athenaeum Club in London and it still exists. For some years Chandler contributed American news to *Chemical News* (perhaps recognizing a kindred spirit in Crookes). He then established a new journal called the *American Chemist* with his brother in July 1870. With the formation of the ACS, the *American Chemist* became the *Journal of the American Chemical Society*, which remains one of the most prestigious chemical journals in the world. Chandler died at his home in central Manhattan in 1925 at the age of 88, having spent his last years working for the Chemical Foundation, which was set up in 1919 to promote American chemistry and chemical industry on the proceeds of German patents seized in the First World War.[25]

Heyday of the Chemical Museum

The chemical museum spread slowly in the first half of the nineteenth century. TCD set up a chemical museum in 1803, when cases were purchased to house the collection of Robert Perceval, the professor of chemistry, possibly with his pending retirement in mind (he retired in 1805). It appears that Perceval had used this collection in his lectures.[26] When a new medical/chemical building was opened in 1887, the museum was located there and contained 'a number of the rarer chemical substances'.[27] Wilhelm August Lampadius set up a teaching museum for the famous mining academy at Freiberg, Saxony, in the nearby village of Halsbrücke in 1815. As he drew up a classification system of classes and genera for this museum, he clearly saw it as being similar to natural history collections. In this respect it was rather different from the later chemical museums. A more typical museum was established at the University of Prague by Adolph Pleischl in 1820.[28] The chemical museum at the University of Edinburgh was apparently established in the new college building completed in 1823.[29] There was an Upper and a Lower Museum, and both were located in the highest parts of the building.[30]

There were several chemical museums in university chemistry departments by the 1850s, notably in Munich and King's College London.[31] At the Universities of Heidelberg and Oslo (and later at

Trondheim), they were called collections of chemical preparations. The preparations collection and mineralogical collections at Oslo were in the lecture-preparation room, so they were evidentially used in lectures.[32] The Chemical Society of London planned to set up a chemical museum after it was founded in 1841, although the attempt was eventually abandoned.[33] William Crookes criticized the society, possibly tongue-in-cheek, for arranging the specimens in order of the size of the containers.[34] He made it clear that museums were still unusual in chemistry departments in 1860 and hence the heyday of the museum was between the 1860s and '90s. Every self-respecting chemical laboratory built in this period had a chemical museum, however small it may have been. I use 'self-respecting' advisedly; the chemical museum was a focus of pride for the chemistry department, and was usually prominently featured in university publications or public descriptions of the laboratory.

The impetus for this surge of museum building was created – as was the case with so many laboratory innovations – in Germany. Wilhelm Hofmann clearly laid great value on chemical and mineralogical museums, perhaps as a result of his familiarity with the Museum of Economic Geology and the South Kensington Museum in London.[35] He continued his association with the South Kensington Museum after his return to Germany, serving as the chairman of the German selection committee for the Special Loans Exhibition held at the museum in 1876. His laboratory building in Bonn had two large halls at the front of the main building devoted to the Mineralogical Museum and the Chemical Museum,[36] which he seems to have regarded as being similar as he combined them in Berlin into the 'Great Museum for the Scientific Collections of the Institution'. Hofmann commented that '[t]he latter is a magnificent hall, 60 feet [19 m] long and 25 feet [nearly 8 m] broad, the arched roof of which is supported by iron columns'.[37] Despite its splendour, it was placed at the back of the ground floor of the laboratories, next to the instrument room. Hofmann was quick to point out, however, that it could be easily reached from the director's residence via a corridor. Hofmann describes the planned layout and position of the museum in the following terms:

> The architecture of the hall suggests a division of the collections into three; minerals, rocks, and metallurgical products

will occupy the section to the left nearest to the colonnade; the middle section, larger than the two others and directly lighted by the three great windows looking into the quadrangle, will be devoted to the chemical collection proper, while in the third section, close to the instrument room and lighted by the Illuminating Shaft, models drawings diagrams &c., will be arranged. The specimens in the museum can be transmitted directly to the lecture theatre by means of a small truck of the same height as the lecture table, running on wheels with india rubber tires. It was in order to facilitate this transmission that three doors connecting collection hall, instrument room. preparation laboratory and theatre, were ranged in the same line.[38]

The position of the museum varied from university to university, although it was usually on the ground floor near a lecture theatre. At Lehigh University in Pennsylvania it was placed next to the recitation room and linked by another door to the preparation room, rather than being a public space near the main entrance or even near the main lecture hall. The Polytechnic Institute in Munich (later the Technische Hochschule), opened in 1868, had a chemical museum on the first floor which was distinct from the 'Museum' on the ground floor.[39] There was also a chemical museum in Greifswald by 1874.[40] Somewhat later a museum of technical chemistry 'illustrating all parts of chemical technology, including the raw products of all countries, series of intermediate and final products, showing the entire course of manufacture' was included as part of the new chemical laboratories at the Eidgenossische Technische Hochschule, Zurich, completed in 1886.[41]

Other countries were quick to set up their own museums. The chemical museum at the Leeds School of Medicine in 1873 contained 'minerals, metallic ores and metals, rare chemical substances, and illustrations of the most important chemical manufactures in their different stages',[42] thus showing a strong similarity to Bergman's museum despite being in a medical school. At Owens College Henry Roscoe set up a chemical museum in the basement underneath the lecture theatre in the new laboratory building of 1873.[43] Alfred Waterhouse then included a chemical museum in the laboratory building he designed at Liverpool in 1878. This museum appears to have been particularly didactic:

This organic [chemistry] portion of the museum is being arranged on an effective educational plan, the fundamental hydro-carbons being placed on the lowest shelf, and each shelf above containing the derivatives in order of derivation, so that the homologous series is in horizontal lines, and the analogous series in perpendicular lines. Thus the specimens graduate in complexity from the lowest shelf upwards, showing a tabular arrangement at a glance.[44]

Subsequently he added a chemical museum, connected to the lecture theatre, in his plans for the laboratory building at Yorkshire College, which was opened in 1885 (illus. 87).[45] Nor were chemical museums in Britain limited to laboratories designed by Waterhouse. The architectural partnership of James Ireland and David Maclaren included a chemical museum in their chemical laboratory at University College Dundee (now the University of Dundee) of 1886, which was connected to a small lecture room, and via a corridor to the main lecture theatre.[46]

Chemical museums were particularly popular in the USA. Charles F. Chandler began to assemble a chemical museum in the mid-1860s, soon after he took up the chair of chemistry at Columbia University, which was moved into Havemeyer Hall when it opened in 1898 (illus. 88).[47] After Chandler's retirement in 1911, it became the Chandler Chemical Museum and survived for many years largely as a memorial to Chandler. Objects from the museum were still on display in 2013.[48] Although Chandler started the museum in the usual way as a repository for chemical samples and objects used in lectures, he had an interest in history, and unusually it also contained historically interesting material such as light bulbs made by Thomas Edison, apparatus used by Joseph Priestley and a sample of mauveine from William Henry Perkin. The collections were arranged by subject areas, usually aspects of applied chemistry such as electrochemistry, explosives, dyes and ceramics. The last of these categories was particularly esoteric, encompassing important examples of early porcelain and a tea set made for a tsar that had been coated with platinum metal from the Ural mountains instead of the silver specified. Presumably the two statues in illus. 88 came under this category. There were also themes closer to pure chemistry, including collections of the elements and their compounds (such as sodium), and a collection of

87 Chemical museum, Yorkshire College, Leeds, 1903.

organic chemicals with 4,200 samples. There was also a significant collection of photographs containing early daguerreotypes.

The University of Michigan at Ann Arbor opened a museum of applied chemistry (also called the museum of chemical industry) in 1882, which was curated by Albert B. Prescott, the professor of organic and applied chemistry and pharmacy.[49] The Museum of Materia Medica in the medical department was even older and dated back almost to the foundation of the department in 1856; in 1862 it was reported that it contained between 500 and 600 specimens of 'crude organic medicinal substances' obtained from Paris.[50] The chemical museum moved into the 1890 extension to the laboratories. In the new chemical building of 1908 the chemistry museum occupied a room next to the lecture theatre on the third floor with the new Pharmacognosy Collection, which may have been set up as a pharmacological alternative to the older Materia Medica Museum.[51] The chemistry museum appears to have moved later to a former classroom nearby. The Chemical Industry Collection contained industrial products from Michigan and elsewhere, and plans and models illustrating industrial processes.[52] In 1890 there was also an educational chemistry section, but this is not mentioned in the later report. Charles Chandler's brother William Henry Chandler set up a chemical museum that was 38 ft x 33.4 ft (11.6 x 10 m) at Lehigh University, in Bethlehem, Pennsylvania, in 1885 (illus. 89).[53]

88 Chandler Chemical Museum in Havemeyer Hall, Columbia University, New York, 1905.

The Kidder chemical laboratory building at Cornell University, erected in 1890, also had a chemical museum.[54] This museum was in fact older than the laboratory, since it was described in some detail in 1886 in words that echo Bergman's description of his museum over a century earlier:

> The Chemical Museum is located in a large room in the eastern end of the Chemical and Physical building, and contains the Silliman collection of minerals, and the collection of applied chemistry. The former comprises about three thousand five hundred specimens, many of them of extreme rarity. The latter consists of materials and products illustrating many of the applications of chemistry to the arts and manufactures, such as the manufacture of soap, sulphuric acid, soda ash, alum, white lead gunpowder, pottery, porcelain, glass, cement, dyes, pigments, oils, the refining of petroleum, etc., etc. These collections are being constantly and rapidly increased by gifts and purchases.[55]

The laboratory building even had a room for the curator. This raises the question of how many of these museums had curators. In some cases it appears that the curator was a freestanding post, but in most cases the curator was also a professor or lecturer. One suspects that the curators were usually young members of staff as they tended to hold such a post early in their careers and for a short period. Somewhat surprisingly, their appointments or deaths were sometimes even recorded in the *Museums Journal* – for example, the death of Ellwood Hendrick, curator of the Chandler Museum, was recorded in 1930.[56] The Catholic University of America (CUA) and the University of Chicago (the Kent Laboratory) had chemical museums by the late 1890s.[57] The chemical museum at CUA contained 'mineralogical specimens ... products of various refineries, and chemicals prepared by students'.[58] The lack of such a museum in the Kidder laboratories at MIT (erected in 1883) was acutely felt. The *President's Report* as late as 1909 pointed out the 'real need' for a chemical museum.[59]

The University of Wisconsin set up a chemical museum in 1901, with donations from no less than twelve companies such as H. J. Heinz and the Babbit Company, and donations expected from seven more companies and two individuals, including the Barrett Manufacturing Co. and Professor Charles Herty.[60] It was reported that:

89 Chemical museum at Lehigh University, 1893.

> These specimens are not now available, except so far as they will illustrate the lectures in organic chemistry; but it is hoped that they will soon be put where they can be readily seen, in order to show the uses of organic-chemistry in the arts and to help to enforce by their easy accessibility the processes used in various lines of manufacture.

Two years later Worcester Polytechnic Institute (WPI) in Massachusetts recorded the donation of

> a full set of samples of coal tar products manufactured by [the Barrett Manufacturing Company of Philadelphia]. This set will be added to the chemical museum, to which it will form an important and valuable addition.[61]

Donations of samples by Standard Oil and Bayer were also noted, and while it was not specifically stated that they would be placed in the museum, this was very likely.[62]

Lyman Newell set up a chemical museum at Boston University in 1906. Although Newell was interested in the history of chemistry, the museum seems to have been purely functional and devoted to applied chemistry, as it soon received donations of material from Royal Baking Powder Company (cream of tartar), the National Lead Company (white lead paint) and the Carborundum Company (silicon carbide).[63] There was already a museum at Harvard in 1909.[64] When Harvard was planning the expansion of its chemistry department in December 1909, it was decided to place the museum next to the library. Although it appears on a ground plan of 1917, this ambitious plan to complete a 'chemical quadrangle' was never completed.[65]

In Australia there seems to have been a focus on chemical collections rather than chemical museums. In Sydney a museum of chemical preparations was part of its museums complex completed in 1865.[66] A plan of the new chemistry laboratory building at the University of Sydney, completed in 1890, shows a large room (32 x 21 ft) allocated to the 'Chemical Collections', which was in a basement, near other storerooms, a similar arrangement to that in Owens College.[67] It appears that this collection was used by Archibald Liversidge for his lectures. The collection was dispersed in the 1960s, when the chemistry department moved to a new building. The

University of Melbourne seems to have had a similar collection, which may have been stored in a museum in the early twentieth century. Joan Radford in her history of the department remarked:

> Technical chemistry had been included as a major subject of the honours course since 1894 in an attempt to keep abreast with European development, and samples of useful technical application had been added to the small museum which [David Orme] Masson tried to build up in the chemistry department.[68]

At the University of Adelaide the collection was stored in a display cupboard in the lecture theatre.[69]

The situation in France is not entirely clear; searches of Google Books on '*musée de (la) chemie*' and '*salles des collections*' did not throw up anything significant, but at the New Sorbonne in 1894, at least, there were collections of samples relating to such famous chemists as Chevreul, Dumas, Wurtz, Friedel, Moissan and Pasteur, which were shown to students.[70] As the Science Museum has collections of chemicals relating to these and other French chemists of the period, one wonders if these collections were eventually donated (or sold) to Henry Wellcome. There was a museum at the Institut de Chimie Appliquée in a building opened in 1923, which was also designed by Paul Nénot. It is perhaps no coincidence that it was an institute of applied chemistry. It became L'Institut de Chimie de Paris in 1932 and the Ecole Nationale Supérieure de Chimie de Paris sixteen years later.[71]

The Crum Brown Museum at the University of Edinburgh was created in its present form for the Joseph Black Building (opened in 1924) by James Walker on the basis of the original museum of 1823, which had been renovated by Lyon Playfair in around 1858. By 1888 the chemical museum was housed in the new medical building in Teviot Place, on the ground floor next to the apparatus room and opposite the preparation room and near the chemistry classroom. It appears to have been a somewhat more historical museum than most chemical museums of this period, containing, for example, the samples of strontia and bartya from the collections of Thomas Charles Hope. It is now called the Crum Brown-Beevers Museum in recognition of the renovation of the museum in the 1980s by the

90 Crum Brown-Beevers Museum at the Chemistry Department of the University of Edinburgh, 2013; note the coffee tables and chairs.

famous X-ray crystallographer Arnold Beevers (illus. 90).[72] As late as 1932 North Carolina State College had a chemical museum that contained 'specimens of the more common minerals, ores, and chemicals, together with many industrial, chemical and allied products'.[73] At the University of Melbourne a room was set aside in the new chemical building in 1936 for a chemical museum, but it appears that it was never occupied for that purpose.[74] This failure to follow through thus marks the end of the drive to establish chemical museums.

The Chemical Museum as a Library of Chemicals

Only a few illustrations of chemical museums when they were actively used have survived. The photograph of the chemical museum in Bonn in 1904 shows a combination of tall, glass-fronted cupboards with shelves with curtains to protect the specimens and desk cases (illus. 91). These arrangements were also typical of public museums in this period. The layout is additionally very similar to rare-book collections, which even today are usually stored in a mixture of open- and glass-fronted bookcases while displaying a rotating display of selected volumes (and other objects) in desk cases. While we must not forget that chemical museums contained other items such as models and diagrams, they can be regarded as libraries of chemical samples, especially the Germanic *Präparatensammlung*. By contrast, the museum at Lehigh had upright display cases that were glazed on all four sides and look remarkably modern (see illus. 89).

In the early nineteenth century the departmental books had often been the personal property of the professor (as indeed was the apparatus in some cases), and were thus shelved as part of his office or home. Other chemical books would be found in the main university library, which was not always convenient for chemists working in the laboratory. In the mid-nineteenth century there was a movement towards setting up departmental libraries. Laboratory buildings built after 1860 usually incorporated a room for such a library. At the same time, at least in Germany, the living quarters of the director (head professor) were incorporated into (or at least next door to) the laboratory building, which facilitated the merging of the professorial collection into the departmental library. As we have seen, the chemical museum became prominent in the same period, and it is quite likely that collections of chemicals and other materials that may have hitherto been the property of the professor were converted into a departmental resource in a similar manner. Hence where museums were lacking, collections of samples may have existed but were kept by the professor. The University of Glasgow may be a case in point as Robert Thomson, who taught chemistry there in the 1840s, called its chemical museum 'private', suggesting that it might have been his own property (or perhaps that of his uncle Thomas Thomson) although students were welcome to use it.[75]

North American universities made major efforts to build up their libraries in the late nineteenth century, often making trips to Europe for this purpose. Similarly, professors of chemistry strove to create and expand their chemical museums. A particularly interesting case is the chemical museum at the University of Virginia. The university decided in 1868 to set up a School of Analytical and Industrial Chemistry and John William Mallet, then at the University of Louisiana, was appointed as professor.[76] Mallet was Irish and had been educated at TCD. Possibly inspired by the museum at TCD, he immediately began to assemble a collection of 'chemicals, minerals, models [and] specimens illustrating the various arts and manufactures as practised on the large scale' from England, France and Germany. He was a judge at the Centennial International Exhibition in Philadelphia in 1876, and used the opportunity to acquire more than 500 specimens from foreign exhibitors.[77] He also obtained

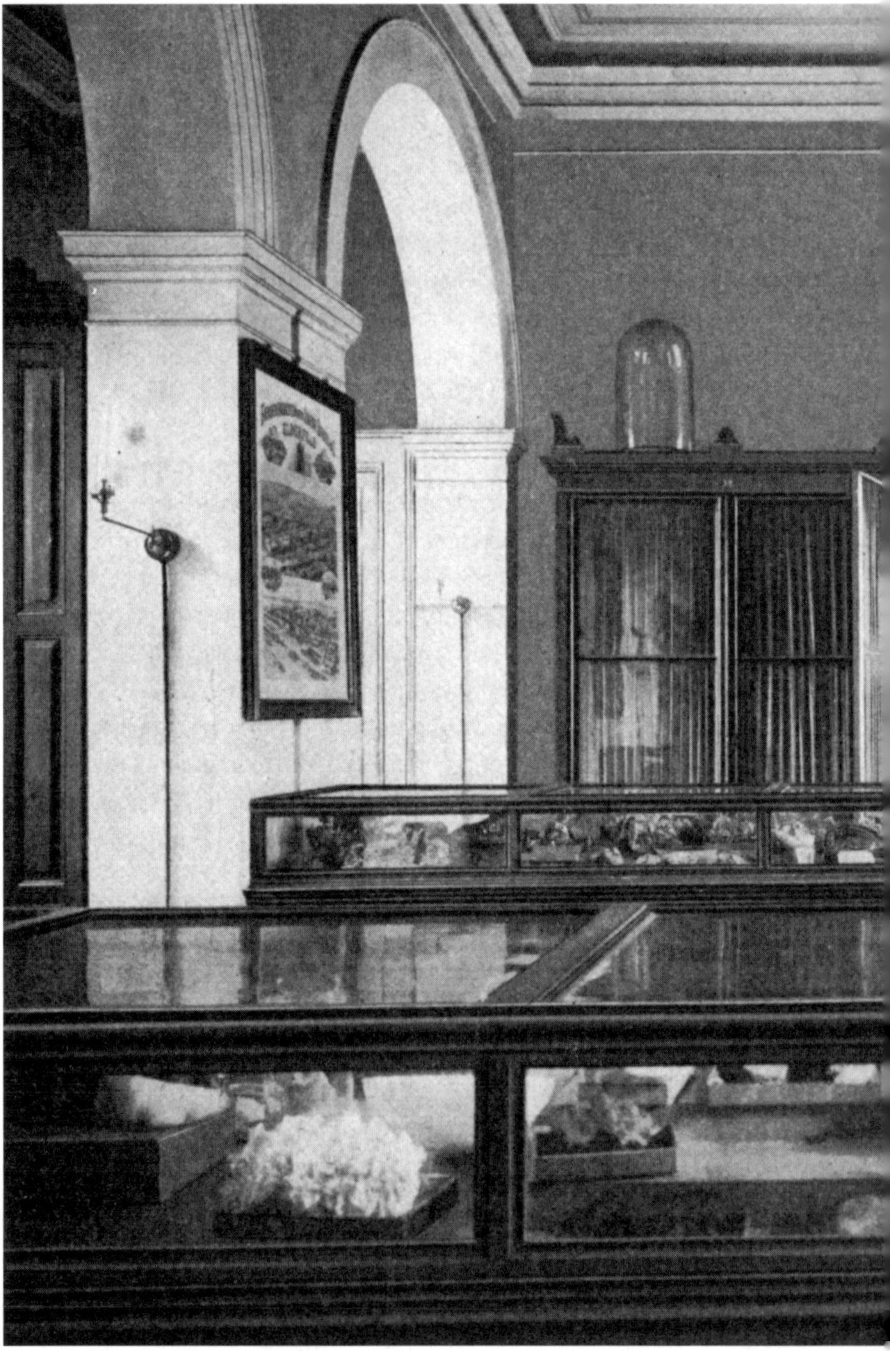

91 Chemical museum at Bonn University, c. 1900.

many exhibits from American firms. The chemistry building and its museum were destroyed by fire on 26 January 1917.[78] As a result of this sad event, Francis Perry Dunnington, Mallet's successor as professor, wrote a list of objects in the museum and their donors as a record of Mallet's service to the university. This is a remarkable book – the only one published on a chemical museum (albeit a reprinted magazine article) – and the list of donors is interesting in its own right.

William Lang, a graduate of the University of Glasgow, set about creating a chemical museum after his appointment to the University of Toronto in 1900, and 'he evidently worked his connections in Britain in the Society of Chemical Industry'.[79] Within three years he had assembled a collection of more than 500 samples illustrating applied chemistry, which had to be displayed in the main balance room because of the lack of space elsewhere. According to the student magazine in January 1903, Lang was expecting another group of specimens soon and planned 'when in Great Britain next summer, "to beg, steal or borrow a great deal more"' despite the lack of space. These samples were used daily in lectures.[80] In a similar manner Irving Fay 'made an extensive trip to Europe in the interests of the Chemical Museum at the Institute' after he retired from Brooklyn Polytechnic Institute in 1932.[81]

Melting Points

Although the evidence is inevitably scarce, it is likely that the chemical samples in the museums were used to check the identity and purity of new samples by comparing the melting points.[82] The use of melting points to characterize organic compounds and their purity was introduced by Michel Chevreul during his study of fatty acids in the 1810s and '20s.[83] The existence of these reference samples would have facilitated the use of mixed melting points to confirm their common identity.[84] The method of mixed melting points was first used by Johann Gottlieb in his study of goose fat in 1846, and it was developed by Wilhelm Heinrich Heintz from 1854 onwards.[85] It seems, however, that this method was not widely used. There is no reference to this technique in the Royal Society of Chemistry journals until a paper by Henry Armstrong and Martin Lowry in 1902,[86] and the phrase 'mixed melting points' does not appear in Google Books before 1906. Of course, at the same time other comparisons could be made using the reference sample, for example the refractive index or the optical activity.

The thin capillary tube, still used today, was first employed for melting point determinations in the 1830s by Robert Bunsen, who closed the tube at both ends, although it is more usually closed at one end only. The most common melting point apparatus used from the days of Bunsen until after the Second World War was simply a heated small beaker half-filled with concentrated sulphuric acid (glycerin or mineral oil were also used), in which was suspended a thermometer and a capillary tube held together by a thin rubber band. The tube was often made by drawing out scrap combustion tubes (used in the Liebig combustion analysis).[87] The observation of the melting point could be improved by using a magnifying glass. Somewhat safer apparatus that enclosed the sulphuric acid was introduced by several chemists in the late nineteenth century, notably by Richard Anschütz and Gustav Schultz (1877),[88] Carl Franz Roth (1886),[89] and Frederick William Streatfeild (1901).[90] Johannes Thiele introduced a b-shaped glass tube filled with oil as the bath for the capillary tube in 1907.[91] The most radical change to melting point apparatus was the introduction of the electrically heated metal block in the 1920s. Frederick Mason developed a version in 1924, which was marketed by Gallenkamp. The now-familiar apparatus with a metal heating block and a magnifying glass incorporated into one piece of equipment only appeared in the early 1960s, when it finally displaced the sulphuric acid bath.

Chemical Storerooms

Chemical museums have been overlooked because of the lack of any connection between them and the development of chemical theories, which has been the major preoccupation of most historians of chemistry. Furthermore, the absence of such museums in modern laboratories makes them appear peripheral to the development of chemistry. Some historians may have assumed that chemical museums were effectively the same as chemical storerooms, although there has not been any significant discussion of the history of storerooms either. It is thus worth examining the history of the storeroom to see if it overlapped with the development of the chemical museum, and to see if institutions with a museum also had a separate storeroom. At the outset it must be stated that there is no obvious overlap between the museum and the storeroom except that they both contained chemicals. However, the status of these chemicals was quite different. Any non-industrial chemical specimens in the museum would have been made in the laboratory and generally preserved, although small amounts may have been used for checking the identity of other samples and possibly even used in chemical syntheses. By contrast, the chemicals in the store room would have been acquired from pharmacists and, by the middle of the nineteenth century, chemical supply houses – such as Allen & Hanbury or Griffin in Britain and Merck in Germany – and were intended to be consumed in the laboratory.[92]

Bunsen had two storerooms in his new laboratory at Heidelberg, and by the 1860s practically all chemical laboratories had a number of storerooms. For example, at both Bonn and Berlin there were storerooms for apparatus, liquid chemicals and solid chemicals. The chemical store rooms were located in the basement and this model was generally copied, although most laboratories simply had a single chemical storeroom, called the 'supply room' in the USA (illus. 92). They were rarely close to the museum and it is clear that there was no connection either in spatial or operational terms between the two. An exception was Edinburgh, where the museum was at one point united with the stores.

92 Store and balance room, Yorkshire College, Leeds, 1908.

93 Lecture Hall, Massachusetts Institute of Technology, 1893, showing the cabinet on the far wall in this picture.

Reference collections (for example of synthetic dyes) were, however, sometimes kept in storerooms, as in the case of Wisconsin and Cincinnati.[93] As already noted, at Oslo the collection of chemicals was stored in the preparation room, presumably so they were readily available for lectures. Indeed, in a few laboratories it appears that the collection was stored in cabinets in the lecture hall itself. The chemical collection at Imperial College was displayed in the main lecture hall in the chemistry department for many years,[94] and there is a cabinet in the lecture hall of MIT (illus. 93) that may well have been used to house the chemical collection (the fact that the doors are open in the photograph lends weight to this argument) and may explain why MIT never had a chemical museum despite its president's best intentions.

The Decline of the Chemical Museum

It is not clear what caused the decline of the chemical museum. On the basis of the scant evidence that survives, it appears that the use of chemical museums for teaching and reference was never properly incorporated into routine chemical practice and teaching. Hence, once the generations of professors who founded the museums retired, the museums were soon overlooked in a lapse of corporate memory and removed to ease the ever-growing need for more teaching space.[95] The decline in the teaching of industrial and applied chemistry in chemistry courses may also have been a factor given the high number of industrial samples in many collections.[96] Furthermore, little effort seems to have been made to keep the collections up to date, and the samples donated in the late nineteenth century would have appeared outdated by the 1940s. Insofar as the museums contained laboratory chemicals, the specimens were probably considered, perhaps unfairly, by a new generation of chemists to be too impure or too ill defined to be of much use as reference samples. Although the chemical museum is still mentioned in a manual of laboratory design as late as 1951, the switch from the teaching of descriptive chemistry based on examples and practice to theory-based teaching had sealed its fate by the end of the Second World War.[97] Another factor was the concurrent change in the general understanding of what a museum was. The definition of a museum was becoming much narrower, usually restricted to museums open to the public and with a dedicated curatorial staff. No longer considered to be museums, the chemical museums became baffling

94 The chemical museum in the foyer of the Baker Laboratory, Cornell University, Ithaca, New York, 1953.

and pointless collections of old chemicals and dusty models. They might have survived if they had become historical museums illustrating the history of chemistry, but the period of their decline was also when the interest of chemists in teaching the history of chemistry to their students also sharply declined. The exact date of their closure is rarely known, but the space formerly allocated to the chemical museum at Illinois was taken over by the library in September 1951.[98] Other museums, such as the one at Cornell, appear to have lasted somewhat longer (illus. 94).

By the 1960s the vast majority of the museums had been forgotten, their contents packed away or simply treated as public displays of little scientific or historical importance. A few historic samples were donated to museums, such as the samples of Hofmann's dyes that were donated to the Science Museum by Imperial College in 1936 (illus. 95), and they were followed by a sample of Perkin's mauve in 1952. It might have been thought that these samples would have become more valuable once they could be investigated using techniques such as infrared, NMR and mass spectrometry, but this was not the case in practice. A generation later most of the samples were disposed of as either a health hazard or at least a waste of space. The display at TCD still existed in the 1970s, containing a significant amount of materia medica, but it was later displaced by offices and much of the material was discarded. It was replaced by a historical display in 1977.[99] Now only a few collections survive and only the Crum Brown-Beevers Museum in its original location, albeit now also serving as a coffee room (illus. 90).[100] The last collection to be used as originally intended was probably the collection at the Norwegian Institute of Technology in Trondheim, which was still in active use by chemists in 2000.[101]

While the disappearance of the chemical museums might have seemed a mark of progress to younger chemists, they left a gap that has never really been filled. If we return to the analogy with libraries, it was as if the library had been closed and chemists expected to keep all the books they needed in their offices or homes. Without a museum to place them in, what could chemists do with the samples they produced in the course of their research? There were sometimes storerooms (or part of a storeroom) for these samples of compounds and other compounds no longer kept in the main chemicals storeroom. They were called 'morgues' and were useful if you only wanted a small

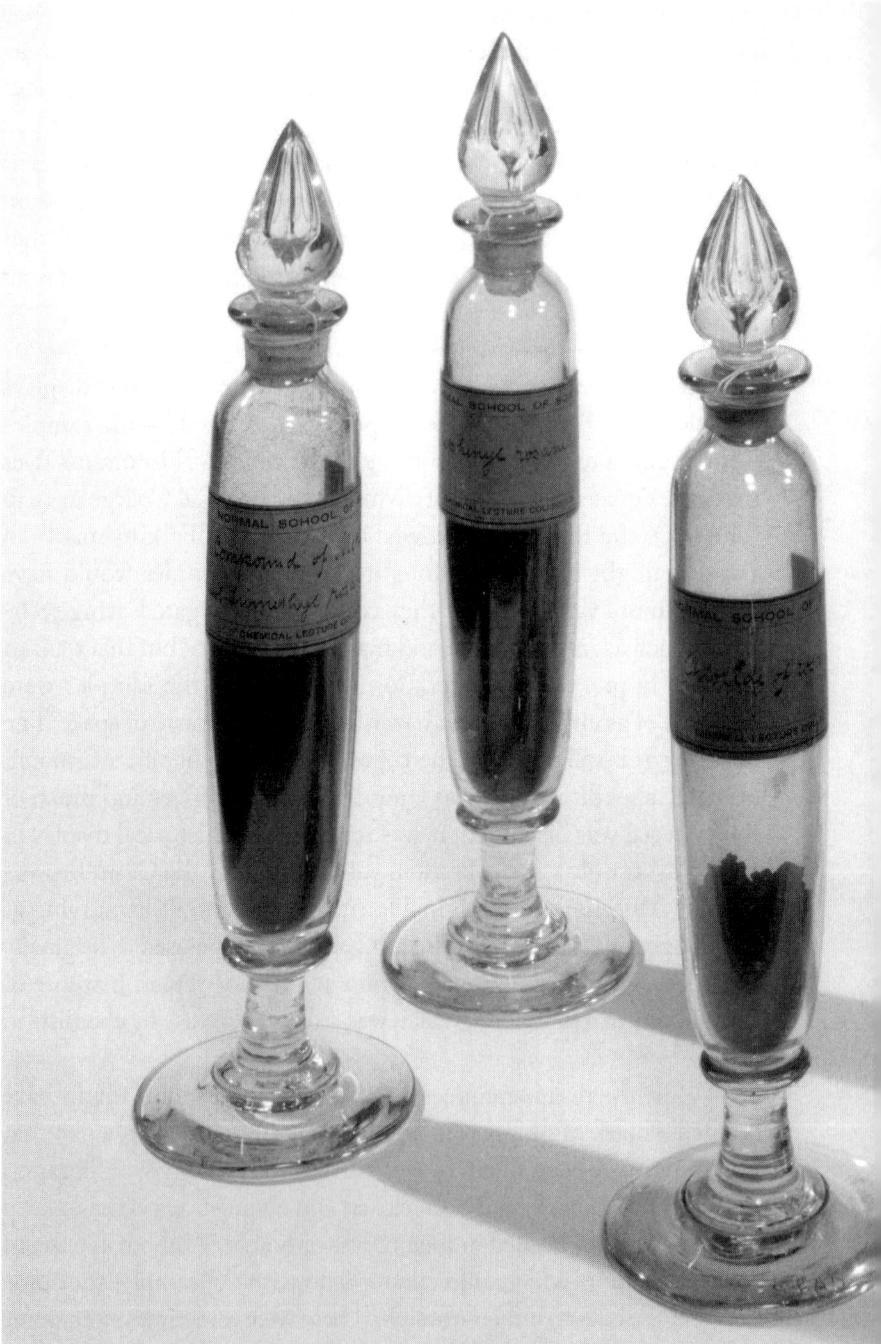
NORMAL SCHOOL OF
CHEMICAL LECTURE

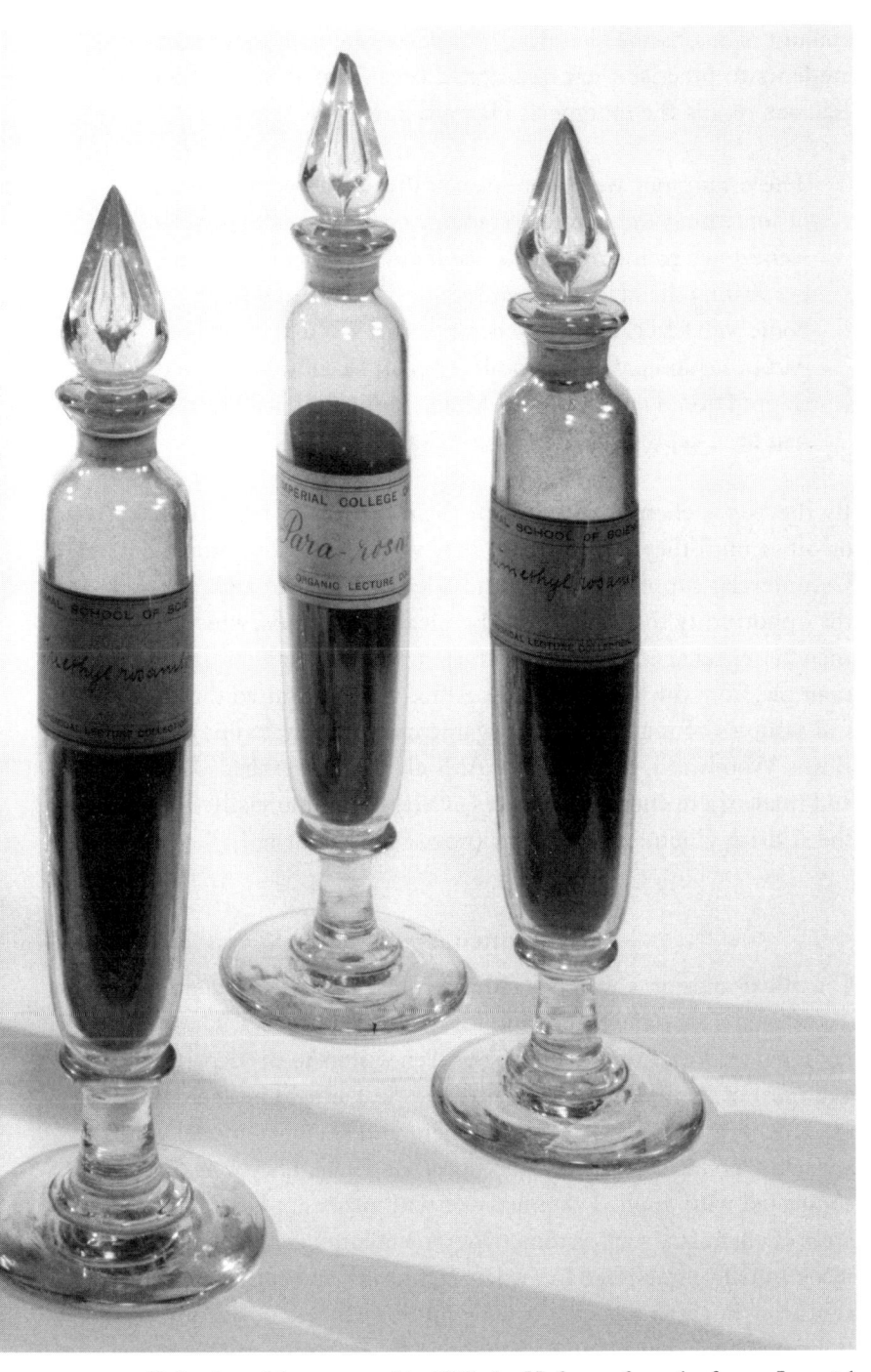

95 Collection of dyes prepared by Wilhelm Hofmann from the former Imperial College 'chemical lecture collection', now held by the Science Museum.

amount of an obscure chemical, or needed samples of unknowns for students to practise their qualitative organic analysis.[102] Richard Laursen recalls the morgue at Harvard in the mid-1960s:

> These 'morgues' were great places: if you wanted a little bit of something and didn't want to buy 500 grams of it, you just went down to the morgue to see if any was there. I remember visiting the Harvard chemical morgue once and finding a bottle with a label handwritten in French. I was told it had been part of a shipment of chemicals (or pharmaceuticals) that was shipped from France to New Orleans during the Civil War and had been captured by the Union Navy.

By the 1970s chemists often kept their samples in the laboratory or office until they retired, when they were frequently thrown out. Commercial suppliers of fine and specialized chemicals spotted the opportunity to acquire rare chemical compounds, which could then be repackaged and sold in small amounts to laboratories. For example, from the 1950s onwards Alfred Bader acquired the chemical samples of many famous organic chemists, including Robert Burns Woodward, to create the Aldrich Rare Chemicals Library, and made these chemical samples available commercially through the Aldrich Chemical Company (now Sigma-Aldrich).[103]

A Forgotten Aspect

Chemical museums were specifically museums within chemistry departments for didactic use and for reference purposes. They were not intended for the general public or even visitors to the department, and most of them did not have any historical aims. They contained specimens, models of factory equipment and chemical models, and apparatus. Some had mineral collections, but not all. They were mainly connected with applied chemistry or with other applied boundary areas of chemistry such as mineralogy/metallurgy, ceramics or pharmacy. Initially established in the late eighteenth and early nineteenth centuries in areas associated with mining (Sweden, Saxony and Bohemia), they only became common in the late nineteenth century. Although the models for these museums in that period were in Germany (Berlin and Bonn), they flourished in the English-speaking world – Britain, the USA and Australia. As the nature of chemistry teaching changed and the exhibits became outdated, they fell out of use. Some museums were moved into public areas, such as an entrance foyer, to become static displays, but most of them were dismantled between the 1950s and '70s. Chemistry departments often only become aware of them when their collections are found packed away in a basement or cupboard.

It is surprising that modern historians of chemistry have overlooked the chemical museum when there is such a strong emphasis on avoiding looking at the past in terms of the present or judging past events by their impact on current science. In realms of theory, historians of chemistry often write about past failures or disappearing qualities, such as phlogiston, N-rays and polywater. Yet the far more tangible concept of the chemical museum has been largely passed over. John Pickstone has shown how the museum, specifically the natural history museum, is characteristic of a certain kind of knowing, namely taxonomies that arose in the mid-nineteenth century in both universities and major cities.[104] The chemical museums, however, stand aside from these developments, although they were doubtless influenced by them. They were not taxonomic in their approach, were intended for teaching rather than research and were not accessible by the general public. While both natural history museums and technology museums have continued to the present day, chemical museums have died out, partly because, *pace* Pickstone, they were 'ways of teaching' rather than 'ways of knowing'. The failure of historians to study chemical museums fully has a parallel in their limited coverage of industrial laboratories. While historians have devoted considerable attention to the development of the research laboratory, relatively little attention has been given to the works laboratory or the manufacturing laboratory. This is the subject of the next chapter, on the laboratory in the chemical industry, which also looks at another hidden aspect of chemistry: the laboratory coat.

9
Cradles of Innovation
Carl Duisberg and Elberfeld, 1890s

Since the pioneering work of the German-American historian John Joseph Beer in the 1950s, there has been considerable interest in the development of the industrial research laboratory in the chemical industry.[1] At first, following Beer, historians concentrated on the laboratories established by the German synthetic dye industry from the 1860s. The research laboratory at Bayer was held up as a model by Beer and later by Georg Meyer-Thurow, but subsequently the laboratory at BASF was shown by Ernst Homburg to have been a more appropriate examplar.[2] Outside the chemical industry, many people have assumed that the laboratory established by Thomas Edison was the first industrial research laboratory, on the dubious basis that it was in effect the research laboratory of American General Electric.[3] Edison's laboratory, however, was really the workshop of an individual inventor. Once we include the laboratories of individuals in our ambit, we have to admit the private laboratories of such pioneers as William Henry Perkin, Ernst Solvay and Ludwig Mond. What is, however, overlooked in all of this is that the laboratory in the chemical industry long predates the industrial research laboratory. If we define an industrial chemical laboratory as an organized space in which a chemical product is analysed as it is being produced, the assaying workshops of the medieval period, as shown so carefully by Agricola and Lazarus Ercker, must surely count as the first industrial laboratories.

Pharmacists and apothecaries have compounded chemical medicines in their workspaces since Paracelsus introduced chemical remedies in the early sixteenth century. We have seen, for example, how the counter-bench was probably introduced into the chemical laboratory from pharmacies (see page 54). Apothecaries' Hall in London, which produced drugs, was part-laboratory and part-factory from the late 1660s, when it was rebuilt after the Great Fire of London.[4] In the early eighteenth century Ambrose Godfrey's laboratory in Southampton Street, Covent Garden, London, was a manufactory for pharmaceuticals on one side and phosphorus on the other (see illus. 16 and the illustration of the other side at wellcomeimages.org). In late eighteenth-century Franconia, at least, the term 'chemical laboratory' covered both a laboratory in the sense that we understand it, and a place for producing speciality chemicals.[5] In the middle of the following century, the firm of William Bailey had a 'chloride of gold laboratory' in Wolverhampton that was probably where the gold chloride was manufactured.[6]

Even if the evidence is thin on the ground, there were clearly works laboratories in the late eighteenth century that performed analysis and what we would now call quality control – in fact not very different from the works laboratory at the pharmaceutical firm I worked for in the mid-1970s. They probably were also used to solve occasional problems thrown up by the manufacturing processes. They were more or less standard in the pharmaceutical industry by the early nineteenth century; the firm of Duncan & Ogilvie in Edinburgh (later Duncan, Flockhart) had a laboratory in the 1820s.[7] The laboratories also seem to have been common in specific sectors such as alum, gunpowder and soap. For example, there was a works laboratory in the soap and candle works of B. T. & W. Hawes in Southwark, London, in 1842:

> Contiguous to offices is a small laboratory fitted up with a furnace, a sand bath, a distilling apparatus, and other conveniences for conducting the chemical analysis of soap, and making experiments incidental to the manufacture [of soap].[8]

By the 1860s at the latest, it would seem that every self-respecting alkali manufacturer in Widnes had a works laboratory, even if most of these laboratories would have been very basic.[9] Chemical laboratories were also starting to appear in other industries. Henry Bottinger set up a laboratory at Allsopp's brewery in Burton-on-Trent when he arrived in 1845, and the London & North Western Railway set up a rudimentary laboratory in a cottage at Crewe in 1864.[10]

The research laboratory began to emerge in the 1860s. The French pharmaceutical firm A. Gélis had a *'laboratoire de recherches'*

[sic] at its works at Villeneuve-la-Garenne, near Saint-Denis in 1862, where it made pyrodextrin (roasted starch) as a colorant for liquors.[11] Charles Girard also had one at the pioneering synthetic dye company La Fuchsine at Saint-Germain in 1865,[12] but we know very little about these laboratories. Then, in the mid-1870s, we encounter the first purpose-built research laboratory, at BASF in Ludwigshafen, followed in 1889 by BASF's central research and development laboratory. Bayer's research laboratory at Elberfeld was opened in 1891. These laboratories, devoted mainly to organic chemistry, and aromatic organic chemistry in particular, were based on the academic laboratories of the 1860s and '70s. They were then copied by other firms in different branches of the chemical industry, such as United Alkali in Runcorn in 1891 and the Du Pont Experimental Station in 1903. Not all industrial research laboratories of this period could match the scales or budgets of these firms. The chemical research laboratories of Burroughs Wellcome, for example, were relatively modest.

Chemical activities in a factory are not necessarily carried out in a laboratory. This is partly because of the scale involved. The Gossage process to remove hydrogen chloride from the smoke of the Leblanc Process was developed in the 1830s in a disused windmill at Stoke Prior, Worcestershire, not in a laboratory.[13] Conversely, in the nineteenth century companies hired chemists without giving them a laboratory. This seems to have been particularly the case in the brewing industry, where there were concerns that the presence of a laboratory implied that the brewery was adulterating its beer.[14] It raises yet again the question of what a laboratory is. The workshops of the early metallurgists or early pharmacies were not called laboratories at the time, although they fall within my definition of a laboratory being a space where chemistry was carried out. The works laboratory seems to have arisen mainly because of concerns about quality control, both of goods coming in and products leaving the factory. This concern was particularly acute in the brewing industry, especially after the arsenic scare of 1900 (see page 278).[15] For a research laboratory to arise there has to be the conviction that chemical research could help a particular branch of industry. For example, there was no point in a brewery having a research laboratory until it was clear that chemistry (and biochemistry) could shed light on the fermentation process. In a similar way, research laboratories in the synthetic dye industry were spurred by the introduction of the azo dyes, which could be permutated to produce an almost infinite number of dyes. However, Homburg concluded that the two key factors in the case of the synthetic dye industry were the introduction of a patent law in the German Reich in 1877, and the existence of an oligopolistic market structure.[16]

This chapter focuses on how the industrial laboratory was designed, and how it was similar to the academic research laboratory, yet different; it also compares the works laboratory to the research laboratory. It is also interesting to consider how pharmaceutical laboratories were different from chemical laboratories, even within the same company. Just as the academic laboratory remained stable in its design for many years, the industrial laboratory remained unchanged for almost a century. However, the impact of new methods and new instrumentation seems to have produced more significant changes in the industrial laboratory than in the academic sector (see chapter Eleven). In order to put these developments into context, a brief history of the chemical industry is given here, as well as a glimpse of one of the earliest industrial laboratories, Apothecaries' Hall.

The Chemical Industry

The question of whether metals are part of the chemical industry – and surely they are if the products of that industry are defined as materials produced by chemical processes – throws into relief the problem of discussing laboratories in an industry that did not exist before the early nineteenth century.[17] It can be argued that there was no single 'chemical industry' up to the 1820s, but a number of different industries that can be considered to be part of the chemical industry as we understand it today. The term can be easily extended to cover a wide range of industrial activities, including the manufacture of soap, glass, paper, sugar, matches and even porcelain.

In the period between 1700 and 1850, the chemical industry was predominantly an endeavour based on the manufacture of inorganic chemicals. Industrial organic chemistry is characterized by its complexity, and only became a mature area of the chemical industry in the latter half of the nineteenth century. Furthermore, natural products provided many of the needs subsequently supplied by the synthetic organic chemical industry. Indeed, the 'chemical industry' in the

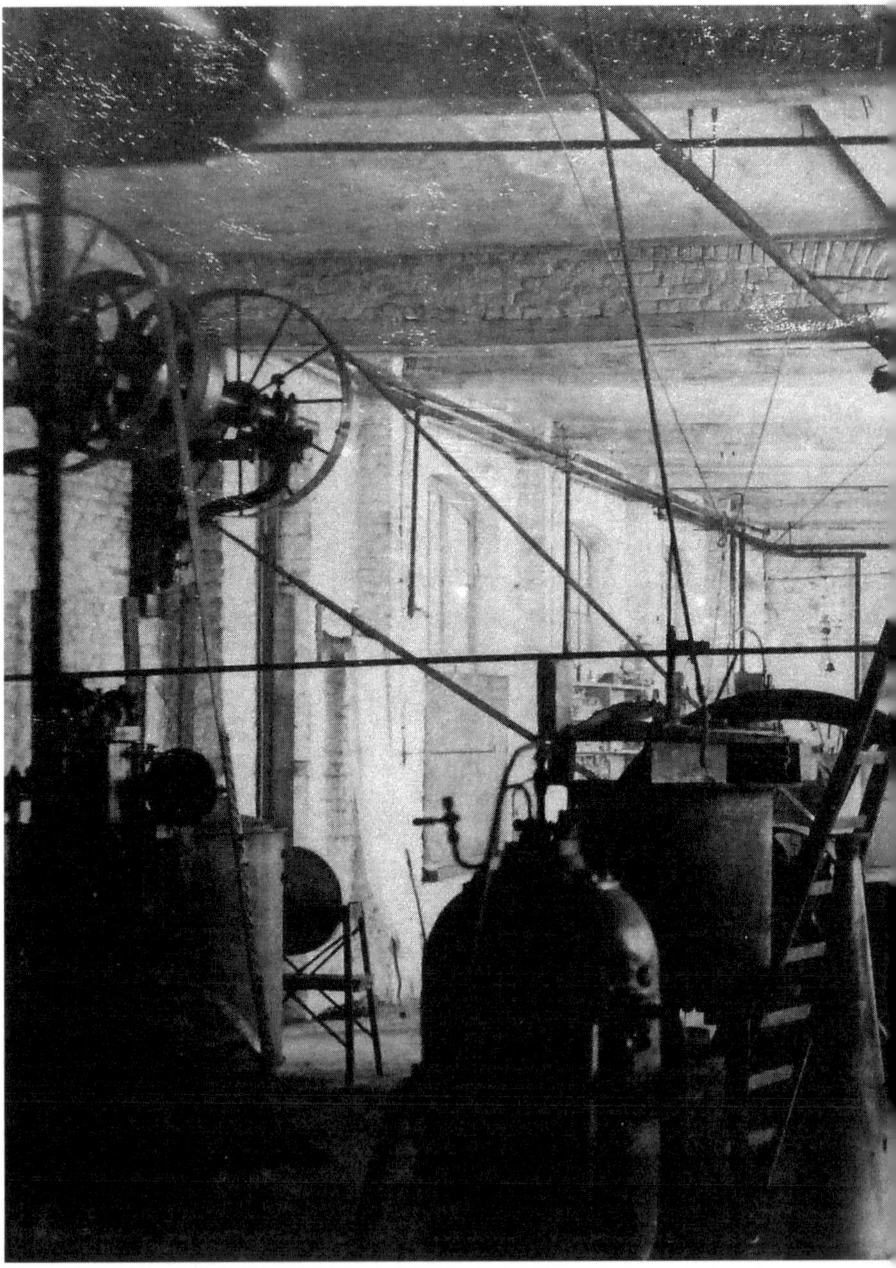

96 *Laboratoire de tests et d'essais*, Solvay, Saint-Gilles, Brussels, 1889. This test laboratory was very different from the standard laboratories of the period, being more like an engineering workshop.

seventeenth and early eighteenth centuries was almost entirely the copperas industry. Copperas, made by weathering pyrites for several weeks in the open air, was iron(II) sulphate. While it was used as a mordant in dyeing with natural dyes and in the manufacture of ink, it was mostly converted into sulphuric acid by heating. In the late eighteenth century this route was displaced by the lead chamber process, which produced a weaker, less pure sulphuric acid, but cheaply and in vast quantities. This abundance of sulphuric acid in turn permitted the development in the early nineteenth century of the Leblanc process for sodium carbonate, which was both technologically complex and fearsomely polluting, fouling the air, water and earth alike.

By 1850 the inorganic chemical industry had achieved a state of maturity and cohesion. Sulphuric acid could be made cheaply using the well-established lead chamber process, which in turn served the Leblanc soda industry and the fertilizer manufacturers. The products of the chemical industry were used by such major branches of industry as the metal producers and the textile industry, as well as soap and glass manufacturers. Yet in little more than a decade two young men, Ernest Solvay and William Henry Perkin (both born in 1838), had established the basis for a completely new chemical industry. Solvay, a Belgian entrepreneur, developed a new way of making soda by reacting brine with carbon dioxide and ammonia, which avoids the energy costs and excessive pollution of the Leblanc process (illus. 96). His process was introduced to Britain, the undisputed leader of the soda industry, by the German chemist Ludwig Mond. Perkin, while an assistant to Wilhelm Hofmann at the Royal College of Chemistry in London, created the first useful synthetic dye, mauve, from the coal-tar product aniline. Mauve soon passed out of use, as it was replaced by other new dyes, but the stage was set for the development of a new industry based on coal tar, hitherto a noxious waste product. A major breakthrough was the laboratory synthesis of the important red dye alizarin by Carl Graebe and Carl Liebermann in 1868, followed by the scaling up of the process to an industrial process by BASF and independently by Perkin. The success of synthetic alizarin enabled the German industry to consolidate its position while both France and Britain fell back. By the 1880s, the German dye companies were already exploring the possible manufacture of pharmaceuticals from coal tar.[18]

Apothecaries' Hall Laboratory

In the Middle Ages and indeed right up to the end of the eighteenth century, there was little or no distinction between a workshop and a laboratory. It is not surprising that some laboratories produced chemicals as well as carrying out chemical operations such as analysis and synthesis.[19] The laboratory of the Society of Apothecaries at Apothecaries' Hall in Blackfriars, London, is an excellent example of such a laboratory.[20] It was originally founded in 1672 and, following a series of developments in the late eighteenth and early nineteenth centuries, a Great Laboratory was constructed in 1822–3. Its main purpose was the manufacture of drugs, which were then sold to a number of customers, principally the Royal Navy and the East India Company. There were two departments, the galenical and the chemical.

The galenical department, which was basically concerned with the manufacture of drugs from natural products by non-chemical means, was merged with the chemical department in 1826 when the operator of the galenical department, Richard Clarke, left. The post of superintending chemical operator had been created in 1812 for William Brande and he held the position until his death in 1866, although his involvement with the laboratory declined after the 1830s. He was an apothecary, a chemist and a Fellow of the Royal Society. From 1813 to 1852, he was also professor of chemistry at the RI. He was closely involved with the Royal Mint and became superintendent of its Coining and Die Department in 1852. Brande did much to improve the scientific and technological standing of the Apothecaries' Hall laboratory including, as noted in chapter Five, introducing steam power. The use of steam for both power and heat, for the ovens and distillation apparatus, was crucial to the effective running of the laboratory's operations, as this description, which appeared in Charles Dickens's magazine *Household Words* in 1856, makes clear:

> Whoever pays a visit to the Hall in Blackfriars will be shown how it is composed of two distinct parts. From a steam-engine room he is taken to where great mill-stones powder rhubarb, rows of steam-pestles pound in iron mortars, steam-rollers mix hills of ointment, enormous stills silently do their work, calomel sublimes in closed ovens, magnesia is made and

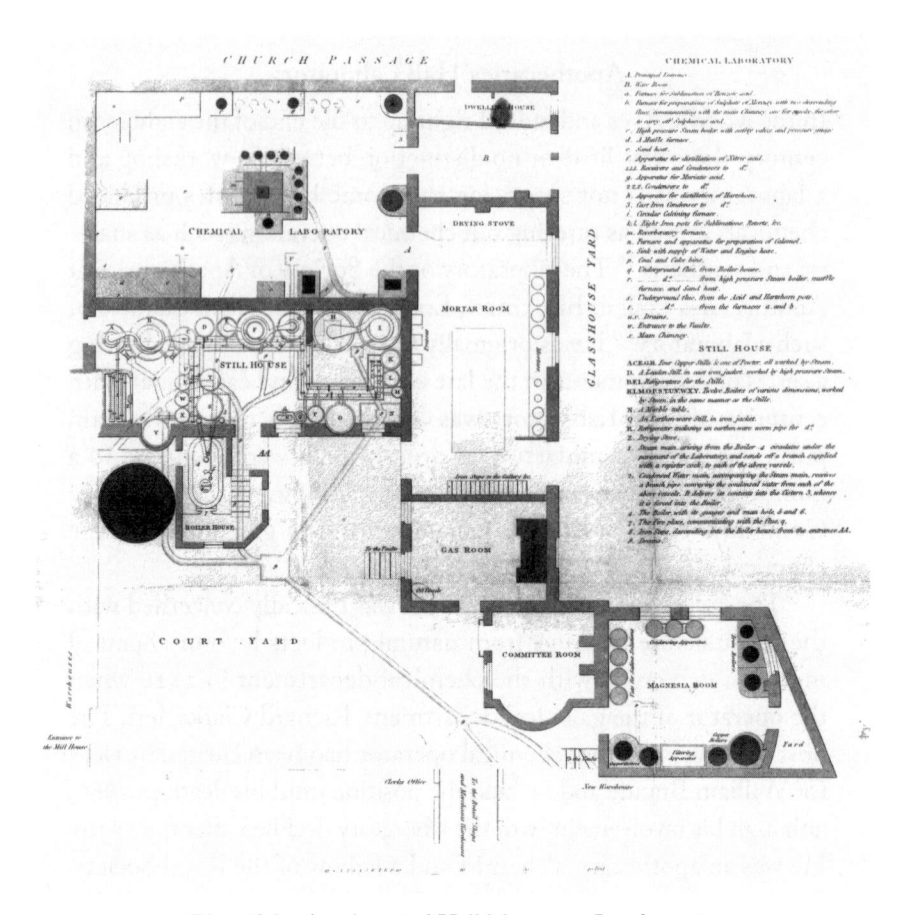

97 Plan of the Apothecaries' Hall laboratory, London, 1823.

evaporated, crucibles are hot and coppers all heated by steam are full of costly juices from all corners of the world. He will find in the cellar barrels fresh tapped of compound tincture of cardamoms, tincture of rhubarb, and such medicated brews; he will find in a private laboratory the most delicate scientific tests and processes employed for the purposes of trade by a skilful chemist; he will find warehouses and packing-rooms, perhaps, heaped up with boxes of drugs to be sent out by the next ship to India, and apparently designed to kill or cure all the inhabitants of Asia.[21]

A much longer but more exact description was given by Brande in 1823 (illus. 97). The main laboratory was large, 50 ft square and 30 ft high. It was divided into the still house, where all the distillations took place in six stills, and the chemical laboratory. Adjacent to the chemical laboratory there was the mortar room, which contained at one end a large drying stove for use when a high temperature was required. At the opposite end a section was divided off for producing gas from whale oil (used mainly for lighting). Above the mortar room there was a storeroom for utensils, and the test room and laboratory

> fitted up with the requisite apparatus for minute and delicate investigations and in which chemical tests and other articles requiring peculiar attention and cleanliness are prepared.[22]

Beyond the gas room there was the magnesia room for the preparation of magnesia and other salts. The laboratory did not make its own magnesium sulphate, but purified crude magnesium sulphate bought from outside suppliers. As Brande remarked,

> In the construction of the new laboratory, safety is insured by the whole being fire proof and it is ventilated by a series of apertures in the roof which may be opened or closed at pleasure.[23]

There were several furnaces in the chemical laboratory, for example one used for the sublimation of benzoic acid, and a sand bath. In addition there was specific apparatus for the production of ammonia, hydrochloric acid and nitric acid. Both hot and cold water was supplied from a well on the premises, and there was even a separate supply for a water engine to fight a fire. The steam was generated by a boiler in a separate steam house, and was supplied to the still house under a pressure of one and a half atmospheres. The lack of flame made this still house particularly useful (and safe) for the distillation of ethyl nitrate and diethyl ether in an earthenware still. As the article in *Household Words* noted, there were rooms and warehouses for the storage of chemicals used in the laboratories and for the final products awaiting shipping.

The testing room and laboratory was presumably used to check that the final products satisfied the criteria laid down by the *Pharmacopoeia Londinensis*, most of which would have been simple visual tests rather than chemical analyses. I carried out similar tests from the *British Pharmacopoeia* in the 1970s, for example dissolving pharmaceutical products in specific solvents and checking whether the solution was turbid. More complex investigations were made by Brande and Henry Hennell, who was the chemical operator from 1821 to 1842. At a time when there was much interest in the discovery of alkaloids and other chemical compounds from plant materials such as opium and Jesuits' bark, they worked on methods of extracting these alkaloids. In 1831 Hennell obtained the crystalline compound elaterin (a triterpenoid, now known as cucurbitacin E, $C_{32}H_{44}O_8$) from tincture of elaterium, derived from the squirting cucumber (*Ecballium elaterium*). In the 1830s Hennell's pupil Frederick Penny worked on the issue of equivalent weights, which arose from the laboratory's routine analysis of potassium nitrate obtained from the East India Company.

Apothecaries' Hall's decision not to continue with the production of quinine (by 1834 at the latest) was a decisive watershed in its production of drugs. It would not thereafter be able to keep up with the increasing pace of innovation in the pharmaceutical industry. Its production range inevitably began to decline, although the demand for its drugs remained high. A related decline occurred in the relevance of its chemical work. Robert Warington, the Chemical Operator from 1842 to 1866, had extensive outside interests, being the founder of the Chemical Society in 1841 and one of the founders of the Royal College of Chemistry in 1845. He was also the chemical referee for four of the metropolitan gas companies, and was interested in the development of aquaria. He used his position at Apothecaries' Hall as a base for his extensive consulting work, rather than making any attempt to build up its scientific status as Brande had done in the 1810s and '20s. Nonetheless, the Hall continued to trade as a pharmaceutical business until 1922.

The Works Laboratory

When I worked briefly as a quality analyst in a British pharmaceutical company during the university vacations in the 1970s, the most sophisticated instrument I used was a UV lamp located in a more advanced laboratory in another part of the factory. The laboratory shown in illus. 98 is not untypical of a works laboratory at the lower end of the scale in the 1930s. The poverty of the laboratory is revealed

98 Works laboratory, 1930s.

by the torn lab coats worn by the chemists, who may well have had only basic training in chemistry obtained in evening classes. The laboratory itself is little more than a garden shed with a table, very basic drawers and a single bottle rack. There is an electric lamp but no sign of any electrical equipment or even gas taps. The chemical operations are the most basic operations of filtration and possibly volumetric analysis. This was quite usual. Even when I worked as a quality analyst four decades later, the bulk of my work was visual inspection of powders and solutions rather than any form of chemical analysis. However, note in the midst of this scientific poverty the presence of a large microscope, which is ostentatiously placed here for the benefit of the photographer — not least because it could not have been used in that position for lack of light. Even more mysteriously it appears to be a polarizing microscope, suggesting that this was a laboratory which handled mineral samples, probably in a different laboratory. Nevertheless, its presence reminds us that the microscope was used regularly in chemical analysis, especially for quality control, up until the 1960s.

A photograph from the same period (illus. 99), by the well-known photographer James Jarché, shows a chemist working in the Llandarcy laboratories of the National Oil Refineries Ltd (a subsidiary of the

99 Researcher conducting a volumetric analysis experiment in the laboratories of National Oil Refineries Ltd, Llandarcy, Wales.

then Anglo-Persian Oil Company, now BP). While the chemist may be better dressed and the laboratory furnishings more luxurious, the chemical operation he is carrying out is still fairly simple, namely the titration of unsaturated hydrocarbons using iodine and a stopclock. A similar setting on a larger scale is provided in the painting by Ernest Wallcousins of an analytical laboratory in ICI in the mid-1950s (see frontispiece). It is a much larger laboratory staffed by chemists in immaculate white lab coats, furnished with many benches, and completely filled with bottles and glassware. Some of the apparatus is clearly specialized, such as the curious long-necked flasks in the bottom left-hand corner of the painting, but much of the work would have been routine analysis. In most respects this industrial laboratory would have been very similar to an academic analytical laboratory in this period. Its most striking difference lies in the number of women working in the laboratory at a time when academic laboratories were still male dominated. The two situations were in fact related; women would often leave university to go into industry, where the hours were shorter and the competition less severe, rather than working for a doctorate or academic position. At the same time, women were finding it easier to gain employment in industrial laboratories after the Second World War, although they were often expected to do routine work.[24] An interesting feature of the painting, perhaps wholly unintended, is the woman chemist on the far right of the picture who looks as if she is offering tea to the seemingly awe-struck male factory worker when she is simply carrying her set-up to the bench.

A Visionary Chemist

Carl Duisberg was born in Barmen (now part of Wuppertal) in the northern Rhineland in 1861, the son of a ribbon manufacturer (illus. 100).[25] Barmen, famous for being the birthplace of Friedrich Engels some 40 years earlier, was an important textile town. Two years after Duisberg's birth, a dye salesman, Friedrich Bayer, and a dyer, Johann Friedrich Weskott, founded a firm in the town to make synthetic dyes. The breakthrough for Bayer was its success in making alizarin in the 1870s, which enabled the firm to take over less successful rivals. Curiously, Duisberg's mother was also a Weskott, but not related to the co-founder of Bayer.[26] She did, however, support her son's desire to study chemistry. He went to Göttingen, then took his PhD at Jena. Finding it difficult to get a job, he was singularly fortunate that Bayer was looking to sponsor university research in 1883 in order to get ahead of its rivals. Carl Rumpff, the chairman of Bayer, offered Duisberg a temporary research post at the University of Strasbourg, then part of Germany. In September 1884 Duisberg joined Bayer permanently and soon became a master of dye patenting, as well as an excellent dye chemist.

Up to this time chemists had concentrated on making new dyes, or copying others, but Duisberg realized – following the enactment of the Reich patent law in 1877 – that having the right patent protection was at least as important, if not more so. His prowess with patent law proved valuable to Bayer in a key patent case over the azo dye Congo Red and its chemical relatives. A small firm called Ewer & Pick argued that as there was nothing novel about using Congo Red as a dye, the patent jointly held by Bayer and Agfa was invalid. Heinrich Caro of BASF, nominally appearing as an independent witness, was able to persuade the German Supreme Court in 1889 that the patent was valid because it showed a novel technical effect, namely direct dyeing on cotton. He also argued, however, that only dyes actually showing this effect (in this case Congo Red) should be patentable. This stance did not suit Bayer as it would allow BASF to produce other related azo dyes. In a final effort, Duisberg convinced the court to accept the technical effect part of Caro's argument, but reject its limitation to Congo Red. Having saved the day for Bayer in court, his position in the firm was secure.[27]

Although most synthetic dye firms had some kind of research laboratory by the mid-1880s – BASF had set up such a laboratory in 1874 – Duisberg had a vision of a much larger laboratory, which he was convinced would soon pay off despite its high cost. By 1889 he was able to persuade Bayer and Rumpff that such a laboratory was essential to the firm's continued growth. Completed in 1891, this laboratory set new standards for the industry. In the same year the Bayer company acquired land from the ultramarine works of Carl Leverkus at nearby Wiesdorf. The area around the factory was known as Leverkusen, and this eventually became its official name. Having completed his dream laboratory, Duisberg now had a vision of an integrated factory at Leverkusen, as set out in a memorandum of 1895. The work of constructing the new factory, with its associated houses and shops, was soon underway. By 1907 the famous chemist Emil Fischer could declare that it was 'the finest chemical works I have ever seen'.[28]

Apart from these physical constructions, Duisberg also pushed the development of pharmaceuticals at Bayer. When the rival firm Kalle had launched acetanilide as the first antipyretic Antifebrin in 1883, Duisberg had the idea of converting a waste product of dye production into a drug that was chemically similar to acetanilide

100 Carl Duisberg as a young man, 1880s.

and was even better at lowering high temperatures in patients. Bayer introduced phenacetin (para-ethoxyacetanilide) in 1888. It remained popular for almost 90 years, until concerns about its impact on the kidneys led to its withdrawal in many countries. Bayer's key breakthrough in this area was the concept of acetylating known pharmaceuticals to produce drugs with fewer side-effects, which stemmed from its success with the acetylation of tannic acid to produce the anti-diarrhoeal drug tannigen in 1894. The product produced in this way by Felix Hoffmann, in August 1897, was acetylsalicylic acid, which was better tolerated than sodium salicylate, and which Bayer promoted under the trade name Aspirin. From 1906 Duisberg also supported the development of synthetic rubber by Felix Hoffmann's near namesake, Fritz Hofmann, and the vista of possible products grew ever larger.

Appointed to the management board of Bayer in 1900, Duisberg now turned to the issue of competition between the German dye firms. The German industry as a whole effectively controlled the world production of synthetic dyes, but the competition between the various firms was fierce. In 1903 Duisberg visited America and came back enthused by the idea of a 'trust' (a formal cartel set up as a holding company) along the lines of the American steel and gunpowder trusts. He was supported by Agfa and BASF, but opposed by Hoechst, which set up a rival grouping with Cassella and Kalle. When the First World War broke out, the industry was faced with the need to finance the expansion of factories for war production, notably at BASF's new Haber-Bosch ammonia plant at Leuna, and to meet the future competition from new foreign dye firms. As a result the two blocs came together with a couple of small firms to form a trust, the *Interessengemeinschaft der deutschen Teerfarbenfabriken* (the trust of German tar dye factories) in 1916. However, this grouping never worked very well, and faced with an increasingly difficult situation in the mid-1920s, most notably the cost of keeping the large Leuna works going and developing synthetic petrol, there was clearly a need to revisit the 1916 agreement. Carl Bosch – the younger and more technocratic leader of BASF – argued for a more complete merger of the industry. His desire for a single firm triumphed over Duisberg's vision of a decentralized grouping, but Duisberg did succeed in ensuring that the new firm remained decentralized in its structure. Bosch dominated the new company, IG Farbenindustrie, as chairman of the management board, and Duisberg moved upstairs to become chairman of the relatively unimportant supervisory board. He died in 1935 just as the combine he helped to form became an integral part of the Third Reich's Four Year Plan. Over the next decade Duisberg's grand vision turned to ashes: IG Farbenindustrie was dissolved by the Allies in 1945.[29]

The Industrial Research Laboratory: Bayer and BASF

The interior of the new Bayer research laboratory resembled the academic laboratories at Leipzig and Berlin, with similar benches and bottle racks, yet it was also very different (illus. 102). Whereas most academic laboratories displayed space and light, despite the large number of students they housed, this was often lacking in the industrial laboratory. The Bayer laboratory was rather like a factory with the benches close together and with a narrow aisle. The fittings were not lavish, there was a corrugated roof and the high ceiling was held up by metal piers, a common feature of French academic laboratories in the period. The company clearly did not wish to spend as much on the building or fittings as a university seeking to attract the best professors or students would have done. The narrowness of the spaces between the benches was also a result of the company's desire to restrict the transfer of information about a given line of research — a result of dye companies suffering from chemists decamping with their industrial secrets. While I would not follow Meyer-Thurow in describing the benches as 'box-like', there was certainly an element of compartmentalization.[30]

Several people in the Bayer picture are dressed like industrial workers in brown aprons, rather than in suits or laboratory coats. The work of the chemists in the industrial laboratory was more routine and closely supervized than that of chemistry students (although the routine nature of academic chemistry in this period should not be underestimated), hence they were often not well qualified or well paid. There was also a hierarchical aspect to the way they were dressed. Those lowest in the hierarchy wore aprons, the middle-ranking chemists wore suits without ties, and the managers (absent in this picture) would have worn suits with ties and probably hats as well. The equivalent laboratory at BASF appears to have been

101 Chemists in the Hauptlaboratorium (main laboratory) at BASF, Ludwigshafen, 1922.

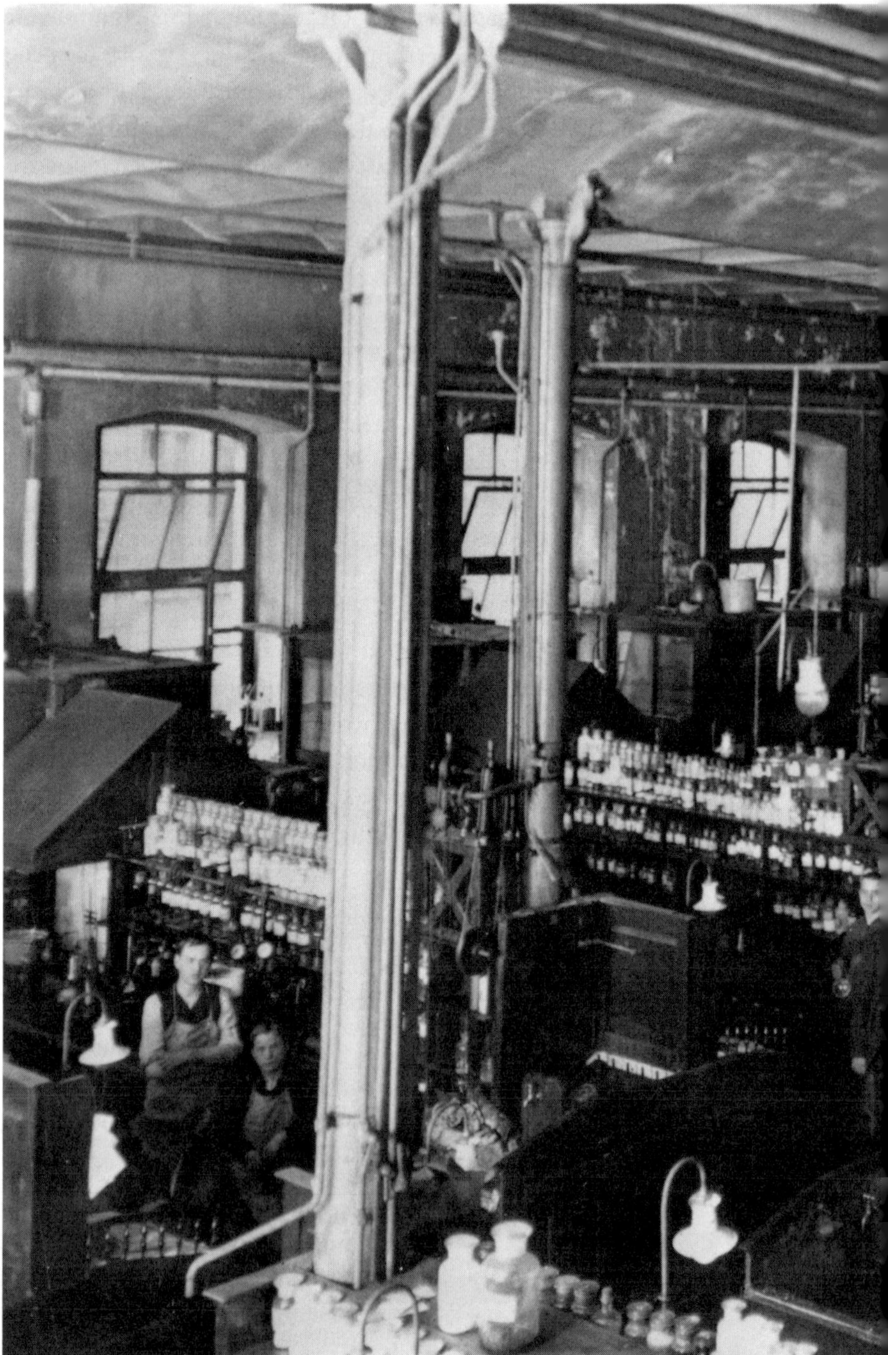

102 Bayer Research Laboratory, Elberfeld, 1890s–1900s.

more like the academic laboratories (and the Government Chemist's Laboratory, see page 281). The laboratory is less crowded and better lit by natural light (illus. 101). The benches are further apart and the aisle is wider. The laboratory fittings appear to be of higher quality and the staff, at least in the photograph, seem better dressed, wearing brown or white laboratory coats. Despite the differences, which hint at differing views of the status of industrial research in the two companies, it can be seen that all chemical laboratories were moving towards a standard design.

The Pharmaceutical Laboratory: Bayer and Wellcome

Pharmaceutical research at Bayer was first established as a field separate from dye chemistry in 1893, when it was given a laboratory in the attic of the main research laboratory. The key year was 1896, when Arthur Eichengrün was hired as a pharmaceutical chemist and a new pharmaceutical laboratory was constructed.[31] It has the laboratory benches with their bottle racks against the wall, leaving an open space in the centre that in illus. 104 is not occupied by a table or desk, although another picture shows a table in the middle with a balance. The layout and fittings are very similar to those of the Crown Contract laboratories at the Government Chemist's Laboratory (see illus. 116). As organic synthesis was the most important activity in pharmaceutical laboratories, it is not surprising that fume cupboards are prominent. The layout of the pharmaceutical laboratory is much more open than that of the main research laboratory. The different layout was not a result of the pharmaceutical laboratory housing a significantly smaller number of chemists. While the number of chemists was practically half that of chemists in the main laboratory in 1902 (eight vs fifteen), it had grown to two-thirds as many by 1912 (twelve vs sixteen).

Henry Wellcome of the pharmaceutical firm Burroughs Wellcome had set up a physiological research laboratory in 1894 mainly to produce the then new anti-diphtheria serum.[32] Encouraged by its success, he then established a parallel chemical research laboratory to be headed by his friend from his student days in Philadelphia, Frederick Power. Initially squeezed into the Burroughs Wellcome head office in Snow Hill, Holborn, it was moved in 1896 to neighbouring King Street.[33] With a Venetian frontage, this laboratory

103 Wellcome pharmaceutical research laboratory, King Street, London, 1899.

building was spread over four rather narrow floors. On the ground floor there was the director's office, linked to the laboratories by telephone, the library and the cabinet of specimens (in effect a 'museum' of the kind discussed in the previous chapter). There were laboratories for the pure and applied chemistry of medicinal substances (to use the terminology of the time) on the other three floors. In the basement there was a combustion furnace room, an electric motor to work the shakers and stirrers, and also the photographic room and the darkroom for polarimetry. As in the case of the Bayer pharmaceutical laboratories, the fume cupboards and the fume hood are very prominent. It is striking how the Bayer pharmaceutical laboratory is closer in appearance to the Wellcome research laboratory (and other laboratories built in 1896; see chapter Seven and illus. 84) than the main research laboratory at Bayer.

The White Laboratory Coat

It is curious that such an iconic feature of the laboratory as the white coat has received so little attention from historians. Insofar as it has ever been considered at all, it has been assumed to be associated with the adoption of the white coat by physicians. I would

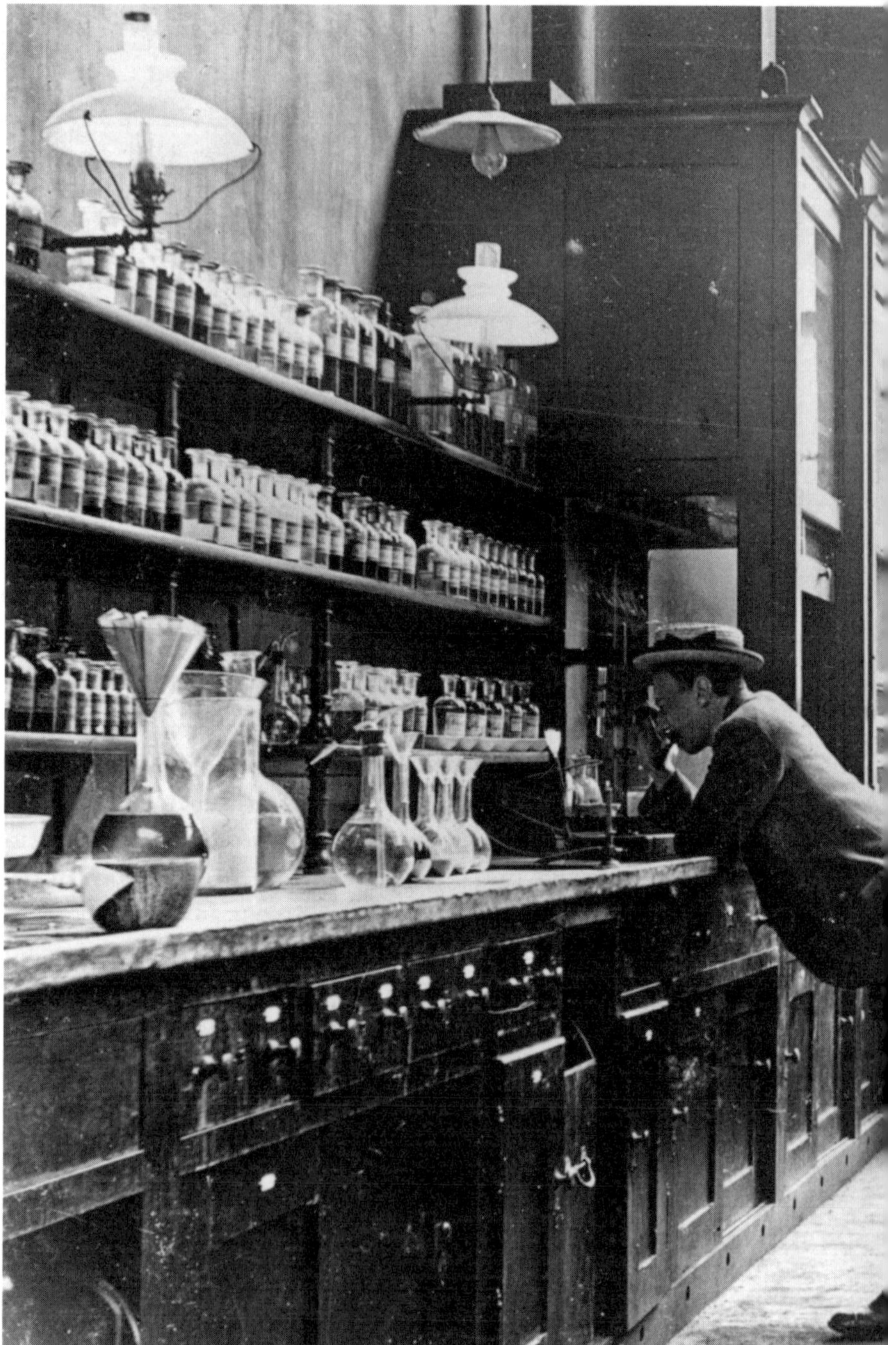

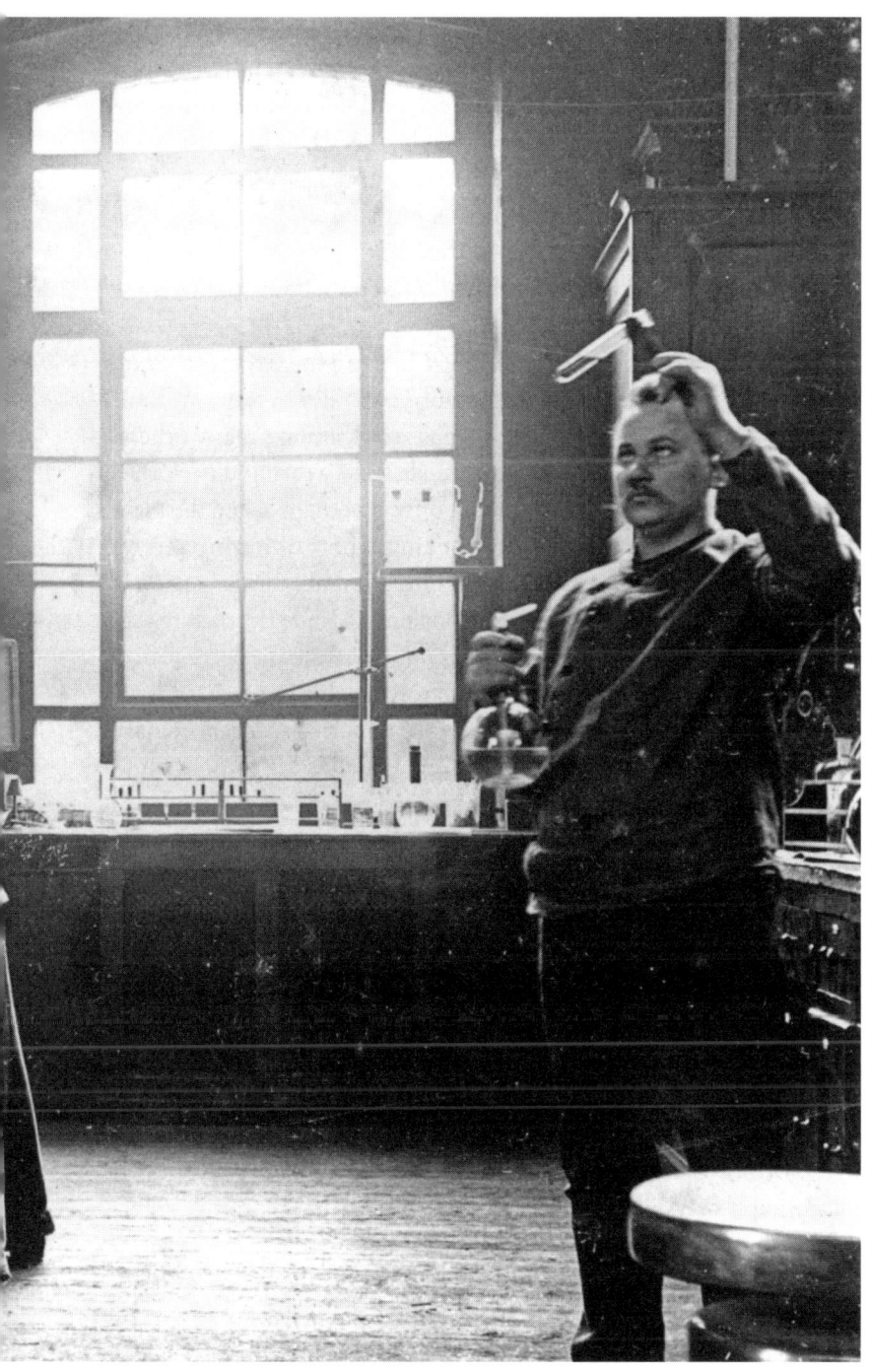

104 Bayer pharmaceutical laboratory, Elberfeld, 1900s.

argue that the white laboratory coat (at least as far as chemists are concerned) is almost certainly a twentieth-century innovation. I have not been able to find any reference in the literature to a white laboratory coat up to 1930. For most of history, chemists have worn ordinary clothes in the laboratory. Few if any chemists wore protective clothing up to the mid-nineteenth century. The claim that alchemists wore protective robes in the laboratory is based on unreliable genre paintings, and in any event is not even borne out by most of these paintings.

Most academic chemists continued to work in street clothes up until the 1930s, as shown in photographs of William Ramsay, Emil Fischer, William Henry Perkin, Jr (illus. 105), among many others. It has been suggested that these chemists were wearing day clothes because this was a posed photograph, or was perhaps taken during a special event such as a laboratory opening. Short of having several photographs of the same chemist on different dates, this assertion is impossible to disprove, but I am not convinced. While the pictures are clearly posed, hardly surprising in a period when cameras were scarce, surely it would have made more sense for the chemist to wear a laboratory coat if it was standard, rather than choosing to wear ordinary clothes? Certainly, modern scientists are often asked to don white coats for photographs, as it is now expected of scientists.

Nonetheless, a few chemists as far back as Johann Becher wore white or brown aprons, and Henri-Etienne Sainte-Claire Deville in France, in a photograph of 1875, wore a large white apron that makes him look remarkably like a weary bartender. A variant of this was a rubberized apron. Such aprons were generally worn by assistants rather than professional chemists, as shown in the picture of Liebig's Giessen laboratory in which only the laboratory assistant is wearing an apron. In one of Harriet Moore's paintings of the RI, Charles Anderson, Faraday's assistant, is wearing a white apron similar to that worn by Sainte-Claire Deville (illus. 106). In the well-known photograph of Wilhelm Hofmann in the Berlin laboratory (illus. 107), it has been assumed that he was wearing a laboratory coat. However, at least in later life, he rarely worked at the bench, and my own impression is that he has just walked in wearing his hat and a lightweight summer raincoat. If it was a laboratory coat, the original was light brown rather than white. As the coat looks lighter in some reproductions of this photograph than in others, it may have been

105 William Henry Perkin, Jr, in the new Dyson Perrins Laboratory, Oxford, *c.* 1916.

retouched on the assumption that he must have been wearing a white laboratory coat. There are other pictures of chemists wearing what look like light brown laboratory coats (or raincoats), for example a picture of chemists at Brown University, Providence, Rhode Island, in 1873, which is almost the same date as the Hofmann photograph. Despite my scepticism, a brown laboratory coat like this may well have been the first one to be worn in academia. One problem with many early pictures of chemists wearing laboratory coats is the lack of certainty about the dates of the photographs.

It does appear that the wearing of specialized laboratory clothing first became commonplace in the industrial laboratory. The clothes worn by the Bayer laboratory workers suggest that the practice originated in the factory rather than in the laboratory. It is very likely that factory workers wore brown aprons with bibs or brown coats, and this kind of clothing was then taken up in the laboratory, not least because in the German chemical industry chemists moved freely between the factory and the laboratory. This type of clothing

Overleaf: 106 Harriet Jane Moore, *Sergeant Anderson in the Basement Laboratory of the Royal Institution*, 1852.

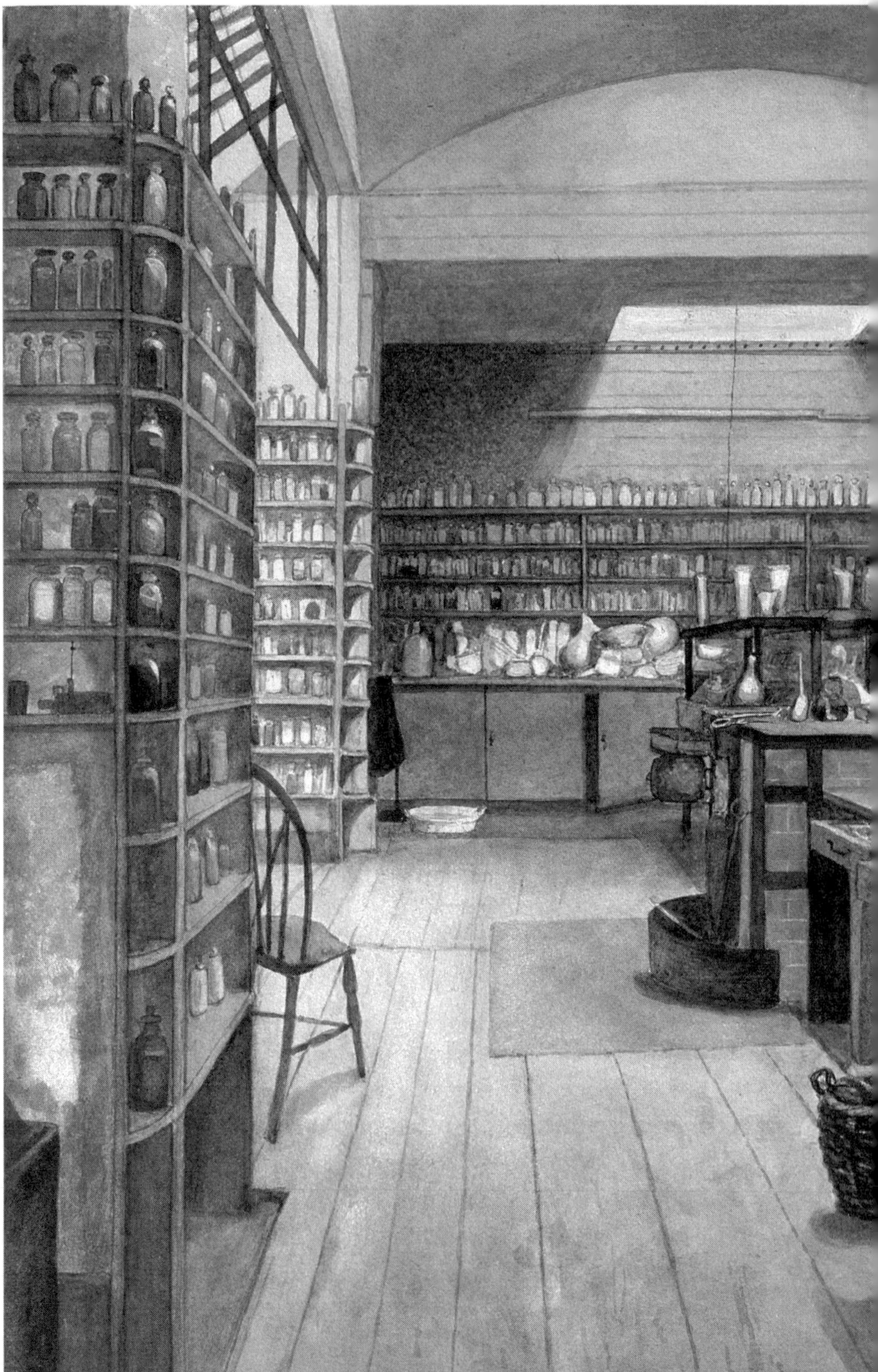

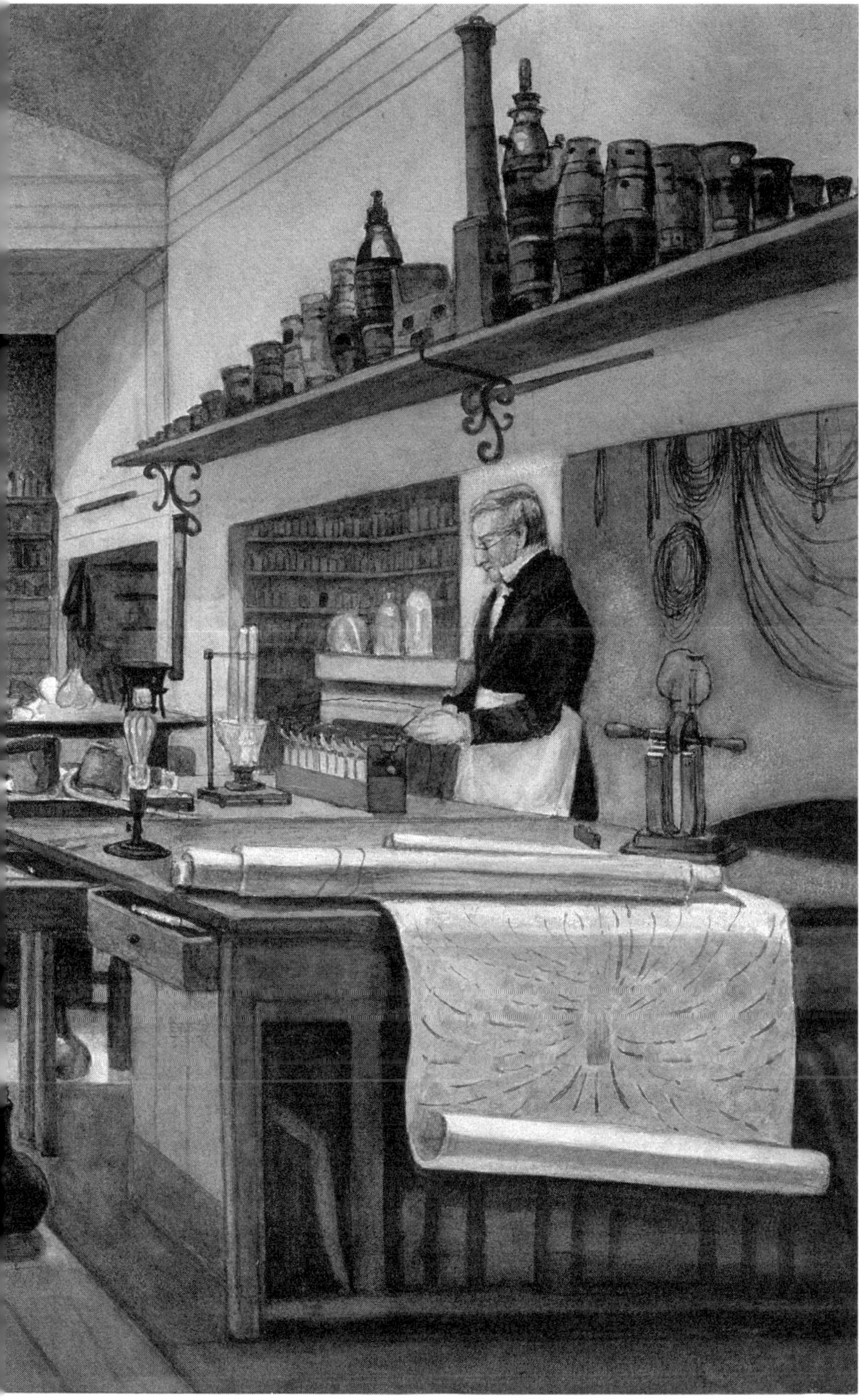

107 Wilhelm Hofmann wearing a coat in his new private laboratory in Berlin University, 1870.

clearly originated in the protective clothing of carpenters and painters. Indeed, the white coat was traditionally associated with the miller; other workmen wore coloured coats, as in the Bayer laboratory. My mother was very fond of a joke which shows that doctors were seen as copying painters in the early twentieth century.[34] Interestingly, the much later picture of the BASF laboratory in 1922 (see illus. 101) does show two chemists in white coats. The chemists of the 1930s in illus. 98 are clearly factory chemists doing some quality-control work in a small works laboratory, not dedicated laboratory chemists.

The custom for chemistry students and academic chemists to wear white coats may nevertheless stem from the wearing of white coats by physicians, if only because this practice made them readily available, cheap and conspicuous. It has been pointed out that most of the physicians in Thomas Eakins's painting dated 1889 of the Agnew Clinic are wearing short white tunics (still worn by medical students in America, as opposed to the standard laboratory coat). In Eakins's earlier painting of the Gross Clinic in 1875, all the medical staff are wearing street clothes (and allegedly black clothes out of respect for the dead).[35] This suggests that physicians started to wear special white clothing as a result of the antisepsis drive of the late nineteenth century. The wearing of protective coats in chemical laboratories possibly predated the 'Agnew Clinic' painting, but the use of white coats (or tunics) by medical staff may explain why the original brown clothing was later changed to white. When working with acids and dyes, a dark-coloured coat makes more sense than a white one, but conversely it is easier to see if a white coat is clean. Furthermore, the 'Agnew Clinic' does not show long laboratory coats, so we are no closer to dating this iconic piece of clothing. Clearly they were in use by the 1930s at the latest, and the 1957 Wallcousins painting shows that they were de rigueur by the 1950s. Additionally, perhaps particularly in the 1950s, there are photographs of schoolboys at my old school wearing white laboratory coats in the 1950s, but we did not wear them in the 1970s. It is possible that the white coat was introduced to inculcate an 'esprit de corps' in students and laboratory workers, and thus to improve morale rather than to protect clothing, but I have found no evidence for this in chemistry as opposed to other fields such as dentistry.[36] Certainly, when I worked in the pharmaceutical industry, the white coat distinguished the laboratory worker, however lowly, from the blue overall-wearing factory worker. Conversely, it was sometimes held that allowing students to wear a laboratory coat encouraged messiness.[37]

A New Kind of Laboratory

As in the case of Stanford University in 1961 (see chapter Eleven and illus. 125), some of the laboratories at Royal Dutch Shell in Amsterdam after the Second World War still looked very traditional (illus. 108).[38] Yet a new type of laboratory was beginning to appear, as shown in illus. 109. The use of electrically powered apparatus and recorders, and the joining together of different pieces of equipment, has created a new laboratory architecture with no benches, sinks or even fume cupboards; in other words a laboratory that is much closer to the academic physics or engineering laboratory, or indeed the factory itself. The apparatus is now supported by simple wooden tables or increasingly by bolted steel racking. Rather than having a dedicated bench space and a stool, the chemists move around the laboratory monitoring and adjusting the equipment, or sit alongside the equipment in ordinary chairs. This photograph shows distillation combined with the analytical technique of gas chromatography.

In illus. 110 it can be seen how a chemical laboratory in the 1950s or early '60s could be dominated by a single suite of apparatus, in this case an early dual-purpose nuclear magnetic resonance (NMR) spectrometer, the Varian DP 40, and an electron spin (ESR) spectrometer, made by Varian, the major American manufacturer of NMR equipment, supplied to the Shell research laboratory in Amsterdam in 1957.[39] As discussed in chapter Eleven, NMR spectrometers were very sensitive and had powerful magnetic fields, which meant that they had to be segregated from other chemical operations in any case. The Varian DP 40 was a high-end version of the HR 40, which was particularly suitable for solids and polymers.[40] Shell had built up a close relationship with Varian in the early 1950s through its American arm, the Shell Development Company, whose research centres were

108 Shell laboratory, Amsterdam, late 1940s.

109 Shell research laboratory, Amsterdam, mid-1950s.

at Emeryville in California (across San Francisco Bay from Varian in Palo Alto) and Houston, Texas (for the history of NMR, see chapter Eleven).

In the early 1950s Shell moved increasingly into the field of pesticides and felt the need to be able to measure pesticide residues on fruit and vegetables in order to ensure that pesticides did not enter the food chain. This objective was shared across three Shell research centres at Amsterdam, Sittingbourne in Kent and Emeryville in California. Both Sittingbourne and Emeryville were close to the fruit-growing industries, which were major consumers of insecticides. At first Shell concentrated on electrochemical methods of analysis developed by Dale Coulson, a former employee of Shell Development; it then switched to the electron-capture detector developed by James Lovelock at the National Institute for Medical Research, Mill Hill, which was partly based on earlier work on a beta-ray detector at Shell Amsterdam and Emeryville.[41] Between 1963 and the late '60s, Lovelock worked as a part-time advisor to Shell and Victor Rothschild, its head of research, in particular.[42] This is a good example of how gas chromatography initially taken up by Shell for its potential use in petroleum refining, and the electron-capture detector originally developed by Lovelock for biochemical use, were combined to tackle

110 NMR laboratory at Shell, Amsterdam, 1959. The physical chemist Cor MacLean is sitting at the console of the Varian DP 40 NMR.

important analytical problems posed by the downstream products of the petrochemical industry such as pesticides.

Writing in Air

By the late 1930s the burgeoning petroleum-refining industry and its offspring, the petrochemical industry, sorely needed accurate but rapid methods of analysing gas streams and the final products of its processes, from petrol (gasoline) to plastics and pesticides. During the Second World War the new field of infrared spectroscopy offered a partial solution.[43] There was, however, a growing need for easy separation of the complex mixtures thrown up by the industry, regardless of whether they were produced in a petroleum refinery or created when a pesticide was sprayed on fruit trees. From an unexpected quarter, namely the National Institute for Medical Research (NIMR) at Mill Hill, a solution to these problems suddenly appeared in the early 1950s.[44] Archer Martin, who had developed liquid-liquid partition chromatography in the early 1940s, was approached by a colleague at the NIMR, George Popják, for a method of separating small amounts of mixed fatty acids. To help Popják, Martin and his co-worker Tony James returned to a concept Martin had first proposed almost a decade earlier but not pursued, namely gas-liquid chromatography. The mobile phase was a gas (usually nitrogen or hydrogen), and the stationary phase was silica (Celite), coated with a suitable liquid and packed in glass tubes. The column was heated with boiling ethylene glycol, the well-known antifreeze. Martin and James demonstrated the new technique at the Oxford meeting of the International Union of Pure and Applied Chemistry in September 1952, by which time they had extended the scope of gas chromatography to mixtures of fatty acids (in the form of volatile esters) and, crucially for industrial laboratories, hydrocarbons. Some important techniques take a decade or more to establish themselves – NMR is a case in point – but gas-liquid chromatography spread dramatically in 1952 and '53, especially after the Oxford conference.

There was, however, a major problem. Chromatography of any variety can only succeed if there is some practicable means of identifying the different fractions as they emerge from the column. The solid chromatographic column demonstrated by Mikhail Tswett in 1907 only worked well with coloured compounds, notably chlorophyll – hence its name, which means 'colour writing'. When Martin introduced paper chromatography in 1944, his proposal to use ninhydrin as a marker for amino acids was crucial to the success of the technique. Initially, Martin and James used automatic titration for gas chromatography, but as well as being rather slow, it could only be applied to acids and bases such as amines. Martin and James then developed the gas-density balance, but it was fiddly to use. Greult Dijkstra, a chemist at Unilever, introduced the katharometer into gas chromatography in 1953. This device measured the thermal conductivity of the carrier gas (or indeed any gas). It had originally been developed by Henk Dijkstra, a physicist at Dutch State Mines (DSM) and no relation, to detect methane in coal mines, and DSM then used it in its ammonia plant in 1948.[45] Addressing the 1956 London Symposium on Gas Chromatography as its chairman, James was critical of existing detectors and concluded that:

> The field is open for the development of more sensitive and simpler detectors, and it is to be hoped that workers in this field will not rest content with the instruments at present available.[46]

The most useful of the second wave of detectors was the flame-ionization detector, which relied on the fact that hydrogen was the carrier gas used in most early gas chromatographs.[47] As it leaves the chromatograph, the carrier gas is burnt and the presence of other substances is detected through changes to the conductivity of the flame caused by the production of ions. The detector was developed independently in 1958 by ICI and the South African Iron & Steel Industrial Corporation, but the patent rights were obtained by ICI. The usefulness of this method lies partly in its general applicability and in its relative simplicity – after it was unveiled by Ian McWilliam of ICI at a conference in 1958, chromatographers rushed back to their labs and made their own versions.

At the same time, a colleague of Martin and James at the NIMR, James Lovelock, was studying other methods of ionizing compounds as they left the gas chromatograph. His main line of research was the use of radioactive elements to ionize these compounds, which ultimately resulted in the development of the electron-capture detector (ECD) in 1959.[48] The ECD is extremely sensitive; it can detect suitable compounds in concentrations of parts per billion or even parts per trillion, but it is only effective with certain types of compound, notably those containing halogens such as chlorine or fluorine. This made it superb for detecting the chlorofluorocarbons (or CFCs) – which escaped from refrigerators and air conditioners – in the south Atlantic, and for the detection of DDT, but otherwise its uses are more limited than those of the flame-ionization detector.

The major breakthrough in the use of the gas chromatograph was the linking of the chromatography to that important analytical instrument, the mass spectrometer. The crucial problem is that a gas chromatograph produces a stream of gas and a mass spectrometer requires a vacuum. The problem had to be solved, however, as the fragments created by a mass spectrometer could be used to work out the identity of the original compounds, making it the ultimate detector for the gas chromatograph. Roland Gohlke and Fred McLafferty at Dow Chemical Company, pioneers in the use of the mass spectrometer for organic chemical analysis, developed the first 'hyphenated' GC-MS in 1957. Despite being commercialized by Bendix, the makers of the mass spectrometer used in this set-up, this version was not very successful. Ragnar Ryhage at the Karolinska Institute in Stockholm then developed a device linking the two instruments, which ejected most of the carrier gas in a jet while allowing the separated compounds to make their way into the mass spectrometer. Space exploration then became a driving force for GC-MS as NASA needed a way of analysing the soil samples on Mars once its two *Viking* landers had reached the planet in 1976. A GC-MS set-up on the *Viking* landers developed by Klaus Biemann showed that there were no organic compounds on the Martian surface at a level of parts per billion.[49] While the GC-MS is by far the most sensitive and versatile detector, it is also very expensive, so other detectors have continued to be employed for routine use.

Similar yet Different

We can only understand the laboratory in the chemical industry properly if we recognize that the research laboratory is not the only possible type, and indeed not even the most common. The industrial laboratories range between manufacturing, quality control and analysis, and process control and research. The standard works laboratory is usually concerned with quality control and process control rather than with research. Quality control (usually simple forms of analysis) is intimately connected to process control, as it checks both the inputs and the outputs of the industrial process. The works chemist ensures that both the raw materials and the end products are of an acceptable standard. Any deviation from these desired norms can then be investigated and often solved, even in a very basic laboratory. Nor were chemical laboratories limited to the chemical industry. We have seen that they arose in the metallurgical industries and pharmacy even before the chemical industry existed as a clear entity. By the mid-nineteenth century these laboratories had spread to other chemically based industries such as soap and brewing – even to the railways. Shortly after 1860 laboratories were first described as research laboratories, and while the concept initially arose in France, Germany soon took the lead.

In their appearance and design, many works laboratories (even in the second half of the twentieth century) were very basic; often they were little better than garden sheds. It was the research laboratories of the late nineteenth century that emulated the type of academic laboratory that had been created in Germany. Even then, the fittings of the laboratory were often less lavish than in the new academic laboratories. Furthermore, industrial laboratories were quicker to change in the face of the new instrumentation introduced in the 1950s, perhaps because they were less bound by tradition and convention, but mainly because they had more money. By the 1990s, as discussed in chapter Twelve, a new generation of pharmaceutical research laboratories set the pace for academic laboratories. The industrial laboratory was thus transformed from being the imitator of the academic laboratory, to the prototype of the latest laboratory designs.

10
Neither Fish nor Fowl
Thomas Thorpe and London, 1890s

Scholarly examination of chemical laboratories in the nineteenth century has largely focused on academic laboratories on the one hand, and industrial research and development laboratories on the other. A third type of laboratory is discussed here: the official or government laboratory, which ranged from the pure research laboratory, through laboratories that sought to measure fundamental data, to analytical laboratories. Many of these were physical laboratories, although often with dedicated chemical sections; or they were at least laboratories that carried out chemistry-related research, such as the National Physical Laboratory in Britain, the Physikalisch-Technische Reichsanstalt in Germany,[1] and the National Bureau of Standards in the United States.[2]

The Government Chemist's Laboratory

A clearly chemical example is discussed here – the Government Chemist's Laboratory in London after it moved in 1897. The GCL (as it was never called) eventually became the Laboratory of the Government Chemist, or LGC, as it *was* usually called. In fact LGC has now become its official name. It was privatized by the government in 1996, and was partly owned by the Royal Society of Chemistry until 2002. The laboratory was founded as a result of a merger of the Excise Laboratory and the Inland Revenue Laboratory in 1894, but it was also a combination of a revenue laboratory that sought to protect the government's income from duties, and a food and drinks laboratory whose mission was to protect the consumer. The overlap arises because both aims involve the detection of adulteration, principally involving padding out (such as the addition of sugar to tobacco or water to milk). The closest counterpart to the GCL was the Laboratoire Municipal de Chimie (Municipal Chemical Laboratory) in Paris, founded in 1878. It was specifically set up to detect adulteration and dilution in food and drink, in particular the dilution and artificial colouring of wine in bistros. As a local body it never covered the protection of revenue, although its director, Charles Girard, was aware of the financial loss to the state as a result of the watering down of wine. There have been a few studies of the GCL and the Paris Municipal Laboratory, and even a comparison of the two laboratories, but they have been mostly written in the context of food history and public health. As such they have focused on the methods used rather than on the laboratories themselves.

Origins of the Government Chemist's Laboratory

The GCL's origins go back to the Laboratory of the Board of Excise founded in October 1842.[3] George Phillips, an excise officer, was appointed to test tobacco under the recently promulgated Pure Tobacco Act. Self-educated in chemistry, he was based at the headquarters of the Board of Excise in the City of London. To the surprise of the tobacco trade, he was able to detect low levels of adulteration in tobacco, mainly using the microscope and some elementary chemistry. The work (and the staff) of the laboratory soon expanded to cover other materials such as pepper, tea and beer, all of which were subject to considerable adulteration. Friedrich Accum had pointed out the dangers of food adulteration as far back as 1820, but the adulteration of food with poisonous colourings and other additives continued unabated, as it was both profitable and rarely detected.[4] Largely thanks to the campaigning of Thomas Wakley, editor of *The Lancet*, and the physician Arthur Hill Hassall, the Act for Preventing the Adulteration of Articles of Food or Drink was passed in 1860, followed by a subsequent act in 1872 and finally the more effective Sale of Food and Drugs Act of 1875.[5] The laboratory had been transferred to Somerset House in 1853 following the formation of Her Majesty's Inland Revenue. Contract work for the testing of goods supplied to other government departments became increasingly important. Phillips retired in 1874 and was succeeded by his deputy, James Bell. Soon afterwards the work of the laboratory was increased even further by the Sale of Food and Drugs

Act, which appointed the Somerset House chemists as referees in disputed cases.

Since at least 1303, customs duty on wine had been charged by volume regardless of a wine's alcoholic strength.[6] In 1860, however, a new law was passed that based the duty on the alcoholic content (proof strength), and the Customs appointed James Johnstone, principal inspector of Gaugers in London, to install a laboratory at Custom House in the City of London, equipped with 'a set of new stills and other instruments for the purpose'.[7] The wine-testing work varied over the years as the regulations were altered. A major change, however, was the obligations laid on the Commissioner of Customs by the Sale of Food and Drugs Act to examine all tea coming into the country in order to combat adulteration, which led to the setting up of a Tea Laboratory.

The amalgamation of the Custom House laboratories with the Inland Revenue laboratories at Somerset House had been proposed in 1890, but was ultimately rejected. The catalyst for their eventual fusion came in the form of another Act of Parliament: the Fertilizers and Feeding Stuffs Act of 1893. Under this act a chief analyst had to be appointed by the Board of Agriculture to analyse disputed samples. Since James Bell was about to retire and the Board wanted to appoint a chemist of some stature, the Treasury set up a committee to study the existing arrangements. In due course the Treasury ignored its own committee and proceeded to follow its own plan to create a Government Laboratory headed by an eminent principal chemist selected by the Treasury. Thomas Edward Thorpe, professor of chemistry at the Royal College of Science (soon to become Imperial College), was appointed to this high office in 1894. All this legislation only produced a moderate increase in the number of samples, but it did greatly increase the variety of work undertaken and thus produced a heavier workload for the GCL.

There is no doubt that new laboratories were required. The existing laboratories at Somerset House were described in 1897 as having 'wretched ventilation and miserable accommodation . . . with the thermometer registering 80 to 90°F' and the

> conditions under which some of the experiments were performed were detrimental to the health of the staff, while the abominable smells which sometimes pervaded the corridors

111 Exterior of the Government Chemist's Laboratory in Clement's Inn, London, 1950s.

of Somerset House were both in variety and number far ahead of those which have been poetically ascribed to [eau de] Cologne.⁸

Perhaps swayed by the evident health risks, the Treasury agreed to the construction of a brand-new laboratory near the Law Courts (illus. 111). The new building was designed largely by Thorpe himself, who had recently designed the new laboratories at the Royal College of Science in South Kensington. The architect was John Taylor, a consulting architect for the Office of Works, best known for Herne Hill railway station and Bow Street Magistrates Court. The limited budget (£20,000, which overran to £25,000) and the confined site produced a fairly mundane building, which does have some similarities to the south range of the much grander Post Office Savings Bank headquarters in Hammersmith built in 1899–1903 to the plans of Taylor's former assistant Henry Tanner (and is now used as a store by three museums, including the Science Museum).[9] The Government Laboratory was opened on 1 October 1897 with a concert by the Whitehall Orchestra and various displays, ranging from historic scientific apparatus from neighbouring King's College (now part of the Science Museum's collections), to examples of lace made by the women of the Azores.

Approving accounts of the laboratories appeared not only in the technical press, but also in the *Strand Magazine* (the new building was just off the Strand, a major London thoroughfare).[10] The magazine article, by the chemist John Mills, is a superb source of photographs and descriptions of the new laboratory building just four years after it opened, and is extensively quoted here.

An Official Chemist

Thomas Edward Thorpe was born on 8 December 1845, in Harpurhey, near Manchester, the son of a cloth and yarn agent (illus. 112).[11] He entered chemistry in 1863, when he worked for Henry Roscoe as a junior assistant at Owens College (now the University of Manchester). As a demonstrator under Roscoe, Thorpe carried out photochemical research, and in 1867 he was awarded the Dalton senior chemical scholarship at Owens College. With Roscoe's support he studied under Bunsen at Heidelberg, where he took his PhD. He then worked with August Kekulé at the University of Bonn. Soon after his return to England in 1870, he was appointed to the chair of chemistry at Anderson's University, Glasgow (now Strathclyde University). Four years later he moved to Yorkshire College in Leeds, where he designed the new laboratories with the architect Alfred Waterhouse (illus. 113), which proved to be valuable experience for his later career at the GCL. Thorpe succeeded Edward Frankland (who had also been at Owens College) as professor of chemistry at the Royal College of Science. When the post of principal chemist of the Government Laboratory was created in 1894, Thorpe was appointed as its first

112 Sir Edward Thorpe, c. 1910.

incumbent. The predecessor of the Government Laboratory, the Inland Revenue Laboratory, enjoyed close links with the Royal College of Science, and Thorpe had published a textbook of chemical analysis. His very broad knowledge of science – he also took part in solar eclipse expeditions – was also doubtless appealing to the appointment board.

Most of Thorpe's chemical research would now be described as physical chemistry. At Leeds he had correlated the molecular weights of compounds to their specific gravities as liquids, but he also studied the chemistry of phosphorus – which led to the elimination of white phosphorus in matches when he was at the Government Laboratory. He was also interested in the determination of accurate atomic weights for metals, and he confirmed the atomic weight of radium that had been determined by Marie Curie. Thorpe was an excellent writer. He wrote a famous multi-volume *Dictionary of Chemistry*.[12] He also wrote an influential history of chemistry, the third edition of which was published after his death.[13] Thorpe was knighted as Sir Edward Thorpe (although he had earlier been known as Tom Thorpe) in 1909 when he retired from the Government Laboratory to return to what was now Imperial College. He finally retired from academia three years

113 General Chemical laboratory, Yorkshire College, Leeds, 1908.

later. Thorpe was a keen yachtsman and moved to Salcombe on the south coast of Devon, where he died in 1925.

What the Laboratory Did

Beer analysis formed a major part of the work of the main laboratory at the new Government Laboratory. Mills commented that:

> Two thousand three hundred and eight-six samples of finished beer, taken from 1,223 publicans were analyzed, and 319 or 13 per cent of the samples were found to have been diluted with water or otherwise adulterated. The practice of diluting beer by publicans is almost entirely confined to London! Beer of a heavy brew has always been regarded as the typical drink of all Englishmen.[14]

The carbon dioxide was removed by whipping the beer with an early type of electric whisk ('a sort of electric screw revolved rapidly in the liquid'), and when the carbon dioxide had escaped the specific gravity of the beer was measured by Sikes's hydrometer – essentially a weighted float of brass suspended in a glass cylinder containing the liquid being tested. At this time, Sikes's hydrometer, introduced in 1816, was the only kind legally recognized in the United Kingdom.[15] In many respects, as Thorpe himself admitted, a 'specific gravity' bottle would have produced better results and the ban on other methods was lifted in 1907.[16] The beer was

> then transferred to a still, by which means the alcohol, under the influence of heat, distils over into a receiver ... the distillate, when weighed in the balance, indicates a different specific gravity, which enables the chemist to compute the percentage of alcohol in the sample of beer under examination.[17]

By contrast, ginger beer and other herbal beers were checked to ensure that the alcoholic content did not exceed 2 per cent; one sample contained as much as 7.6 per cent![18]

Tea was the other beverage dear to the British heart. Mills was emphatic on this point, remarking that:

> It is estimated that the Anglo-Saxons are by far the biggest tea-drinkers in the whole world ... All Europe treats tea with disdain. ... Coffee, in fact, holds the same position in England that tea does in Germany.[19]

It was therefore entirely natural for the Government Laboratory to be heavily involved in the testing of tea. Customs inspectors based at the ports would send samples of tea to the laboratory. It was then subjected to a number of chemical and microscopical tests for adulterants such as tea dust and sand:

> In the Customs Department during the last year 226 samples of tea, representing 3,322 packages were found to contain exhausted leaves or to be mixed with sand, or other substances, and were refused admission for home consumption. Of these packages, 2,274 were exported[!] and 1,048 destroyed.[20]

In a period when milk was usually sold in open churns or jugs, it was easily adulterated by the addition of water, and it was not very difficult to skim some of the cream off milk that had stood overnight. The analysis of milk took place under the provisions of the Sale of Food and Drugs Act. The food inspector took pint samples of milk from churns, then divided each sample between three similar clean, dry medicine bottles, which were sealed with a cork and a wax seal. One was handed to the seller of the milk for him to have analysed, another was handed to the Public Analyst, and the inspector would keep the third, the reference sample, in a cool, safe place. If the Public Analyst found that his sample contained less than 3 per cent fat or less than 8.5 per cent of solids other than fat, it was presumed that the seller of the milk had committed an offence under the provisions of the Sale of Food and Drugs Act. If his analytical results were disputed at the court hearing, the reference sample was sent to the GCL for analysis.[21]

Butter was also the subject of fraud in the nineteenth century, as Mills pointed out:

> In the analysis of foodstuffs the object aimed at is protection against fraud in, for example, the sale of margarine under the name of butter. Margarine may be a wholesome and palatable form of food for those who can only afford to pay a moderate price and who are not given to inquire too curiously whether they are consuming animal fats ingeniously manipulated or the product of legitimate dairy produce. The ordinary farmer makes real butter, and he has to confront the competition of the manufacturer of what looks like butter, and is sold as such, though it is quite a different thing – an artificial product which may deceive the eye and even the taste.[22]

Even genuine butter contained adulterants:

> A large number of butters contained boric preservative, and were artificially coloured ... it was found that the use of boric acid is most prevalent in France, Belgium and Australia, and is very common also in Holland. The most frequent colouring-matter is annatto, but the use of coal-tar yellow appears to be on the increase. ... The bulk of the margarine imported comes from Holland, and it is usually made with cottonseed oil, contains boric preservative, and is artificially coloured with a coal-tar yellow.[23]

In contrast to many continental countries, Britain did not levy a duty on sugar until 1901. This meant that sugar analysis did not play a major role in the work of the Government Laboratory, although James Bell had covered the subject in the first volume of his book *The Analysis and Adulteration of Foods*.[24]

The laboratory had no supervisory role over regional or municipal laboratories; it carried out all the analyses required by the state or statute, rather than checking the carrying out of analyses elsewhere. It was thus a highly centralized system that arose from the desire to protect the government's revenue rather than the public's health, although the two aims eventually coalesced. The Act for Preventing the Adulteration of Articles of Food or Drink of 1860 had permitted the appointment of part-time public analysts by local authorities. This was, however, a voluntary scheme that was never widely adopted, although the Society of Public Analysts, founded in 1874 to promote the interests and standing of the public analysts, became a respected body. The public analysts were either private analysts/consultants or medical men. The government laboratory held the public analysts at arm's length, and the public analysts complained that they could not obtain assistance or information (for example about analytical methods or standards) from it. After his appointment as the government chemist, Thorpe made an effort to bridge the gap between the two sides and relations eventually became more cordial.

An In-between Laboratory

The GCL is important for any examination of the laboratory in the nineteenth century, as it is neither an academic laboratory nor an industrial research laboratory. Furthermore, I would argue that it did not carry out research – although that depends on what we call 'research', an issue in itself. Neither was it a straightforward analytical laboratory. To be sure, the laboratory carried out a vast number of analyses of a great variety of products. A celebrated case soon after the new laboratory opened was the testing of arsenic in beer, following an outbreak of arsenic poisoning in Manchester in 1900 that was traced back to the sugar used in brewing. As Mills reported:

> The recent epidemic of arsenical poisoning attributed to beer caused a thrilling sensation throughout the country. At the

time I visited the Government Laboratory this grave subject was under the consideration of one section of the department, and many samples of the condemned or suspect beverage were under examination. I was permitted to look at the arsenic extracted from beer which had been submitted to the most searching chemical analysis [in the form of the yellow sulphide.] ... I also saw the naked arsenic itself in the form of black lustrous mirror which had been deposited inside a glass tube in the process known as Marsh's test; by means of this test the most minute traces of arsenic can be detected.[25]

Most of these products were, however, complex natural materials that had usually undergone some form of industrial processing. They were not pure (or even impure) chemical compounds, and the examination of these materials required a range of tests. Some of these were clearly chemical, but others were more pragmatic and based on the analysts' familiarity with the materials. The tests were not available in any manual or taught in any chemistry courses (although the analysts at the laboratory were expected to take chemistry classes). Nearly all of them were developed by Phillips and his successors. In developing these methods, the analysts drew on their experience as excise officers as much as chemists (indeed, Phillips himself was a professional excise officer and only an amateur chemist). Phillips had first developed a series of test methods, notably microscopy, for tobacco between the 1840s and '60s. Many of the tests were indeed chemical, since the main adulterants of tobacco (especially snuff) were chemicals, including red iron oxide, lime – lime snuff was particularly popular with the mill girls of Ulster – and red lead oxide.[26] Yet taken as a whole the character of these test methods was as much forensic as chemical, for the government analysts were investigating crimes that would end up in court. There is more than a touch of Sherlock Holmes in some of their tests. For example, in 1858 Phillips was asked by the General Post Office to find a way of checking whether cancellation marks had been removed from postage stamps, and a way of preventing this fraud.[27]

Hence in the course of their work, the Government Chemist and his analysts developed new methods and new apparatus for carrying out the tests. For example, the laboratory created a tea-testing set (illus. 114). The tea was infused in a special mug with a serrated edge.

114 Tea-testing set used at the Government Chemist's Laboratory, now in the Science Museum's collections (inv. no. 1994-1325).

A lid was then put on top and the infusion was poured through the serrations into a bowl. Both the infusion and the spent leaves were examined – for flavour and mould in the former, and foreign leaves and stalks in the latter. Another important area for the laboratory was milk analysis, where there was a need to know the composition of genuine milk and to develop accurate tests to detect the watering down of milk. In the wake of the arsenical beer scare, the laboratory developed a sensitive test for small amounts of arsenic.[28]

This takes no account of the ad hoc work carried out by the laboratory in the period up to the First World War, including the preservation of medieval wax seals held by the Public Record Office; the best way of eliminating death watch beetles in Westminster Hall (which showed the need to develop better insecticides – the best the laboratory could come up with was sulphur dioxide); analysis of pesticide residues on apples, and whether white lead paint posed a health hazard. Continual innovation was therefore taking place without formal research. This was perhaps less unusual than we might now expect. Works laboratories would have made similar innovations in methods, apparatus and processes to 'meet the need', if perhaps less methodically and less often (see chapter Nine). The question here is how this combination of analysis in the broadest sense and innovation of methods shaped the new laboratory when it was set up in 1897.

Layout and Fittings of the New Laboratory Building

According to Mills, the work of the Government Laboratory was

> divided into four distinct departments: (1) the main laboratory, wholly reserved for the analysis of alcoholic products – beer, wine, tinctures, rum, brandy etc.; (2) the tobacco-rooms, fitted with appliances for the examination of manufactured and the so-called 'offal' tobacco, for the determination of fraudulent or improper admixtures; (3) the Board of Agriculture Department, where all cases of disputed analyses of fertilizers, etc., are referred here, and on which the decision of the principal chemist is final; (4) the Crown Contracts laboratories, in which all manner of substances may from time to time be examined, from the gilt buttons and gold lace on the uniforms of our naval and military grandees to the steel rails of a railway.[29]

The rooms in the basement were mainly what one would expect in this period. The boiler house was there, as were the workshop, the storekeeper and the store rooms for the old samples and for reagents, rooms for the housemaids and porters, and a room for hydrometers (presumably for storing rather than using hydrometers). Somewhat more surprisingly, the sample reception and the Crown Contracts for India laboratories were also in the basement. However, it has to be realized that the basement area had windows and was open to the outside; it was not a cellar. As was usual in this period, the library, and the offices of the principal (the Government Chemist) and the deputy principal, were on the ground floor. Here there were also a reference sample laboratory and a research laboratory, which was used by the principal for his private research. The main Crown Contract laboratories, three in all, were also on the ground floor.

The main laboratory on the first floor was the largest room in the new building (illus. 115). According to a report in *Nature*, it was '49 feet long by 43 feet wide, lighted by eight lofty mullioned windows, and a flat-roofed dormer lantern, the open roof being carried

115 The main laboratory in Clement's Inn, London, *c.* 1902.

on light iron principals'.³⁰ Mills was eager to portray the drama of this laboratory:

> The main laboratory presents a scene of extreme activity, and one is almost bewildered by the variety of operations in which the many chemists are engaged. There is a profuse distribution of bottles of all kinds of alcoholic drinks, tinctures, etc., on the top shelves of the benches – the work set out for the day.³¹

Adjoining the main laboratory there was 'a dark room for polarimetric work and a refrigerated room for storing samples'. The refrigeration plant, supplied by the firm of J. and E. Hall, was a wonder of the age. In addition to cooling the cold room, it made ice, and supplied very cold water to the laboratories for cooling purposes and operations requiring water at a cold temperature. Mills noted:

> The refrigeration apparatus employed is in the basement, but outside the main building. Liquid carbonic acid [carbon

dioxide] is evaporated to cool brine, which in turn reduces the temperature of the tank containing water.³²

Both the superintending analysts were on this floor, much as the laboratory demonstrators would have been on the same floor as the main teaching laboratory in a university. The tobacco laboratory and the tobacco furnaces (to dry out the tobacco in order to check its moisture content) were also on this floor. Finally, on the second floor there was the photographic room (and presumably a photographic darkroom, although this is not listed), the typists' room, the supervisors' room and the museum, which may have been similar to the Black Museum at Scotland Yard, rather than the typical chemical museum of the period. The photographic room was presumably used to take photographs of evidence to be used in court.

We now turn from the layout to the fittings of the laboratories. The view of the main laboratory is very similar to that of Kolbe's laboratory in Leipzig, with long benches and bottle racks. The only major difference is that the fume cupboards were not at the ends of the laboratories as was often the case in academic laboratories; somewhat unusually, they appear to have been outside the laboratories, perhaps reflecting the relatively low usage of fume cupboards in the absence of organic chemistry. The washbasins that were attached to the ends of the benches in Berlin and Leipzig were inset into the benches here, which made it easier to place items such as dirty glassware next to a basin. The centrifuge was prominent, but that could also have been the case in academic laboratories in around 1900. A period feature of the laboratory is the ceramic tiled walls, which can be found in several British laboratories of the 1890s (for example the Davy-Faraday Laboratory at the RI and the new laboratories at the RCS), and at the Post Office Savings Bank headquarters, as well as in public conveniences.

The Crown Contracts laboratories were fairly small rooms with a bench along the window for the microscope and the balance (illus. 116). The bottle rack was fixed to the wall. There was probably a table and chair in the middle of the room. A laboratory mock-up like this, based on the recollections of retired LGC chemists, was displayed in the Science Museum for many years using fittings and furniture from the Clement's Inn building. While the microscope was particularly associated with the forensic work of the GCL, it was a common

116 Crown Contracts Laboratory, c. 1902.

feature of industrial laboratories up to the 1960s. There were also steam baths and steam ovens, reflecting the need for gentle heat and the widespread use of steam in this period.

Checking Out the Bistros

How does this compare with the most similar institution to the GCL and its predecessors in the late nineteenth century, namely the Municipal Laboratory in Paris?[33] The Paris laboratory was founded in September 1878, following a proposal by Jean-Baptiste Dumas in 1876, to detect wines that had been dyed by bistros to produce diluted and inferior wine. This concern about the artificial coloration of wine arose from the development of the aniline dyes – used to mask the adulteration of wine with water – and the destruction of the French vineyards by the *Oidium* fungus (1850s) and the *Phylloxera* insect (1870s), which made wine more expensive and hence more prone to adulteration. It also has to be seen as part of the broader civic-hygiene movement of the period. As a municipal laboratory, it reflected the return of power to local communities in the early Third Republic after the highly centralized imperial system of Napoleon III. Until 1911 the laboratory was directed by Charles Girard, the former manager of the ill-fated French dye monopoly, La Fuchsine, perhaps on the basis of his knowledge of aniline dyes. Most of the attention on the laboratory both in the late nineteenth century and more recently has focused either on the laboratory in the context of public health or on the methods it used.

The Paris laboratory operated in a very different way from the GCL.[34] It was concerned with adulteration, but it also dealt with fire safety in theatres, water quality and gas supply. In this respect it was more like New York's Metropolitan Board of Health. In contrast to the GCL, it dealt directly with the public, which it served from March 1881. It was mostly concerned with wine and milk, but also analysed other foodstuffs such as butter and pepper. When members of the public suspected adulteration, they could request that two inspectors from the laboratory (which worked under the Paris police) accompany them to the institution in question, and that preliminary checks be made. If adulteration was suspected by the inspectors, samples would be taken. One would be analysed by the laboratory, the other being set aside in case of dispute. If adulteration was found the matter was usually settled informally, the offending shop or bistro being advised to destroy the remaining stocks of the adulterated item. In its early days the laboratory was criticized publically for the quality of its analyses and, to be fair, the analysis of wine – especially distinguishing natural wine from artificial wine made from raisins – was not easy. After a change in the law in 1905 that effectively centralized the detection of adulteration, the Municipal Laboratory refused to collaborate with the new Service de Répression des Fraudes, formed in July 1906, and thus lost its legal standing in court cases.[35] In 1911, probably after Girard's retirement, the Municipal Laboratory began offering merchants a quick quality-control service to check the quality of the products they were buying in the market, such as milk and wine.[36] This had the intended effect of winning the merchants' support for the laboratory, as well as extra business. By 1914 the Municipal Laboratory was carrying out over 20,000 analyses a year.

Fortunately, there are a few illustrations from the period that shed some light on the laboratory itself (illus. 117 and 118). The laboratory rooms are large but do not appear to have been constructed for the purpose. There appear to be two rooms that both contain wide benches

117 Main laboratory, Laboratoire Municipal de Paris, 1883.

118 Laboratory at the Laboratoire Municipal de Paris, 1889.

119 The Bureau de Présentation, Laboratoire Municipal de Paris, 1883.

with bottle racks and tables. In general the apparatus is similar to that of the GCL, including, for example, glassware for the determination of the alcohol content of beverages. The laboratory shown in illus. 118 appears to have a steam oven similar to that of the GCL. Both the Municipal Laboratory and the GCL had darkrooms for polarimetry and spectroscopy. Interestingly, it appears that the French also used the microscope in the darkroom. In common with the GCL, there was a photographic room to take photographs of evidence. There are also differences between the laboratories. Apparatus such as balances, glassware and furnaces was kept on shelves on the walls until needed, an arrangement that echoes William Lewis's laboratory of the 1760s. Crucially, because the Municipal Laboratory dealt with the public, in contrast to the GCL there was also a public office, the 'Bureau de Présentation' (illus. 119).

Echoes of Leipzig

The GCL had a specific mission, to detect adulteration of all kinds, which could not be completely fulfilled using the chemistry of the day. To be sure, much of its work was standard 'wet' analysis but this was not sufficient, for example, to detect all types of adulteration of tobacco or the watering down of milk. In common with other laboratories of a similar nature, the GCL had to develop new methods and tests to detect all the adulteration and other underhand activities it might encounter. This meant that the analysts of the GCL had to draw on their expertise as excise officers, which in turn implied a familiarity both with legitimate manufacturing methods and the 'tricks of the trade'. Yet this by itself was not enough; the analysts had to use common sense and ingenuity to come up with new ways of finding adulteration. Manufacturers and chemists testifying on their behalf in court would complain bitterly that the methods used by the GCL and its predecessors were 'unscientific'. Their sense of grievance may have been genuine, but their complaints overlook the fact that the aim was to detect adulteration, not to carry out scientific experiments. At the same time the government analysts had to use the 'scientific' nature of their work to justify their procedures. So there was a tension between what would work and what could be shown to work on scientific grounds. A major problem was the variability of the natural materials they were working with.

This type of scientific work also falls outside the usual understanding of 'research and development'. New methods (and apparatus) were innovated as the need arose, either to detect a new form of adulteration or in response to a request from a government department. The importance of tackling new varieties of adulteration cannot be emphasized sufficiently. Unscrupulous manufacturers were very quick to take advantage of legal loopholes. This led to a kind of 'innovation race' in which manufacturers would develop new ways of avoiding duty (as well as fleecing the consumer), and the laboratory would scramble to detect the new type of adulteration. There is a similarity here to the use of performance-enhancing drugs in sport, to give a more recent example of such a race. However, these innovations took place within the framework of ongoing work rather than as a separate process of research and development. In order to stop the adulteration or fraud as soon as possible, the work often had to be done very quickly. To give an admittedly early example, Phillips and his manager, Thomas Dobson, worked out how much sugar was equivalent to the standard amount of malt or barley used in brewing beer in just four months in the winter of 1846, thus enabling the government to permit the use of sugar in brewing. Phillips and Dobson then just as quickly established in the summer of 1847 how to determine the original specific gravity of the wort if sugar was used.

120 Kolbe's teaching laboratory, Leipzig University, 1872.

The layout and fittings of the GCL do show evidence of being adapted to the specific type of analysis and innovation it carried out. There is the high profile of steam ovens and cold rooms, and the relative unimportance of fume cupboards. The Crown Contracts laboratories are the most distinctive element of the GCL, being relatively small, lacking fume cupboards and having a window bench for microscopic work and other activities requiring good light. To a limited extent these features were shared by the Municipal Laboratory in Paris. Yet the rooms in Paris look different and small laboratories appear to be lacking.

The Municipal Laboratory looks very much like the laboratories of Henri-Etienne Sainte-Claire Deville, Marcellin Berthelot and Louis Pasteur – in other words like a standard French chemical laboratory rather than an anti-adulteration laboratory. In the case of the GCL, the most striking similarity is between the main laboratory (see illus. 115) and the main organic chemical laboratory in Leipzig (illus. 120), which had also been the basis for Thorpe's earlier laboratory at Yorkshire College in 1885. As noted in chapter Seven, this shows how a specific type of chemical laboratory that arose in German organic chemistry in the 1860s became the model for the whole discipline even in a sector such as adulteration detection, although it was dominated by pragmatic concerns and sector-specific methods of working. It also reveals how this model was, at least up to the 1890s, more widely adopted in Britain than in France.

11
Chemistry in Silicon Valley
Bill Johnson and Stanford, 1960s

In the mid-twentieth century organic chemistry was transformed by the introduction of electronic instrumentation. Between 1940 and 1970, working out the structure of a complex compound such as chlorophyll was reduced from being a lifetime's work for a professor to a few weeks' work for a graduate student aided by a technician.[1] Robert Burns Woodward remarked in 1963 that 'structures may now be determined in a small fraction of the time once required'.[2]

The Instrumental Revolution

The instrumental revolution was a marriage of ambitious instrument firms and chemists eager to find new ways of doing organic chemistry, not only in research but also in teaching and textbook writing. It was also a reflection of the growing influence of physical chemistry from the 1920s onwards, and the concomitant development of electronics technology that made relatively cheap, desk-sized instruments possible. Indeed, the revolution was only completed in the 1970s, after electronics manufacture had become a mature industry. As most of the early instruments were developed and constructed in the USA, they were first made available to American chemists. Furthermore, only American universities had the American currency to buy them. This gave a great boost to American organic chemistry and, with the help of increasing federal funding, ensured its supremacy over European organic chemistry, thus displacing the former great power in this field, Germany.

The revolution did not happen overnight; it took nearly three decades to become embedded, and while it revolutionized what happened in the laboratory, the layout of the laboratory did not change so dramatically. This was because the new technology was not put into the laboratories – they were too dusty and corrosive an environment for the sensitive (and expensive) electronic equipment, and ordinary chemists were not trusted to operate it. The classical laboratory had created specialized rooms for the then new-fangled spectroscope and polarimeter, and it was simply a matter of putting aside similar rooms for the NMR machine and mass spectrometer, albeit with stronger floors to bear the weight of the magnets they contained. So here we have a paradox of a major shift in laboratory practice not affecting laboratory design. The apparent contradiction is solved when we bear in mind that activities such as synthesis, which remained in the province of the organic chemist, were mostly unaffected by this revolution. To be sure, organic synthesis underwent a revolution in this period as well, but it was a change of strategy, with new reagents and new reactions brought about by the use of organic reaction mechanisms, not of chemical practice. Such change in practice that did occur, like the introduction of the rotary evaporator, was on too small a scale to affect the laboratory.

A Departmental Chemist

William Summer Johnson was born in February 1913 in New Rochelle, New York State, the son of a business journalist (illus. 121).[3] He went to Amherst College, Massachusetts, in 1932 to study chemistry and, as he was interested in organic chemistry, he took his PhD at Harvard under the famous organic chemist Louis Fieser, thus following a typically American academic training of a first degree at a small liberal arts college, followed by a PhD at a major research university. His thesis, completed in 1940, was on the synthesis of substituted benzathracenes and chrysenes for screening as potential carcinogens – chrysene was to play an important role in his later career. However, in common with many organic chemists of that era, including Fieser himself, his main interest was the synthesis of steroids. He joined the University of Wisconsin at Madison, then one of the best organic chemistry groups in the USA, in 1940. Like Fieser he was interested in synthesis methodology – developing new chemical methods to achieve desirable outcomes – rather than organic synthesis as an end in itself. One key goal of his early work was the development of reactions that would produce compounds with the correct stereochemistry. Many natural products, such as the steroid Vitamin D, are optically active and thus have two (or more) stereoisomers, only

121 William S. Johnson, c. 1970s.

one of which is found in nature. If a chemical compound has several optically active atoms, as many as sixteen different isomers can exist and it would clearly be very wasteful to produce all of them. Johnson began with the introduction of one particular methyl group into the steroid structure. This led to the use of hydrochrysene as a starting material with one stereochemical centre already in place. The final outcome of this line of research was his synthesis of the steroid aldosterone.

At Madison, Johnson gained a reputation of being a good administrator as well as an excellent organic chemist. Anxious to retain him, Wisconsin created its first research-only chair in chemistry, the Homer B. Adkins Professorship, in 1954 when it seemed possible that he might move to Harvard University. When the president of Stanford University, Wallace Sterling, and the provost, Fred Terman, sought to improve the standing of the university near Palo Alto in the late 1950s, it was natural that they would seek to attract Johnson to California.

Happy at the University of Wisconsin, Johnson gave Terman a list of what he considered to be impossible demands. When Stanford called his bluff by agreeing to all of them, Johnson had little option but to move to Palo Alto. In any case, California was an excellent place to develop academic chemistry in the 1960s, with several important instrument firms located in the state. He would also have the opportunity to build up an outstanding chemistry faculty backed by the great wealth of Stanford University. Effectively, Stanford was applying to chemistry what had long been standard in college football: hiring a good manager, then giving him the funds to attract star players. Within the space of four years Johnson had recruited such stellar chemists as Carl Djerassi, Paul Flory, Henry Taube, Eugene van Tamelen and Harden McConnell, and by the time he retired as department head in 1969, Stanford had become one of the finest chemistry departments in the world. The arrival of Paul Flory was particularly important partly because of his eminence (he won the Nobel Prize in 1974) and partly because he was a polymer scientist, in what was then a comparatively new field with a need for very specific kinds of instrumentation. Flory had worked in industry on and off during his career, and had retained strong links with it. This was also true for Johnson, who consulted for Du Pont and Sterling Winthrop. As a result, consulting for industry became one of the hallmarks of Stanford chemistry.

After stepping down as head of the department, Johnson was made Jackson-Wood Professor of Chemistry and he finally retired in 1979. At Stanford he developed his conversion of polyenes (long hydrocarbon chains with several double bonds similar to the polysaturated acids in margarine) into polycyclic steroids. Eventually he could convert his chosen polyene into the desired stereoisomer of the target steroid with a very high yield. He then concentrated on the mimicking of enzymatic synthesis of steroids, a continuation of his earlier interest in reactions that took place under mild conditions. In all this synthetic work, he was very much aware of the importance of the new physical instrumentation for organic chemistry, especially NMR. Johnson had a huge number of students in his long career, over 100 of them predoctoral and 200 postdoctoral. He died in August 1995 at the age of 82.

Plus ça change

The advancement of the Stanford chemistry department to the first rank sprang not from within the chemistry department or even from a new professor, as is often the case, but from Stanford's provost between 1955 and 1965, Fred Terman.[4] He was born on a farm in Indiana in 1900, but his father trained as an educational psychologist and moved to California for health reasons. He then found an academic position at Stanford, which was at that time in a rural setting (it was sometimes called 'the Farm'), and Fred took engineering and chemistry there with the aim of becoming a chemical engineer. Like many young men of his time, however, he was interested in radio and decided to take his master's degree in electrical engineering. He then took his PhD at MIT under Vannevar Bush. Thereafter he was much influenced by east-coast institutions such as MIT and neighbouring Harvard, and by Bush and Harvard's president, James Bryant Conant.

As a pioneer of electronics himself, Terman had a vision that electronics would dominate the post-war world, California could dominate electronics and that Stanford could play a crucial role. He was thus one of the fathers of Silicon Valley.[5] When he became provost, President Sterling had already identified chemistry as one of the problem areas, but finding the right person to reorganize it was not proving easy. The physical chemist Willard Libby was convinced by Sterling and Terman of the benefits of moving to California, but chose in the end to go to the University of California, Los Angeles (UCLA). They then transferred their attention to Johnson and persuaded him to move to Stanford rather than to his alma mater Harvard. From the outset Terman was determined to construct new laboratories as well as to find a new head of department, thereby bringing chemistry up to the level of physics or mathematics. Indeed, the issue of a new laboratory had been raised in 1954, long before Johnson and Djerassi were appointed. An important factor was the move of the medical faculty from San Francisco to the main campus near Palo Alto, with a concomitant shift in focus from clinical care to basic medical sciences. A revamped chemistry department could play an important role in the development of the biomedical research at the Farm. Furthermore, Terman had been the driving force behind the development of the Stanford Industrial Park (which soon became the

122 Carl Djerassi, c. 1970s.

Stanford Research Park). The instrument firm Varian had already moved there, and with the presence at Stanford of Johnson and especially Djerassi (illus. 122), with their strong links with industry, chemical and pharmaceutical firms might set up business in the industrial park as well. Hence the arrival of the new chemists at Palo Alto was not simply a change of guard in a fractious chemistry department, or merely a major element in Terman's strategy to build up Stanford, but part of his overall strategy to make the area around Stanford a breeding ground for entirely new kinds of industrial development.[6] A good example of this was Syva, a joint company between Varian and the Syntex Corporation set up to develop plastic superconductors, thanks to Djerassi's connections with Syntex. While plastic superconductors did not come to fruition at the time, Syva moved into the marketing of kits for testing military personnel for illegal drugs.[7]

Although the large size of Djerassi's research group made the construction of a new laboratory imperative, the Stauffer chemical laboratory was not simply the result of a demand for good laboratories by Johnson and Djerassi, or even of the prospect of a large number of graduate students arriving.[8] Terman provided Sterling with the arguments to put before the Board of Trustees for a new chemical laboratory: the strengthening of the chemistry department, the great influx of graduate students, a probable threefold increase in research funding and the prospect of industrial support. Fortunately, John Stauffer of the Stauffer Chemical Company and his niece Mitzi Briggs offered to fund a new chemical laboratory in memory of Stauffer's parents. It was also a good time to obtain funds from federal institutions (in this case the National Science Foundation and National Institutes for Health) for new research facilities.

The architect was Birge Clark, a Stanford graduate whose father had been a professor of drawing at Stanford. He had already designed buildings on the campus, and in the city of Palo Alto he had drawn up the plans for a post office and the fire and police station, as well as many houses.[9] His signature style was the Spanish colonial revival style with tiled roofs and white adobe walls, but the new chemistry building was different, although a Spanish influence can perhaps be seen in the tiled overhanging roof. Its design probably owes more to Clark's work for the Stanford Research Park. Indeed, it could be argued that the Stauffer building was a complete break from the past. This was no longer the imposing traditional architecture of the previous century – be it neoclassical, Romanesque or Edwardian Baroque – but a standard office building, at least from the outside. As the historian of the department remarks, Clark had 'an extraordinary ability *to listen to his clients* and to take careful note of their needs'.[10] As a result of his discussions with Johnson and Djerassi, he designed a building that was ideally suited for modern organic

123 Exterior of the Stauffer Chemistry Building (Stauffer One), Stanford University, California, 1961.

chemistry and was also flexible for future needs (illus. 123). Terman played a key role, as he insisted on a complete basement being constructed, although there was no need for one at the time.

This was indeed prophetic, as Flory was appointed soon afterwards (partly to stop him moving from the Mellon Institute to Cornell University), and he initially occupied the basement. This was clearly unsuitable (and given Flory's personality he would have lost no time in saying so), and equally prophetically the original Stauffer building was called Stauffer One from the outset. With the arrival of Henry Taube in 1962, the need for a new laboratory for physical chemistry was now compelling and, once again, John Stauffer agreed to sponsor the building of Stauffer Two, also designed by Clark. In fact the two buildings are identical from the outside. However, a new space requirement had arisen. NMR was becoming increasingly important in organic chemistry, and Djerassi was pioneering the use of mass spectrometry in the study of natural products. Even at the time of the opening of Stauffer One, NMR instrumentation featured prominently in the brochure produced for the opening symposium. In fact, Johnson had demanded the upgrading of the department's instrumentation, in particular the newly launched Varian HR 60 NMR spectrometer (illus. 124), as part of the deal for his acceptance of the chair at Stanford.[11] The instruments required were both heavy and bulky. The solution was the construction of underground laboratories between the two Stauffer buildings (which presumably incorporated the original basement), probably the first laboratories specifically constructed for the use of NMR and organic chemical mass spectrometry. Djerassi argued eloquently for the erection of a small octagonal building, rather like a gazebo, between the two laboratories for seminars, viva voce examinations, coffee breaks and the like, which ran into opposition from the university authorities. He overcame this internal resistance by marshalling the corporate funds required to build it.

Given the dramatic changes in structure determination in this period, you might think that the new laboratories at Stanford would be completely different from older laboratories, in much the same way that organic chemistry and inorganic group analysis had revolutionized the laboratory between the 1830s and '60s. As illus. 125 shows, the actual laboratories at Stanford remained much the same and in fact have a remarkable similarity to an analytical laboratory at ICI

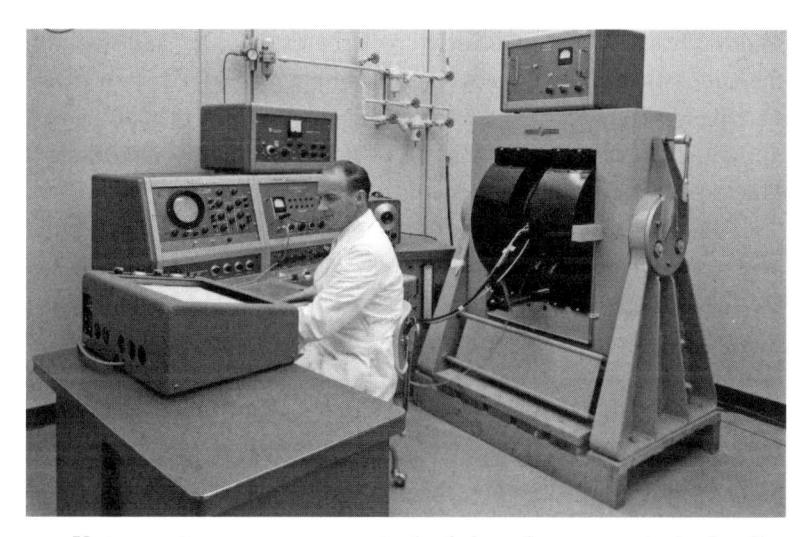

124 Varian HR 60 NMR spectrometer in the dedicated NMR room in the Stauffer Laboratory, Stanford University, 1961.

125 Main Stauffer One laboratory, 1961. Afanasii A. Akhrem (Moscow University) on the left, and Neville Stanton Crossley (Imperial College).

(see frontispiece) four years earlier. This similarity should really not surprise us. In the mid-nineteenth century the practice of chemistry changed and the laboratories were modified to meet the needs of this changing practice. In the 1950s and '60s the practice of chemistry changed dramatically. However, some aspects of organic chemistry, specifically the operation of electronic instrumentation, were taken away from chemists and placed in the hands of specialized technicians.

In other respects chemists carried on as before, making and studying the same compounds. There is a parallel here with combustion analysis. Liebig had tried to make his system of combustion analysis part of mainstream chemistry and thus a skill acquired by all chemists. However, in the new large organic chemical laboratory buildings of the late nineteenth century, the complex, hot and dangerous business of combustion analysis was carried out by trained technicians in special rooms, usually located in the basements. Similarly, the determination of an NMR spectrum was taken out of the hands of organic chemists and given to a technician to carry out. The operators had to be trained to use these instruments, and careful preparation of samples was required. The machine was therefore housed in a special room in keeping with its specific function. There were other reasons why the NMR machine was housed in a separate space rather than being placed in the working laboratory. NMR instruments (and mass spectrometers) produced powerful magnetic fields, and they were very susceptible to vibration and the unwelcome presence of moisture, chemicals and dust. The new machines were both very expensive and delicate, so it was clearly undesirable to have ordinary chemists near them and the room had to be locked out of hours.

Whereas Johnson went for four small laboratories designed for five researchers each, Djerassi wanted a large laboratory for twenty researchers. He had a vision of a 'quasi-socialist enterprise presided over by a benevolent dictator' in which there would be no barriers to communication and all the equipment was shared (see illus. 125).[12] The centrality of the instrument room was unusual for the period, but the separation of the new instruments from the main laboratory and ordinary chemists was commonplace in most academic laboratories. When it was completed, the Stauffer building had six instrument rooms. They provided space respectively for NMR, spectropolarimetry (for optical rotatory dispersion, a technique developed by Djerassi), microcalorimetry, microwave spectroscopy, microanalysis and, somewhat surprisingly, paper chromatography. Just as the NMR room can be compared with the combustion analysis room in older laboratories, the spectropolarimetry room has its parallel with the darkroom used for conventional polarimetry.

Why America?

Why did the United States become the world leader in chemistry after the Second World War?[13] As in the case of Germany in the 1860s, the major factors were money, fear and competition. During the war the U.S. government had bankrolled specific programmes such as the anti-malarial drug programme and the synthetic rubber programme, which had effectively funded many leading American chemists. After the war the funding came from a number of government and military agencies, including the National Institutes for Health (which took over this funding from the wartime Office of Scientific Research and Development in 1946), Office of Rubber Reserve (founded 1945), Office of Naval Research (founded 1946), Atomic Energy Commission (founded 1947) and eventually the National Science Foundation (founded in 1950).[14] In addition to this wave of federal funding, which was mainly for research programmes, there was considerable philanthropy, largely directed towards the construction of laboratories and other buildings that naturally bore the donors' names. Companies also funded the building of laboratories and research projects. This support was often linked to the consulting activities of academic chemists with those firms. Indeed, firms sometimes set up research laboratories independent of both the university and the company for their leading consultants' research, as in the case of CIBA and Woodward. Further support for chemical research was provided from 1955 onwards by the Petroleum Research Fund of the American Chemical Society.[15] The main drawback of this system of financial support was the lack of funding for the day-to-day expenses of the academic laboratories, which had to be met by the universities themselves. This was usually sustainable as long as the department attracted large numbers of students.

Much of the driving force behind the funding was fear.[16] Initially it was the fear of falling behind Nazi Germany during the Second World War, and then the Soviet Union, especially during and after the Korean War. The impact of the Cold War was strongly reinforced by the 'Sputnik Crisis' following the launch of the first orbiting satellite by the Soviet Union in October 1957 (illus. 126).[17] While American concern about the Russians winning the Space Race was probably misplaced, the shock that the Sputnik delivered to the self-belief in American technological superiority spurred the U.S. to redouble its

„С новым годом, с новым счастьем"

126 Soviet New Year card showing (from left to right) *Sputnik 3*, *Sputnik 2* and *Sputnik 1* floating above the USSR, 1958. The launch of the *Sputnik* satellites was a cause for celebration in the Soviet Union, but caused alarm and introspection in the USA.

efforts, culminating in the Moon landing in 1969. However, the impact of *Sputnik* was not limited to space exploration. Scientific and technical education in the USA was reformed, and the funding of scientific research was increased. The research funding obligations of the National Science Foundation almost doubled from U.S. $33.6 million to $60.4 million between 1958 and 1959.[18]

Another important factor was the competition between different universities.[19] Just as there was competition between the German states in the 1860s, there was a more direct competition between American research universities in the 1950s. At one level it was a competition to attract the best and brightest students, partly to obtain their fees and partly as a matter of reputation. That in turn meant that the universities had to attract the best teachers and research leaders. On another level, the competition was for the funding mentioned in the previous paragraph. The elite universities received a disproportionate share of research funding.[20] Private donors were keen to give their funds to top universities or to universities that might become elite institutions as a result of their donations. It was crucial to this competition that most universities were free from federal or even state control, and that students had a free choice of universities subject to making the necessary entrance grades.

Nor should the role played by American scientific instrument firms be overlooked. Most of the important firms in the 1940s and '50s were based in the U.S., including Beckman Instruments, Perkin-Elmer, Varian and Consolidated Engineering Corporation. This meant that their products were priced in dollars, which was a major problem for European universities, which were not only lacking funds even in their own currency, but were also unable to obtain dollars because of currency restrictions. Furthermore, the instrument firms liaised closely with American chemists, giving them advice and access to the latest models.[21] Needless to say, the instrument makers benefited from the feedback from chemists and the publicity from leading chemists using their instruments. The American instrument makers also had a competitive advantage from being based in the world's leading developer of electronics, stemming from the pioneering research at AT&T Bell Laboratories in the 1940s. Perkin-Elmer was on the east coast, in Connecticut, but Beckman and Varian were based in California close to Silicon Valley. Indeed, Arnold Beckman and Russell Varian were involved in the creation and development of Silicon Valley.[22]

America also benefited from the open nature of its higher education system. As well as being able to choose from a wide range of universities, American students could switch to chemistry after taking their first degree. Many leading American chemists in the first half of the twentieth century took their first degree at a small liberal arts college, often in the Midwest, before moving to one of the major research universities to take their PhD in chemistry. This has similarities to the German model of the *Wanderjahre*, but was completely different from the British system in which the decision to specialize in chemistry had to be taken at around the age of fourteen. The American chemical and petroleum industries were growing rapidly in the 1940s and '50s, and some of the largest (and wealthiest) companies in the world, such as Du Pont, Dow and Exxon, were based in the United States. These firms were looking for good graduates to recruit, and the chemical industry was seen to provide an attractive career for a bright student (as illustrated by the famous exchange between Benjamin and Mr McGuire in the 1967 film *The Graduate*).[23] Having taken their degree, some of these students then opted for an academic career, thereby maintaining high standards in chemistry departments.

Communicating with Atoms by Radio

A modern chemist has a clear idea of what chemical structure is. It is the way in which the different atoms in a molecule are connected to each other, and how these atoms are arranged in three-dimensional space. The idea that there is a clear connection between a molecular model (however it is created) and the actual structure – to the extent that the chemist often speaks about the model as if it were the molecule – is a relatively new one dating from the 1930s. In the nineteenth century chemists were much more ambiguous about the status of the early molecular models. Mostly, they saw these models as representations of what was known about a molecule rather than its physical reality.[24] In that world view, the chemical structures of compounds constitute stores of chemical information – what we know chemically about that compound. Hence working out the structure of the molecule involved building up a library of information about a given molecule from reactions carried out on it or on the chemical fragments produced by destructive reactions. The sum of that knowledge was embedded in the proposed structure, which could be 'proved' by synthesizing the original molecule from simple compounds.[25]

By contrast, in the modern physical approach to molecular structure it is assumed that we are determining the physical arrangement of the molecules. How this arrangement is mentally constructed and understood depends on the actual technique used. X-ray crystallography, infrared spectroscopy and ultraviolet spectroscopy were all in regular use by the 1950s, and they all helped to determine the structure of chemical compounds, penicillin being a notable example of the use of all three techniques.[26] However, they all had their drawbacks. X-ray crystallography had to be done by specialists and required a crystal or at least a crystalline powder. The determination of the structure also needed a good deal of number crunching, so the process could take several years if the molecule was a large and complex one such as Vitamin B_{12}. By contrast, the spectroscopic methods were easy to use and produced quick results, although at the cost of only providing a limited amount of information, usually about the type of bonds a molecule contained. Simply because it was so easy to carry out and readily available by the late 1950s, infrared spectroscopy has always been the most common physical technique and has remained so to the present day.[27] What was sorely needed was a technique that would tell the chemist what type of atoms a given atom was attached to, and how it was placed relative to the other atoms in the molecule. To get that kind of detail, you would have to find a way of 'interrogating' the atom. Remarkably, a couple of American physicists independently discovered a way of doing this almost by accident after the Second World War.

If I were to go into the details of NMR, this chapter would become far too technical and for most readers very boring, hence this brief overview.[28] Felix Bloch at Stanford University and Edward Purcell at Harvard University discovered the phenomenon of NMR (nuclear magnetic resonance) in 1946.[29] If atoms are placed in a very strong magnetic field and bombarded with radio waves, they give off their own tiny radio signals; they begin to 'speak'. Only certain atoms give off these signals. Fortunately, both ordinary hydrogen (^{1}H) and a minor isotope of carbon (^{13}C) fall into this group, which makes NMR useful in organic chemistry and molecular biology. In 1949–50, physicists began to realize that the position of these signals in the NMR spectrum was a result of the chemical environment I have mentioned.[30] This meant that it was now possible to identify the kinds of group that existed in the compound. In 1951 a postgraduate assistant of Bloch, James Arnold, produced the first NMR spectrum that showed different peaks for the different kinds of hydrogen atom in the molecule. By itself this was only of limited use. At this point NMR offered little or no advantage over infrared spectroscopy, given its expense and complexity. The breakthrough came with the discovery that the hydrogen atoms were interacting electronically with each other (called spin-spin coupling), and hence it was possible to work out their positions in the molecule from this interaction. So you could now tell what kind of hydrogen atom was present and where (in relation to the other hydrogen atoms) it was located in the molecule. This effect was first studied by Warren Proctor and Fu Chun Yu at Stanford, and by Herbert Gutowsky and David McCall at Illinois during 1952–3. As a chemist, Gutowsky was the pioneer in the use of NMR in organic chemistry. NMR now allowed a chemist to work out the position of a given hydrogen atom relative to other hydrogen atoms. However, it did not give the chemist the position of the given atom relative to all the hydrogen atoms in the molecules – only to its neighbouring atoms. Moreover, of course not all atoms in a molecule – assuming there are more elements present than hydrogen and carbon – are NMR-active. Taken together, these two limitations mean that NMR could not be used on its own in the 1960s to determine the complete structure. With the advent of more sophisticated NMR techniques from the 1970s onwards, this uncertainty has now been eliminated and NMR spectroscopy can determine structures as least as well as X-ray crystallography and for a wider range of substances.

To use NMR, chemists had to both understand how it worked and have access to an NMR spectrometer (usually just called an 'NMR machine'). Many chemists with only a basic grounding in quantum physics and mathematics struggled to come to grips with its theoretical aspects. This meant that only a few determined groups explored this new field in the 1950s. Furthermore, the chemists had to either make the machine themselves – as Rex Richards did at Oxford University and Raymond Andrew at the University of St Andrews – or to purchase what was in the early 1950s a very large and expensive instrument. The companies making these machines had to achieve several goals. It was crucial to make the machines cheaper, smaller, more robust and easier to use. To increase the ability of the machines to produce good results – to interrogate the atoms more thoroughly – the magnetic fields had to be made stronger and the radio frequencies increased. For this reason progress in the development of NMR machines in the 1950s and '60s can be tracked by their frequency rating. The stronger magnetic fields produced a broader spectrum that allowed chemists to distinguish interactions with a hydrogen atom's immediate neighbours (spin-spin coupling) and those with more distant atoms. This reduced the amount of guesswork required to deduce a molecule's structure.

The firm of Varian Associates was founded in 1948 by Russell and Sigurd Varian with several members of the Stanford University faculty.[31] Fred Terman joined the board a few months later and did much to support the young firm. The expertise in NMR was provided by Martin Packard, who had been a student of Bloch. Packard and a chemist, Jim Shoolery, went around chemistry laboratories in America and Europe to persuade chemists to buy Varian's machines by offering technical support (illus. 127). Varian produced its first commercial NMR machine rated at 30 MHZ in 1952, but it was very expensive and was only bought by large firms such as Du Pont, and Shell Development at nearby Emeryville.[32] Even before the arrival of Johnson, Stanford hosted the Varian factory in its new industrial park in 1953, and played an important missionary role. Jack Roberts, then a physical organic chemist at MIT, first heard about NMR from Richard Ogg of Stanford University:

> One day in late 1949 or 1950, he [Ogg] was at MIT, and I invited him to lunch. He was really wound up and proceeded to tell me about the wonderful new magnetic resonance spectroscopy, with such promise for chemistry. I wish I could say that I could understand even 5% of what he told me, but I had too little knowledge of magnetism and absorption of radio frequency radiation – indeed, hardly any knowledge of other radiation ... It was clear there were applications to chemistry, even if I didn't understand what they were.[33]

Roberts in turn introduced Johnson to NMR during a visit to Wisconsin in 1957. Even more importantly, Roberts also obtained the Varian HR 40 [MHZ] machine for Caltech in 1955 with the help of Linus

127 Jim Shoolery and Virginia Royden with an early 30 MHZ NMR system at Varian Associates lab, 1952.

Pauling; he has claimed that this was the first commercial NMR machine to be located in a university.[34] It was certainly the first Varian NMR machine to be purchased by British universities, including Cambridge University and, thanks to a grant from the Wellcome Trust, University College London. This model was followed by the HR 60 in 1958, but it was the 60 MHZ Varian A60 NMR spectrometer that finally made NMR routine (illus. 128). Robust and magnetically stable, it was almost half the cost of earlier machines despite its higher frequency, which meant that almost any university could afford to buy one. Around the same time Perkin-Elmer, best known for its infrared and ultraviolet spectrometers, brought out the R-10 NMR instrument. While this machine was popular, Perkin-Elmer never became a major force in this field.

An important technical breakthrough was the use of superconducting magnets in NMR. In 1964 Varian marketed the first commercial instrument to incorporate a superconducting magnet, the HR 220. At this point a new player entered the field. Bruker, a German firm founded by the physicist Günther Laukien in 1960, collaborated with the University of Zurich to develop the concept of pulsed NMR, then adopted a powerful mathematical technique called a Fourier transform.[35] It brought out a 90-MHZ machine in 1969, which was followed a year later by a superconducting Fourier transform model operating at 270 MHz. The hitherto dominant Varian was now scrambling to catch up and it launched the XL 200 superconducting FT machine in 1978.

Mass spectrometry followed a similar path.[36] Although it was first developed as a technique in 1919, it was largely confined to physics and physical chemistry until the mid-1950s. The use of mass spectrometry in organic chemistry began in 1956, when Fred McLafferty at Dow began his work on the fragmentation patterns formed when a complex organic compound breaks up in a mass spectrometer. Klaus Biemann, an Austrian chemist working at MIT, entered this field soon afterwards. By the mid-1960s it was possible to display a complete mass spectrum on a single photographic plate, then use a mainframe computer to calculate the exact molecular mass of each fragment. With this method Biemann was able to determine the hitherto unknown structure of the marine sex hormone anthediriol in 1968, using the mass spectrum alongside its ultraviolet, infrared and NMR spectra. Its mass spectrum was obtained from a sample of 'a few micrograms'.[37] Again Stanford played an important role. Soon after he joined the Stanford faculty, Carl Djerassi drew on the earlier work of Ivor Reed at the University of Glasgow to use mass spectrometry to determine the structure of the steroids, terpenoids

128 Varian A60 NMR spectrometer, formerly used by ICI.

and alkaloids he was already studying. He then used the technique of isotope labelling, already employed in chemical kinetic studies, to label different parts of a molecule with specific isotopes (such as ^{2}H) in order to explore how complex molecules broke up in the mass spectrometer.

The Final Hurrah

With hindsight, the Stauffer laboratory buildings represented the final stage of the classical laboratory. Major changes were occurring in chemistry, but the laboratory design developed in the 1860s was still able to accommodate the new equipment and new ways of working. The ghost of William Henry Chandler could have walked into Stauffer One and been at home there, even if he thought the fume cupboards were much better, and some of the equipment would have inevitably taken him by surprise. With the building of Stauffer Two a few years later, the new equipment was beginning to alter the design of the laboratories, since basements were specially constructed to house this heavy instrumentation (although nineteenth-century laboratories also had basements for heavier machinery).

Nonetheless, other changes that were evident at Stanford in the 1960s were about to change the chemical laboratory fundamentally. Health and safety was playing an increasingly important role in the design of chemical laboratories. Pharmaceutical companies were increasingly influential in academic chemistry. As we have seen, Djerassi was also involved in spinning out chemical research into commercial ventures with Terman's blessing. This semi-commercialization of academic chemistry did not go unnoticed by a Fulbright Fellow from Oxford University at the department in 1975–6. Graham Richards returned home determined to develop a similar spin-out strategy, which was only possible once Margaret Thatcher became Prime Minister in 1979 and universities were allowed to make profit from their research. Taken together, these different strands made a completely new laboratory possible in Oxford, which is the subject of the final chapter.

12

Innovation on the Isis
Graham Richards and Oxford, 2000s

Hitherto this book has emphasized the positive aspects of the new laboratories, especially the classical laboratory. With the benefit of hindsight, however, the classical laboratory was far from perfect.

Problems with the Classical Laboratory

The construction of the laboratory was inflexible. Once it was in use, the fittings and layouts of the rooms could not be altered easily, and the island benches took up considerable space. To make one of these laboratory buildings suitable for modern work would be extremely complex in terms of the temporary rehousing of the staff and remediation – probably about as expensive as building a new laboratory building.[1] One of the advantages of the classical format was that it allowed a laboratory to be used for a range of scientific activities, especially if the bottle racks were removable. Furthermore some fittings, such as the fume cupboards, could be updated when necessary. However, the utility supplies (water, gas and electricity) were difficult to maintain, as they were usually laid under floors that were not always easily accessible.

Above all, while it was a considerable improvement on its predecessors, the classical laboratory was not completely safe. The fume cupboards were usually at the ends of the laboratory (or at best on the sides), and hence some distance from the majority of the benches. There were often insufficient fume cupboards for every student or researcher, and they had to be shared. The fume cupboards themselves were far from ideal. The airflow was usually low (I can recall several fume cupboards with hardly any detectable airflow at all).[2] The water and gas taps were located inside the fume cupboard. As it was not easy to clean, the inside of the glass pane was often very dirty and even obscured. It is therefore not surprising that there was little inclination to use the fume cupboards, the harmful effects of doing experiments on the bench being alleviated by a forced draught through the laboratory in most cases. Certainly, any experiment informally deemed to be harmless (the vast majority even in an organic chemical laboratory) was carried out at the open bench.

Between the 1950s and '90s a number of books, mostly American, considered the subject of laboratory design and laboratory safety. In 1951 a monograph on laboratory design was equivocal about the large teaching laboratory, saying that larger laboratories used floor space more effectively, but 'smaller laboratories were better under most circumstances'.[3] By 1962 a similar monograph argued that 'the large laboratory is an administrative nuisance'.[4] There was much concern about space, both in the laboratory itself and around the outside of the laboratory building for further expansion.[5] It was stated in 1973 that 'the use of peninsula benches at right angles to the windows has become almost mandatory'. They were preferred to the classical island bench because less space was taken up; 'the installation of services is easier and less costly' and the shadow of the bottle racks was reduced.[6] Hence the laboratory designs of the 1960s and '70s placed the benches next to the windows and at a right angle to them, with the fume cupboards usually along the opposite wall, as can be seen in the picture of the Stanford laboratory in the last chapter (see illus. 125). The need for the correct storage of flammable and hazardous chemicals was emphasized, a theme also taken up by safety manuals.[7] The discussion about safety did not, however, suggest individual fume cupboards, but only an increase in their width. One book suggested that there should be one fume cupboard between two students in the organic research laboratory.[8] The safety manuals concentrated on emergencies and hence on procedures, automated safety systems and alarms.[9] For example, eating in the laboratory and lone working were identified as potential hazards.[10] They did not try to deal with safety issues by altering laboratory design.[11]

The biggest change in laboratory design in this period was the introduction of the C-frame laboratory bench – a C-shaped steel frame from which the cupboards were hung. This has been dated to the 1960s, although the first mention of the C-frame bench in Google Books was in connection to its introduction by the British laboratory furniture firm Morgan & Grundy (now APMG) in 1980, when it was presented as being new.[12] William Ferguson mentioned the use of suspended cupboards in 1973, but noted that the loss of storage space outweighed their practice's advantages.[13] The next stage in the development of the bench was the cantilevered casework system, which enabled storage to be suspended over the bench.[14] Perhaps partly because of the introduction of the C-frame bench, there was a recognition by the 1990s that a rationally designed chemical laboratory would have to be more flexible, have utility supplies that could be easily accessed for maintenance and be much safer.[15] The subsequent revolution in design was driven by industrial laboratories, especially pharmaceutical laboratories, where concerns about safety were high and the construction budgets were large. Major new laboratory buildings were opened in 1995 by Fisons Pharmaceuticals (now Astra Zeneca) at Charnwood, near Loughborough, and by Glaxo Wellcome (now GlaxoSmithKline) at Stevenage. Both firms were happy to pass on what they had learned to universities. Stevenage was particularly influential on the design of the laboratories at Oxford, including the glass walls of the laboratories and the concept of clean and dirty areas.

Back to Medicine

In the seventeenth and eighteenth centuries chemistry was closely connected to medicine and even anatomy (see page 53). In the eighteenth century chemistry was also linked to mining and porcelain manufacture. For a relatively brief period in the nineteenth century it seemed to be breaking free of these links to become an academic discipline in its own right, while still being an important service science to industries such as soap, brewing, steel-making and railways. In many respects it has succeeded in establishing its independence. From the 1860s onwards, however, organic chemistry became closely allied with the growing synthetic dye industry. As the chemical industry moved into new areas, such as high-pressure chemistry, and the petrochemical industry grew in size, industry's links with physical chemistry developed. An important watershed was the rise of chemotherapy and steroid chemistry in the 1930s, which strengthened the connection between chemistry and the pharmaceutical industry.

One result of this renewed alliance with pharmacy was the increasing focus of synthetic organic chemists on natural products of interest to medicine. Robert Burns Woodward was one of the pioneers of this shift, synthesizing cholesterol, reserpine, prostaglandins and antibiotics, as well as non-medical targets such as chlorophyll. This focus on medically important compounds has been maintained by more recent chemists such as Samuel J. Danishefsky (taxol), K. C. Nicolaou (taxol, vancomycin) and Steven Ley (thapsigargin, rapamycin).[16] At the same time, the pharmaceutical industry has replaced the chemical industry, at least in the Western world, as the main arena of industrial chemical activity. The industry has developed new techniques such as computational chemistry and robotic synthesis. It has also encouraged the development of new techniques in academia, most notably molecular graphics programs such as the ones developed by Graham Richards.

The pharmaceutical industry has, moreover, been in the vanguard of the design of a new generation of chemical laboratories that has been largely emulated in the new Oxford laboratory building. In the last two decades there has been an even greater shift. Biochemistry and molecular biology, which had moved away from chemistry (and arguably been pushed away) between 1930 and 1970, are now finding common ground with chemistry again. As biochemistry becomes more cellular in approach, the molecular aspects are returning to chemistry, and new disciplines including chemical biology and synthetic biology flourish. Biochemists can be found, for example, in the new Oxford laboratory building. For their part, chemists have become increasingly interested in proteins and other large biomolecules, rather than the smaller (if still large) natural products that were the focus of their attention between the 1950s and '80s. In the early twenty-first century it seems that organic chemistry, at least, is returning to medicine.

Chemistry at Oxford

Oxford University's first chemistry laboratory was opened in 1683, but the regular teaching of chemistry at Oxford did not begin until the 1780s.[17] Even then it was an extra-curricular activity, mainly for medical students, and it was not possible to take a degree in chemistry until 1850. For the next 90 years much of the chemical research at

Oxford was carried out in the college laboratories.[18] The movement back to the central university began with the arrival of W. H. Perkin, Jnr in 1913, and the opening of the Dyson Perrins Laboratory for organic chemistry three years later.[19] The DP, as it was usually called, was designed by Paul Waterhouse, the son of Alfred Waterhouse, who had just designed the Morley Laboratories at Manchester with Perkin (illus. 129). Two decades earlier Perkin had designed Manchester's Schorlemmer laboratories with Paul's father.[20] The style of the laboratory is best described as neoclassical with red and cream bricks, which is similar to Aston Webb's design at South Kensington a few years earlier, but less ornamented. It also has a touch of the revived Queen Anne style used for the Meteorological Office building in South Kensington (now used by the Science Museum), which was completed by the Office of Works in 1910. The crucial turning point in Oxford's chemical history was the plan for a new physical chemical laboratory next to the Dyson Perrins Laboratory, which the automobile magnate Lord Nuffield agreed to fund in 1937. When it opened in 1941, the Dr Lee's Professor, Cyril Hinshelwood, moved from the Balliol-Trinity Laboratory, and the era of the college laboratories was almost over, the last one (at Jesus College) closing in 1947. Inorganic chemistry and physical chemistry were under the same chemistry professor, but the Old Chemistry Building, first built in 1860, became the de facto inorganic chemistry laboratory building after 1941. An extension to the Old Chemistry Building, in effect a new laboratory

129 Exterior of the former Dyson Perrins Laboratory, Oxford, 2013.

building, was completed in 1957 – despite being designed in the 1930s.²¹ This is a rubble-faced building that resembles a Crusader fortress. The inorganic chemistry department was formally established in 1963 with the creation of a new chair. For the next four decades chemistry at Oxford would be dominated by the 'big three' professors and their laboratory buildings. By the 1990s, however, the DP was becoming inadequate for modern chemical research and change was in the air.

The New Alchemy

One major change in academic chemistry in the 1980s and '90s was the growth of spin-out (also called spin-off) companies. Of course, the commercialization of academic research had taken place since the mid-nineteenth century at least. The famous example is William Henry Perkin's discovery of mauve dye when he was still a student at the Royal College of Chemistry, London, in 1856. After his return to Germany, Wilhelm Hofmann (Perkin's former professor) was involved with the firm of Haarmann & Reimer in the essences field and Agfa in dyestuffs, both founded by his students. Many famous chemists have been consultants to industrial firms, for example Roger Adams at the University of Illinois and Du Pont.²² None of these commercializations, however, directly benefited the universities concerned. Furthermore, the connection between academic chemistry and industry was not universally popular with academic chemists, some of whom regarded such collaboration as a betrayal of 'pure' science. As a result, many chemical advances were not patented either through principle or less often by oversight. It is hard to say whether more patenting would have benefited either the chemist or the university. For example, when Othmar Zeidler made DDT in 1873, he was not aware of its insecticidal properties and even if he had been, the patent would have run out long before Geigy started to make it as an insecticide in 1940.

After the Second World War and Britain's alleged failure to profit from the discovery of penicillin, the Labour government set up the National Research and Development Corporation (NRDC) in 1948. One of the earliest products exploited by the NRDC was the cephalosporin antibiotics developed by Edward Abraham at Oxford. While the NRDC held the rights, Abraham was given an income from the patents which he used to set up the Edward Penley Abraham Research Fund and the EPA Cephalosporin Fund. However, this route was not a money-spinner for most chemists or even the state. Meanwhile in America, universities sought to generate an income from their advances through patent agents or hired patent lawyers themselves, but generally with little financial success. Furthermore, if the research was sponsored (for example by a federal agency), the sponsoring body's permission had to be obtained.

As described in chapter Eleven, Stanford University was a pioneer in building links with industry. Largely through the driving force of Fred Terman and Alf Brandin, vice-president for business affairs, the university set up the Stanford Industrial Park (later the Stanford Research Park) in 1951. The first tenant, Varian Associates, was soon followed by Eastman Kodak, Hewlett Packard and Lockheed Research Laboratories, but the research park still did not exploit the university's intellectual property directly. In 1968 Niels Reimers came to Stanford as the associate director of its Sponsored Projects Office.[23] In this position he was responsible for getting the sponsor's permission for patenting inventions. He soon realized that the crucial thing to do was to market the invention rather than the patenting itself, which could be left to external patent lawyers. He persuaded the university to allow him to set up a pilot scheme with a sinking fund of $125,000. Within a year he had made $55,000, which was ten times more than the licensing income for the previous fifteen years. The university then made him director of the new Office of Technology Licensing (OTL). An early breakthrough was the licensing of 'gene splicing' in 1974, after Reimers persuaded its inventors, Stanley Cohen of Stanford and Herbert Boyer of the University of California, San Francisco, that it could be patented.

It was exactly at this point that the young Graham Richards arrived from Oxford for a year's sabbatical as a Fulbright Fellow. Soon after his return to England, in 1978, Richards became chairman of the University and Industry Committee of Oxford University just before Margaret Thatcher came to power.[24] The first step in spinning off Oxford science was a collaboration with Monsanto, which had set up a venture capital fund. The real shift came in 1986, when the government allowed universities to exploit research council-funded research, having already abolished the NRDC's right of first refusal on university research. Oxford University – at the urging of David

130 Graham Richards.

Cooksey, the head of the Oxford-Monsanto venture fund Advent Eurofund, and Richards – set up Oxford University Research & Development Ltd in November 1987, which (perhaps mercifully) was renamed Isis Innovation eight months later. Like Stanford's OTL, it initially concentrated on licensing and improving links between academics and industry, setting up the Isis Innovation Society for this purpose in 1990. Meanwhile the very first spin-out company, Oxford Glycosystems (which became Oxford Glycosciences in 1996), was created in October 1988 by the Oxford biochemist Raymond Dwek with the help of Monsanto.[25] Soon afterwards Richards set up Oxford Molecular (now Accelrys), a computer-modelling company, with his former student Tony Marchington.[26] In 1994, however, with only one other company set up – Oxford Asymmetry by Oxford chemistry professor Steve Davies – and disappointing revenues for the university, a review was carried out.[27] As a result of this review, the university asserted its rights over the intellectual property of its staff and students, reinvigorated the leadership of Isis Innovation with the appointment of Tim Cook, an experienced executive from industry, and increased its level of support. Subsequently, the Isis Angels Network of potential funders was created in 1999.[28] It was in this context that the innovative funding arrangements for the new Oxford laboratory building were put in place (see next page). The rate of creation of new spin-out companies increased rapidly, with 74 set up between 1998 and 2012.[29] In its annual report for 2012, the company reported that:

> Isis revenues increased by 21% to £10.2m with growth across all areas of the business. Over the last ten years Isis revenues have shown a compound annual growth rate of 20%.

A similar process to that at Stanford and Oxford has occurred across the world between the 1970s and the present day. It has completely transformed the relationship between academia and industry, and has made some academic chemists very wealthy. It has also potentially changed how the construction of laboratory buildings is funded.

Creating a New Laboratory Building

After ten years of inaction on the part of the chemistry departments, in 1996 Oxford University insisted that a unitary chemistry department should be set up. It was also agreed that the new head should not be one of the heads of the existing departments. Graham Richards was appointed to the role, perhaps aided by the fact that his research did not easily fall within the usual disciplinary boundaries. One of his first actions was to build a new central building for chemical research, named rather prosaically, the Chemistry Research Laboratory (CRL).[30]

It is not often realized that Oxford has the largest chemistry department in the Western world, and the new laboratory was capable of housing 440 researchers.[31] Apart from any technical problems in refurbishing one of the existing laboratory buildings, a new building was needed to cement the identity of the centralized department. Richards could have gone for a greenfield site outside the city, but he was keen to keep the laboratory building in the centre of town to encourage researchers to work in the laboratory in the evening.

Fortunately a suitable site, a car park and former vicarage, was available in the middle of the so-called Science Area close to the existing chemistry buildings.

The financing of the laboratory building shows the diversity and complexity of modern academic research facility funding. Oxford received a £30 million award from the newly founded Joint Infrastructure Fund (JIF) – a partnership between the Wellcome Trust, the Office of Science and Technology, and the Higher Education Funding Council for England – having caused some alarm by applying for the whole cost of £60 million. Nonetheless, the grant it received is the largest grant so far given by JIF. In the face of some internal opposition, an additional £20 million was raised through a novel scheme with the investment bank Beeson Gregory. In return, Oxford University transferred a share of its future rights to equity in spin-out companies arising from the chemistry department to Beeson Gregory for fifteen years, with the firm helping Isis Innovation, the university's subsidiary technology-transfer company, with the formation of these spin-outs. The deal was attractive to Beeson Gregory as the chemistry department had a good record of setting up spin-out companies, having set up four of the university's 23 spin-out companies. According to the department website:

> This funding model, labelled by the Financial Times 'a blueprint for universities everywhere', has since been copied by other institutions, and three new spin-out companies have already been created since its inception.[32]

The score is now fifteen companies, with five having had initial public offerings (IPOs). More traditional donations were received from the Wolfson Foundation, the EP Abraham Research Fund, Thomas Swan & Co. and the Salters' Company and, most recently, the family of Landon T. Clay. Thomas Swan is a privately owned chemical company in County Durham founded in 1926 as a tarmacadam manufacturer. The owner was a friend of Graham Richards. The Salters' Company is a City guild that has traditionally supported chemistry. Landon Clay is an American financier who founded the Clay Mathematics Institute, currently based in Oxford, in 1998 and funded a new telescope at the Harvard University Observatory in 2007. Surprisingly, there was no financial support from the pharmaceutical

131 Exterior of the Chemistry Research Laboratory, Oxford, 2013.

industry despite its use of Richards's software and the industry's high profile in Britain.

The building's architect, RMJM, was founded by Robert Matthew and Stirrat Johnson-Marshall in 1956.[33] One of its early iconic buildings was the Commonwealth Institute in North Kensington. The firm quickly obtained experience in building 'glass-plate' university campuses (Stirling and York), and more recently laboratories such as the new Laboratory of Molecular Biology at Cambridge and Genzyme's research facility in Beijing.[34] This is in keeping with the trend for laboratory design to be carried out by large practices with broad remits, rather than by architects who specialized in laboratory design in the tradition of Alfred Waterhouse.[35] The laboratory is a glass-plate building like many modern office buildings, reflecting the tendency since the 1960s for laboratories to be like standard business buildings (illus. 131). Similarly, the construction was done by Laing Construction, a large, well-known international firm, rather than by a local builder as would have been the case a couple of generations ago. However, Laing had to sell its construction company during the building of the laboratory, and it was completed by the new firm of Laing O'Rourke, founded by Ray O'Rourke, which was relatively unknown at that time, but has since become an important builder, constructing Terminal 5 at Heathrow, and in Oxford the new Biochemistry, Earth Sciences and Mathematics buildings.

The overall management of the project was run by the Chemistry Management Committee, with Richard Jones as client representative and project manager for building occupation, while the details were worked out by the departmental building committee.[36] There was close cooperation between the building committee and the architects, with visits to pharmaceutical company laboratories. One member of the building committee, the Waynflete Professor of Chemistry Steve Davies, even travelled to a quarry in Italy to choose the stone used for the cladding and the flooring of the foyer. The fact that the department had raised all the money required for the building permitted the best solution to be employed rather than the least expensive – for example installation of a Trend 963 Building Management System to manage air movement, and pressure regimes required for the control of the 325 fume cupboards. The building was formally opened by HM The Queen and HRH The Duke of Edinburgh in February 2004.

A Modern Laboratory

To illustrate the changes in the laboratory building since the 1960s, let us compare a visit to a classical laboratory building (such as the former Dyson Perrins Laboratory building across the road) with a visit to the new chemical laboratory building in Oxford.[37] Previously anyone could just enter a laboratory building and walk through echoing corridors to the laboratory, but security today (as in most buildings, including the older laboratories) is strict. You enter a reception area and either have to swipe a security card or sign in at the main desk. To enter the laboratory area you then have to use the swipe card again. Whereas the offices in the classical laboratory were few in number and scattered throughout the building, one side of the new Oxford laboratory building is entirely devoted to offices for the professors and administrators. Although stairs still remain, lifts have become the predominant way of moving between floors. The continually moving lifts ('paternosters') common in laboratories in other countries are now rare in Britain, having enjoyed a brief spell of popularity in the 1960s and '70s, including one in the Biochemistry building (Hans Krebs Tower), which has now been removed.[38] The laboratory is

132 Atrium of the Chemistry Research Laboratory from the first-floor bridge, 2004.

133 Writing area in the grey-carpeted 'clean' laboratory area; the whiteboard can be seen on the right in front of the laboratories, CRL, Oxford, 2004.

purely a research building and there are no teaching facilities, but there are a few seminar rooms. There is then a covered atrium between the office block and the laboratory block (illus. 132). This has been designed deliberately as a space for socializing and peer interaction between different laboratory groups, and between academics and technicians. Hence the cafe is located here.[39]

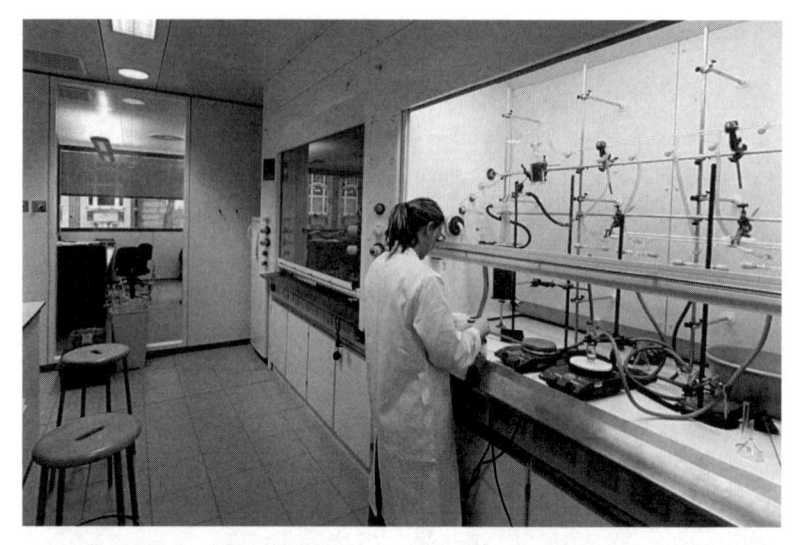

134 Fume cupboard in the laboratory – the writing area can be seen outside, CRL, Oxford, 2004.

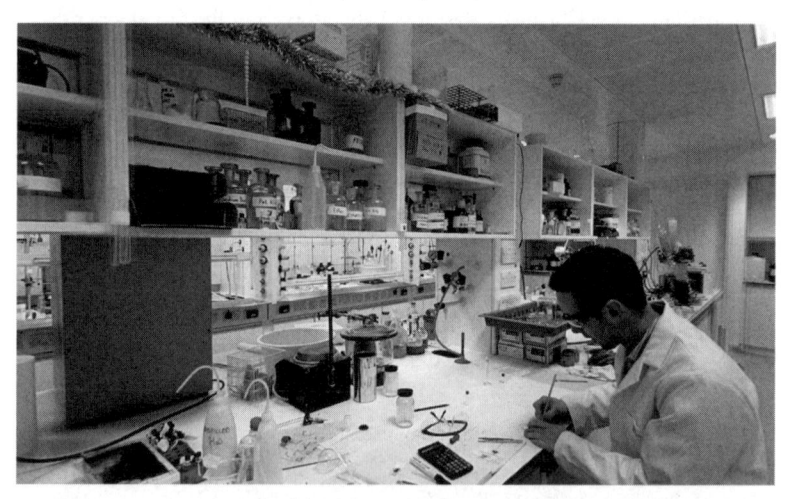

135 Laboratory bench in the laboratory – the fume cupboards can be seen on the left, CRL, Oxford, 2004.

There are four floors and a basement in the laboratory block; two are given over to organic chemistry, the third one to inorganic chemistry and the lower ground floor (first storey) to biochemistry. The central facilities and larger apparatus are located in the basement. Walking into the laboratory side of the building, you are again struck by the contrast with older laboratories. The flooring in the corridors and the laboratory offices is carpeted rather than bare wood or stone. This is deliberate as the different types of flooring indicate whether an area is 'clean' (carpet) or 'dirty' (linoleum flooring). Instead of a classical laboratory with benches, you initially enter a carpeted open-plan office, or 'writing area' as it is called (illus. 133). This office space is a result of a shift from paper-based laboratory records to computers, and a strategy of separating the 'dirty' experimental work from the 'clean' office-based tasks. It is open-plan, partly for flexibility, but mainly on safety grounds as it provides a clear line of sight across the office and into the laboratories.[40]

The laboratories themselves are much smaller than traditional laboratories, and are dominated by eight fume cupboards, four on each side (illus. 134). There is a bench in the middle (illus. 135); the successor to the bottle rack is a couple of shelves above the bench, now filled with a variety of bottles and boxes. This is in contrast to a proposal in a 1993 monograph on laboratory design that the fume cupboards should be in the middle and the benches on the sides.[41] The side facing the office is completely glazed, including the doors, to allow researchers to scrutinize the labs from the desk area in case of accidents. Between the glazed panels are 'Optiwrite' full-height white glazed panels, which can be written on to facilitate development of formulæ and discussion. As a 'dirty' area, the laboratories on the upper three floors are floored with ceramic tiles that are resistant to stains and damage, while the biological chemistry and support areas have proprietary vinyl flooring. As is customary nowadays, workers in the laboratory wear plastic safety glasses, white coats and nitrile gloves. There is a bench in the middle of the laboratory, but much of the experimental work is carried out inside the fume cupboards, which are now completely white and have internal lighting, with all the controls located outside the cupboard. The glass on the front is treated to enable the researcher to write on it – it can perhaps be the reaction being carried out, a message to another researcher or a warning. Efforts are being made to optimize the flow of air through the fume cupboards to reduce the energy consumption while maintaining containment. Each researcher should have their own fume cupboard, but the rising number of researchers has forced a degree of sharing of cupboards. The laboratories are planned to be as flexible as possible. In the inorganic chemical laboratories, fume cupboards are sometimes replaced by glove boxes. In the biochemical laboratories, the fume cupboards are often completely absent and the partitioning walls are taken down to create a larger space, which is superficially more like the classical laboratory, but with flat-topped benches rather than a bench with bottle racks. This means that laboratories can be reallocated to different groups each year according to their needs. Indeed, it would be possible to convert laboratories into open-plan offices if this was ever deemed necessary.

What is striking in this laboratory building compared with the chemical palaces of the late nineteenth century is the absence of the leading professors. Of course the Oxford building does not have the palatial residences of Hofmann or Kolbe – professorial residences have never been a feature of British laboratories – but now it even lacks any private laboratories for the professors. For a few decades now, professors and other leading researchers have not carried out benchwork themselves, but achieve their research goals by directing the laboratory work of their associates and students. The professors (and other leading academics) have a fairly modest office in the office block of the CRL, usually lined with their students' theses, but no laboratory to call their own. Libavius's secret space for the philosopher, which survived for four centuries, no longer exists in the CRL at Oxford.

Behind the laboratories there is a 'dirty' corridor that is used to move chemicals and equipment to and from the laboratory. There is also a 'dirty' lift to move things from floor to floor. On the other side

136 Solvent room across the corridor from the laboratories, showing the nonthermal purification system, CRL, Oxford, 2004.

137 NMR room in the laboratories area, showing the NMR spectrometer with its automatic sample loading belt, CRL, Oxford, 2013.

of the 'dirty' corridor there is a suite of rooms on each floor, which contain the equipment shared by each of the laboratories on that floor. Precisely what they contain depends on the research in their vicinity. Chemical Biology has cold rooms and freezer morgues; other floors have routine NMR or mass spectrometry instrumentation and the potentially hazardous solvent-purification room. In the solvents room solvents are dried by forcing them through a drying column under pressure, rather than by the traditional method of distillation, thereby removing the risk of fire (illus. 136). In another room is a high-performance liquid chromatography (HPLC) machine that is linked ('hyphenated') to a mass spectrometer, allowing routine analysis of reaction. A third room contains an NMR machine that enables the study of the structure of compounds produced in the neighbouring laboratories (illus. 137). In the 1960s, as noted in chapter Eleven, NMR and mass spectrometry were conducted in specialized central facilities. At one level this is still true and the central facilities are in the basement of the building, but standard NMR and mass spectrometry are now carried out next to the laboratory, and they have become routine in the same way that infrared spectroscopy was in the 1960s. This has been made possible by the automation of sample submission and insertion. The samples are placed in a queue in a belt and automatically loaded into the HPLC or NMR machine, and the results are stored on a computer for later retrieval. This avoids any waiting to use the machines, and prevents mishandling of the equipment. While the costs of NMR and mass spectrometry machines have come down over the years (in real terms), the provision of these techniques at the laboratory level remains a hugely expensive business.

The basement of the laboratory building houses the main central facilities, including NMR, mass spectrometry, X-ray crystallography and surface analysis.[42] Some of the machines can be used directly by any researchers and others by trained researchers, rather than the conventional route of submitting samples to the staff operating the equipment. The NMR room contains eleven solution-state and two solid-state FT NMR instruments, with proton-operating frequencies ranging between 200 and 700 MHz, mostly made by Bruker in contrast to the dominance of Varian in the 1950s and '60s.[43] Given the presence of thirteen superconducting magnets, a quarter-inch thick steel plate was put into the ground floor to stop any magnetic field going up the building and interfering with PCs. The mass spectrometry facility is one of the largest and best equipped in the UK, with sixteen mass spectrometer systems and a comprehensive range of ionization techniques and inlet systems.[44] Most of these machines are hyphenated to a liquid or gas chromatograph. There is also

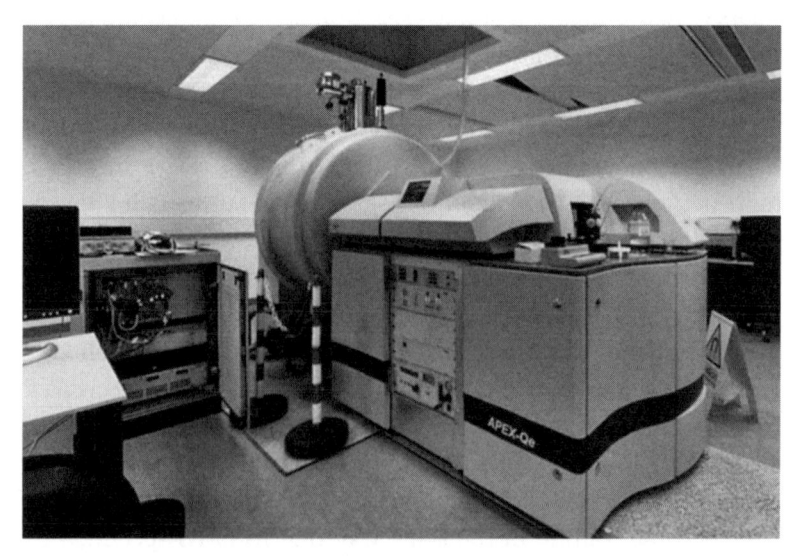

138 Bruker Daltonics Apex QE-QH-FTMS Fourier Transform Ion Cyclotron Resonance Mass spectrometer in the mass-spectroscopy research area in the basement, CRL, Oxford, 2004. This very high-resolution instrument is used for proteomics, the study of proteins by identifying the peptides they contain.

a MALDI TOF (Matrix-Assisted Laser Desorption/Ionization Time-Of-Flight) mass spectrometer for large molecules, either polymers or biological molecules. The substance is absorbed on a solid organic material ('matrix'), then blasted off the material ('desorbed') by a UV laser. The hot plume thereby generated is then ionized and passed into the TOF mass spectrometer. This mass spectrometer uses an electric field to give all the ions the same kinetic energy. The time taken to pass through the mass spectrometer then gives the mass-charge ratio and hence the mass. Biological molecules, for example proteins, are an increasingly important field of research for both NMR and mass spectrometry (illus. 138). Carol Robinson, the Dr Lee's Professor of Chemistry, uses an ion mobility mass spectrometer to study the three-dimensional architecture of protein complexes.[45]

The X-ray crystallography room has four machines: two Nonius KCCD Diffractometers and two dual-wavelength Oxford Diffraction/Agilent SuperNova A systems.[46] The switch from photographic film to cooled charge-coupled detectors (CCDs) and automation has revolutionized X-ray crystallography just as it has astronomical imaging. A wider range of reflections can be captured by the large detectors and they can be run effectively all the time, allowing the capture of more data and a higher throughput of samples. To complement these specific approaches, the surface analysis section employs a wide range of techniques to study the surfaces of solids, including scanning tunnelling microscopy, atomic force microscopy, Fourier transform infra-red spectroscopy and X-ray photoelectron spectroscopy.

The basement also houses six laboratories ideal for sensitive experiments that might be jeopardized by vibration. Indeed, two of these labs have in the floor a vibration-isolation zone achieved by having an isolated thick concrete plinth that has the capability to ride on air jacks. The floor is also useful for the operation of research equipment that must be precisely controlled or measured. A good example of this is Professor Mark Brouard's research on the mechanisms of simple gas phase reactions using polarized laser pump and probe (flash photolysis) techniques.[47] Polarized light from a laser is aimed at a sample of a simple molecule such as nitrogen dioxide (NO_2), which then breaks up and its fragments are studied. The advantage of using polarized light is that it allows the direction in which the fragments move to be determined. Clearly such a sensitive measurement requires complete stability. The equipment is heavy and is placed on an optical table, which in turn rests on the specially constructed anti-vibration floor.

An Enterprising Chemist

Graham Richards was born in Hoylake, Cheshire, on 1 October 1939, the son of a Welsh printer and a Welsh domestic servant.[48] He was educated at Birkenhead School, a direct grant school,[49] then took chemistry at Oxford. He took his Part II (the fourth undergraduate year of the Oxford chemistry degree) and his DPhil under Richard Barrow on the spectroscopy of the chlorine molecule. In 1964 he became a fellow of his undergraduate college, Brasenose. Rex Richards, then the head of the physical chemistry department and later vice-chancellor, told his unrelated namesake to spend a couple of years thinking about the kind of research he wanted to do. He decided to work on quantum mechanical calculations of small molecules on a computer. For several years he had to use unreliable British computers before a press leak that he engineered led to the shake-up of the computing service and the introduction of American computers. Invited in 1968 to be a consultant on the project led by James Black at Smith, Kline & French to develop H2 blockers for the treatment of stomach ulcers, Richards switched from the quantum chemistry of small molecules to the larger molecules of interest to the pharmaceutical industry. This move had an important impact on his career. Pharmacological molecules needed a graphic interface and much more computer power. Moving into an area of commercial interest also led to his later involvement with spin-out companies.

Richards (illus. 130) was a rather unusual figure in Oxford in the 1960s and '70s. He contributed to a collection of essays, *The Conservative Opportunity*,[50] at a time when most chemistry dons were not overtly political (or if they were, they tended to be Liberal or Labour), and was a supporter of Margaret Thatcher when she was not popular with Oxford dons. Furthermore, he supported the commercialization of academic chemistry – partly as a result of being a regular visitor to the chemistry department at Stanford – when this was still anathema to most dons. Most of his fellow academics (and I was at Oxford at the time) regarded him with a mixture of bewilderment, consternation and a certain admiration for going against the current. It was universally agreed, however, that he was a humane laboratory demonstrator who could see merit even in the weakest student reports of laboratory work. At the same time, Richards was instrumental in the introduction of women into hitherto all-male colleges, and his own college was one of the five colleges to 'go mixed' in 1973.

As computer-aided molecular design hardly existed when Richards began working with the pharmaceutical companies, he had to develop all the techniques required – even the colour graphic interface. From its rudimentary beginnings in the early 1970s, it soon became a mature technology routinely used by chemical and pharmaceutical companies to make new compounds. As it was so useful to industry, Richards wondered if his research could be commercialized. He was a chemist rather than a businessman, and needed someone to run the commercial side of any start-up business. Even if he wanted to set up a business, however, the NRDC had an automatic first refusal on any patents he might take out. It was only possible for Richards to exploit his research after Margaret Thatcher's Conservative government gave the rights to patents taken out by their staff back to the universities in 1986. Richards already had a potential partner in his former DPhil student Tony Marchington, who was working on molecular graphics for ICI, but also had a profitable sideline in promoting steam-traction engine rallies. The final spur to action was the death of Richards's wife in November 1988, leaving a void that he aimed to fill with work. He got together with Marchington, who had left ICI. They raised £350,000 and after eighteen months their company, Oxford Molecular, had fifteen employees working in a prefabricated building outside Richards's laboratory. The company then moved to the Oxford Science Park, was floated on the Stock Exchange in 1994, was valued at £30 million and began to take over other similar companies. At its peak in 1997 the firm was valued at £450 million, employed 400 people and held 25 per cent of the world's bioinformatics market. It was eventually sold for £70 million in 2001, yielding a £10 million profit for the university.

Taking inspiration from the screensaver program distributed by Project SETI, the search for extraterrestrial life, Richards developed a similar program to harness idle PCs to work out how small molecules were bound to a given protein, with the aim of developing new drugs between 2001 and 2007. The main targets were for anti-cancer drugs, but following the terrorist attacks of 2001 on the World Trade Center, the search was extended to cover defences against anthrax and smallpox. The project grew to involve over 3.5 million personal computers in more than 200 countries, and led to another spin-out company.

Under Pressure

While gas-liquid chromatography (see chapter Nine) was a major breakthrough in the 1950s, a different form of chromatography had come to dominate chemistry by the 1980s.[51] Gas chromatography, as its name implies, works well with gases and volatile liquids. However, many organic compounds and biological molecules are high-boiling liquids or solids. The problem can be circumvented to some extent by converting the compound into a volatile derivative, for example making the ester of an acid. This is, however, a limited solution and many compounds are either too sensitive to heat or lack a suitable functional group for derivatization. Many compounds can be converted into a solution and separated by conventional liquid-liquid chromatography, but the process is slow, the separation is often poor, and there is no easy method of detecting the bands as they are produced in the solid column unless they are coloured (or fluoresce in UV light). Using smaller particles in the column produces better separation, but slows the flow of solvent through the column even more.

The very first high-pressure liquid chromatograph was developed specifically for amino acids by William Stein and Stanford Moore at the Rockefeller Institute in New York in 1958. Meanwhile Lloyd Snyder, working for Union Oil, was trying to find a way to analyse high-boiling petroleum components. His work was unusual insofar as he started with a mass spectrometer and sought a way of separating compounds before they entered the instrument. By the mid-1960s he had settled on liquid-solid chromatography as the best way forwards. In 1965 Csaba Horváth was a Hungarian chemist at Yale University working with Seymour Lipsky, who had collaborated with James Lovelock on the development of the electron-capture detector. Coming from gas chromatography, Horváth realized that switching to liquid-solid chromatography would allow the use of narrow reusable columns that could also be heated – features that already existed in gas chromatography. To speed up the process, he then applied pressure of 1000 psi (70 bar). The detector was an ultraviolet spectrometer.

Similar work was being carried out at Eindhoven University of Technology in the Netherlands by the Austrian chemist Josef Huber, who benefited from the advice of Lou Keulemans and Archer Martin in the same faculty. Martin had actually forecasted the development of HPLC in his paper on liquid-liquid chromatography in 1941. The process was initially called high-pressure liquid chromatography, but Horváth soon preferred the superior-sounding term high-performance liquid chromatography. (Some wags suggested that it should be called high-price liquid chromatography.) The two terms were then used side by side, and to reduce the confusion Horváth introduced the abbreviation HPLC in 1970.

Within a few years in the early 1970s, the pressures used had increased to 6,000 psi (420 bar) and the technique was commercialized. A major player in this field was Waters Associates, who had taken up a technique called gel-permeation chromatography, developed in 1962 by John Moore of Dow Chemicals. This was similar in many ways to HPLC, which gave Waters an easy entry into the field. Jim Waters, the founder of the firm, persuaded the famous organic chemist Robert Burns Woodward in 1972 to use a Waters machine to separate isomers during the synthesis of Vitamin B_{12}, and to allow the firm to publicize his use of the instrument in a flyer.[52] Waters's great rival was Du Pont, where Jack Kirkland had taken up Huber's work in order to analyse pesticide residues. He was aided by the firm's expertise in on-stream process analysis using a UV photometer. Du Pont subsequently concentrated on the development of packing materials for the columns. Varian also entered the field, as it was already making Aerograph gas chromatographs. However, Waters soon dominated the field and has continued to do so. The particle size continued to fall and the pressures employed rose. By the late 1970s the speed and sensitivity of the technique was comparable to gas chromatography; namely a sample size of picogrammes and a detection limit of parts per trillion.

Nowadays HPLC is often used in conjunction with mass spectrometry as the detector, and this is the standard set-up in the Oxford laboratory. The mass spectrometer is by far the most sensitive detector for either gas chromatography or HPLC and it is also very rapid, with a complete detection time of around a minute. The drawback is that the sample is destroyed within the mass spectrometer. As in the case of gas chromatography, the carrier substance (a solvent in this case) has to be removed before the fractions enter the mass spectrometer. The first way of doing this was the thermospray technique introduced in the early 1980s by Marvin Vestal and Calvin Blakley at the University of Houston. This was soon displaced by the electrospray method developed by John Fenn at Yale, who won the Nobel Prize in 2002 for this innovation.[53] In the electrospray technique an atomiser sprays the emerging liquid from the chromatograph into a blast of hot nitrogen gas in an electrically charged field. This removes the solvent and ionizes the compounds that have been separated by the chromatograph. The ions are then swept into the mass spectrometer to be analysed. HPLC is often used with tandem mass spectrometry-mass spectrometry, whereby the molecules that have been separated by their mass (strictly speaking their mass/charge ratios) in the first mass spectrometer are then broken up into fragments in a second mass spectrometer. The characteristic mass spectrum of the 'daughter' fragments is then used to identify the compounds, usually with the help of a computer-based library of mass spectrometry spectra. A UV spectrometer is, however, still commonly used as it is a non-destructive technique and can be used in conjunction with mass spectrometry by splitting the outlet stream as it comes out of the HPLC. Another common technique is the Evaporative Light Scattering Detector (ELSD), in which the outlet stream is atomized, the carrier solvent evaporates and the sample is analysed using light scattering.

A New Path?

The CRL at Oxford was novel in several respects. Over the centuries the construction of laboratory buildings and laboratories had been funded in various ways, ranging from state support, local authorities, public subscription (in the case of the RI), private philanthropy and companies. The CRL is highly unusual in being partly funded through the creation of spin-out companies. It remains to be seen if this is a model that will be copied by other universities. The relative lack of corporate support is also striking.

Even more importantly for this book, the design of the laboratory has changed completely and in a way not anticipated by the laboratory designers of the 1960s and '70s. The laboratory space is now sealed off from the outside by glass walls, and most of the experimental work (at least in organic chemistry) is carried out in the fume cupboard with the ideal of one cupboard per researcher (but not achieved even at Oxford). While sealed off, the laboratory is also overlooked from outside through the glass walls for safety reasons. The laboratory is now much smaller than the classical laboratory, but the island bench has returned. The experimental working space as a whole is now clearly demarcated and separated from offices and other facilities. What you can do in the laboratory is increasingly circumscribed; you cannot eat or drink any more in the laboratory (no more tea brewed using a glass beaker, a glass rod and a Bunsen burner!), and in some laboratories (but not at Oxford) you have to use special sealed wipeable laptops to make notes instead of a paper notebook, which can easily become contaminated.

One result of this separation is a growing divide between the leading professors and the laboratory. Originally the top chemists had their own laboratory (indeed, it was a sign that they were top chemists); then, by the 1950s, they led teams of researchers and hence needed a bigger laboratory. At the same time chemists such as Woodward ceased to work in the laboratory themselves, but had an office next to the laboratory. One particular anecdote about his practical skills relates how Woodward was able to overhear a remark in the laboratory from his office. In the CRL, however, the professors have their offices in the administrative block away from the laboratories. For this reason, and to encourage different research groups (and sub-disciplines) to interact socially, there are now central spaces in laboratory buildings, such as Oxford's atrium cafe, for mingling. In terms of health and safety, the focus is now on creating inherently safe design rather than the development of facilities that respond to incidents such as emergency showers (although the showers still exist, of course). An example of this is the solvent-purification room, which does away with the hazardous distillation of solvents and removes the handling of solvents in bulk as far as possible away from the laboratory itself.

Originating in pharmaceutical laboratories, this is a wholly new kind of laboratory that represented a clean break with the classical laboratory in a way that laboratories in the 1960s and '70s, such as Stanford, did not. Will this new laboratory type replace the classical laboratory, and how far will it spread? On one level this is a difficult question to answer — it is rather like asking in 1866 if the classical laboratory would catch on. There are, however, two problems with this kind of laboratory which suggest that its diffusion may be limited, at least in the medium term. It is not convenient for teaching, which means that there may soon be a clear distinction between teaching laboratories and research laboratories, something that did not exist in the classical laboratory period. This separation had already started to appear in the 1970s as research laboratories became smaller and more elaborately equipped, but now the difference is evident. Given ever-growing health and safety concerns, it is hard to see how teaching laboratories will survive in their present form. One possibility is that a larger version of the new laboratory design for teaching will appear (as has been proposed for Oxford), but it will be difficult for several people to work simultaneously in such a confined space, even if it is a larger space.

This brings us to the other problem with the new laboratory design: that it is very expensive to erect and operate. I cannot see many less-developed countries being able to afford such laboratories for their universities (or indeed many universities in the developed world). I surmise that a modified version of the classical laboratory, with C-frame benches and more fume cupboards, will continue to be built (or installed in refurbished laboratory buildings) for many years to come. The organic chemistry teaching laboratory in the new chemistry buildings at Leipzig University (opened in 2000) is exactly of this type.[54] On the other hand, the refurbished teaching laboratories at the University of Bristol are very similar to the CRL laboratories, but have a different layout, partly to avoid the crowding problem just mentioned.[55] Given our earlier discussion of the development of the lab coat, it is interesting to note in passing that the different types of student and staff in the laboratories are distinguished by different coloured laboratory coats.

Conclusion

The most striking aspect of the history of the chemical laboratory is the predominance of two types of laboratory in the last 400 years. The original concept of a workspace dominated by furnaces and distillation apparatus was well established by the late sixteenth century and was very long lived, surviving until the early nineteenth century. The period between 1770 and 1860 was a period of transition. Lavoisier's laboratory marked a major break, but did not create the template for the laboratories of the mid-nineteenth century. Between the 1830s and '60s a new type of laboratory arose, largely in Germany but also in Britain, which was geared to housing large numbers of students and researchers.

By the mid-1860s a standard design ('the classical laboratory'), dominated by benches, bottle racks and fume cupboards, had been created and adopted across the world. This layout proved to be remarkably robust, partly because it could be adapted without fundamentally changing the template; with changes in the position of the benches it was still used in new laboratories up to the 1980s at least. By the 1990s, however, the pharmaceutical industry was developing a new kind of chemical laboratory, which was centred on the fume cupboard. This design had reached academia by the early twenty-first century and will probably dominate laboratory design for the foreseeable future. It seems unlikely, however, given the pace of technological and scientific change, that it will endure for as long as the classical laboratory. It would be foolhardy to predict the future, but it is possible that robots (illus. 137) may take over the laboratory by the middle of the twenty-first century, if only because of increasingly stringent health and safety regulations. The robots can be envisaged working in sealed rooms that are monitored remotely by webcams and have a constant

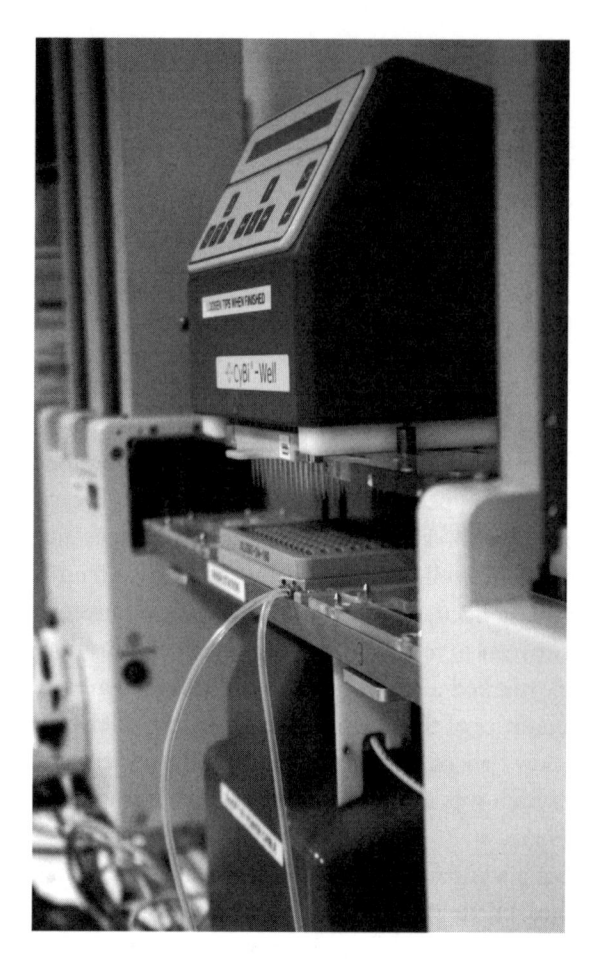

139 Robotic synthesiser at the Chemical Research Laboratory, Oxford, 2004.

flow of air in the room rather than a fume cupboard. In effect the fume cupboard will have become the room.

Why are Laboratories Changed?

It might be thought that laboratories are changed because of scientific progress, the improving status of chemistry and increased funding. These factors undoubtedly play some role in the improvement of the laboratory, but I would argue that they are not the most important factors. The crucial factor in my view is competition; it has changed over time, but it has always played an important role.[1]

The competition can be on different levels that can co-exist at the same time. On one level it is the competition between states, be it Saxony and Prussia, Germany and Britain, or the USA and the Soviet Union. This state competition usually exists because chemistry is perceived as adding to a nation's clout – in terms of agricultural improvement, better health, economic growth or military capability. On the next level down there is competition between cities or universities (or firms). You can see this at work in the rivalry between London, Manchester and Birmingham, and between Stanford and the east-coast universities. The rivalry is both economic – a popular university boosts the economy of a town – and a matter of local pride. A top chemistry department can also help to promote the economy of a region, whether it involves Lancastrian textiles in the nineteenth century, or Californian electronics in the twentieth. Universities need top-rate laboratories to attract the best staff and students. Even after Stanford had attracted Bill Johnson and Carl Djerassi to California with the promise of a new laboratory, it had to build another new laboratory to keep Paul Flory. Finally, there is the personal rivalry between chemists which spurs them to create the best possible laboratory when the opportunity arises.

If competition explains to a large extent why new laboratories are built, it does not elucidate why these laboratories are different from the old ones. Two factors are involved in the creation of a different laboratory: chemical progress, and new or better utilities. A continuing theme in this book has been 'form and function'. As the function of chemistry changes, so does the form of the laboratory. The development of a new field of chemistry has often produced important alterations in laboratory design. The first major change was a result of the new gas chemistry of the late eighteenth century, but the kind of laboratory it created did not endure, as gas chemistry did not became the dominant aspect of chemistry. The changes in the early nineteenth century were partly the result of new kinds of chemistry – organic analysis and group analysis – which required rigorous training. Organic analysis (as combustion analysis) became a specialized technique and was hived off into its own special room. By contrast, group analysis had a major impact on the laboratory with its need for various aqueous solutions, hence the bottle rack. It also required the use of the extremely toxic gas hydrogen sulphide, and the concomitant need to remove it safely (see page 101). The fume cupboard arose from both group analysis and organic synthesis.

By the mid-nineteenth century the growing number of specialized techniques – gas analysis, spectroscopy, polarimetry – resulted in an increase in the number of rooms needed in a laboratory building (and indeed the need for a building rather than just a laboratory). A successful template for a laboratory building having been created in this way, later chemical developments had less impact. Physical chemistry did not need bottle racks, and in some universities tables made a return, but most laboratories reflected the usual 'classical' template. As explained in the Introduction, the leading chemist (usually an institute director or professor) plays a key role in the development of the laboratory, firstly as the decisive element of competition – the winner is the institution that acquires the best professor (or professors) – then by the professor's demands for a better laboratory. He (and historically it was always a he) then designs the laboratory to incorporate the latest developments, but also introduces new features to accommodate the chemistry of the future.

Modifications in the utilities provided to a laboratory can sometimes bring about equally fundamental changes. The furnace did not disappear because chemistry moved beyond the furnace; rather it was the introduction of piped steam and mains gas that allowed the furnace to be dropped. Changing utilities also have a profound influence on chemistry itself. The development of atomic spectroscopy was dependent on the availability of mains gas; recrystallization, a mainstay of organic chemistry, was revolutionized by the introduction of pressurized mains water and the fume cupboard was improved by the use of mains electricity. It seems self-evident that new technologies will change the design of the chemical laboratory, yet the evidence is sparse. Such a major technological revolution as the introduction of physical instrumentation only wrought a fairly minor change in the laboratories at Stanford. In fact the ergonomic study of space and service provision in laboratories produced a more significant change in the 1960s, resulting in the movement of the benches towards the windows and eliminating the central aisle.

Health and safety has had an even more profound effect on the laboratory – above all, in the development of the fume cupboard and laboratory ventilation. In the nineteenth century this was linked to the widespread use of hydrogen sulphide. In the twentieth century the focus was more on the dangers of organic chemistry and flammable solvents, although the changes were more subtle, involving, for example, segregated storage of solvents, automatic sprinkler systems, eyewash stations and emergency showers.

Finally, it has to be acknowledged that growing numbers, originally of students and increasingly of researchers, have also shaped the laboratory by increasing its size and changing its layout. Even today, space is still a major concern, and Oxford's chemistry research laboratory already has more researchers than its planned capacity.

Other authors dealing with chemical laboratories have stressed the need to explain what type of chemical laboratory is being discussed – that one cannot discuss chemical laboratories in the abstract.[2] I have tried to follow this advice, but am struck by the similarities between laboratories with different functions rather than by their differences. Clearly the equipment in a teaching laboratory is different from that in an organic research laboratory, and it is usually larger. Similarly, an industrial research laboratory is not the same as an academic laboratory, and it is very different from a works laboratory. However, the basic layout and furniture of any two laboratories in a given laboratory building are very similar. Furthermore, the similarities are often greater for laboratories of the same period with different functions, than for laboratories with similar functions built in different periods. A good example of this is the similarity of the Bayer pharmaceutical laboratory to other laboratories built in the same year (1896), in contrast to the differences between the Bayer research laboratory and the Bayer pharmaceutical laboratory, despite being built only five years apart.

The Laboratory Building

One important interaction in this book has been the development of the laboratory on the one hand and the rise of the laboratory building on the other. I have been careful to make it clear when I am talking about the laboratory building rather than the laboratory, but it is very easy to mentally elide the two concepts together. To unite them in this way is to ignore the richness of the laboratory building as a design for scientific work – the building is far more than just a group of laboratories.[3] In the late nineteenth-century classical laboratory building (and even in the latest buildings) there are numerous specialized rooms, including darkrooms for spectroscopy and polarimetry, balance rooms, gas-analysis rooms and above all the very hot and dangerous combustion analysis room, usually in the basement. This division by function not only made laboratory work more efficient and safer; it also made it more flexible, enabling the basic design to accommodate the rise of physical instrumentation in the 1950s. Then there were the spaces associated with teaching: lecture theatres of various sizes, demonstration rooms (recitation rooms in the USA) and libraries. Another aspect of teaching is often overlooked, namely the chemical museum. There were many chemical museums by the end of the nineteenth century, notably in the USA, but also in Britain and Ireland, Australia, Germany and Japan. That such a prominent part of the laboratory building has been ignored suggests that more detailed studies of the rooms in the laboratory building are sorely needed.

It was often said in both the nineteenth and twentieth centuries that buildings make a statement and that laboratory buildings in particular are making a statement about the status of chemistry and academic teaching and research. In some cases this is undoubtedly true. No one could deny, for example, that Hofmann's Bonn laboratory made a powerful statement about the growing status of chemistry on the one hand, and the growing power of Prussia on the other. Similarly, the Normal School in South Kensington made strong claims for science (rather than chemistry), and the Department of Science and Art. When I worked at a pharmaceutical firm in the 1970s, the modern multi-storey research block towering over the single-storey industrial buildings of the 1930s subtly hinted at a more elevated sphere of activity than mere manufacture. On the other hand, due to its cramped position Hofmann's Berlin building was mundane, more like an office block or even a warehouse – albeit a heavily ornamented one. Similarly, the Government Chemist's Laboratory, a victim of budget overruns and a restricted site, looked like a set of rather inexpensive lawyers' chambers, and appropriately so given its position near the law courts. Indeed, I suspect that most university laboratory buildings were built cheaply, which is hardly conducive to making a statement. In addition, many academic laboratories were erected outside the centre of the town (or university) in a 'science area', and they soon became a huddle of different architectural styles (Oxford and Uppsala are good examples of this), which inevitably detract from any assertion their architects (or professors) wanted to make. Since the 1960s architects accustomed to designing many different types of building have tended to make chemical laboratories look like standard office buildings, as Stanford and Oxford illustrate. While they may be impressive in their own way, if only because of their size, it is hard to see them as making any specific statement about chemistry or learning. In reality, laboratory buildings have always tried to look beautiful to attract a better class of student and/or researcher, and increasingly to impress current or potential donors, rather than to make a statement to the world at large. It is worth mentioning that the surge in the building of laboratories between 1850 and 1910, and even the increase in their size, was not unique to chemical laboratories. The same process can easily be traced in the building of museums and lecture halls (both closely related to laboratories), as well as government buildings, town halls, secondary schools, public workhouses and factories.[4]

The Future of the Chemical Laboratory

It is clear that the chemical laboratory has changed more in the last twenty years than in the previous 100 years, but to what extent is it still a *chemical* laboratory? Due to the move to more interdisciplinary research and the introduction of modularity and greater flexibility, the modern chemical laboratory can accommodate research that goes beyond what is normally considered to be chemistry. For example, the Oxford laboratory building has two cold rooms on the lower ground floor – not unlike the walk-in chill room at the back of a supermarket – which are used to carry out biological experiments. What is the future of the chemical laboratory? Will this new design result in laboratories converging towards a common model that will not be specifically chemical?

This question is closely linked to another question I have asked elsewhere – namely, does chemistry have a future? In 50 years' time will there be a discipline that is recognizable as chemistry, or will it split into the material sciences on the one hand and the biomolecular sciences on the other?[5] When I discussed this issue in a presentation back in 2007 I was pessimistic, but the Oxford laboratory building gives me fresh optimism for the future of chemistry. Its construction shows how chemistry, by renouncing its old sub-disciplinary boundaries, can gain a renewed confidence. Furthermore, it houses both the biochemical and biomolecular sciences alongside the study of the solid state, creating a marriage between the solid state and the living state rather than the divorce I feared. Above all, it shows that only chemistry is capable of bringing these disparate fields of study together harmoniously.

If we can perhaps be optimistic about the survival of chemistry, can we be equally confident about the fate of the chemical laboratory? Obviously there will always be chemical laboratories as long as there are experimental chemists, but will the chemical laboratory remain distinctive? I would argue that the chemical laboratory through history shows distinguishing features that are specific to chemistry and its neighbouring fields, such as metallurgy and biochemistry. Up to the early nineteenth century the feature was the furnace and its associated distillation apparatus. In the early nineteenth century the bench became specific to the chemical laboratory. It was soon joined by the fume cupboard, which remains the chief characteristic of the chemical laboratory, although of course not all chemical laboratories have fume cupboards – they might have glove boxes, for example. To put it another way, chemical laboratories can look like other laboratories – especially when carrying out work on the boundary with other disciplines – but other laboratories rarely look like archetypical chemical laboratories. Furthermore, the fume cupboard is now more important than ever before. Whereas it was largely a marginal feature of laboratories even 40 years ago, it is now central to chemical experiments, at least in organic and inorganic chemistry. So we can say that as long as fume cupboards are central to chemistry, the chemical laboratory will remain distinctive. As it is now inconceivable for many experiments to be done outside a fume cupboard, the future of the chemical laboratory is secure. Even if chemical experiments become completely automated on the grounds of health and safety, Robert Burns Woodward's vision of organic synthesis carried out by robots, the space in which the robots will operate will still be specifically chemical.[6]

Could chemistry become a completely theoretical activity – with reactions carried out in a virtual world? The development of molecular graphics by Richards and others has brought this possibility closer. However, theoretical chemistry is based on experimental data, and chemists cannot assume that what happens within a computer will happen in the real world. For at least a few centuries longer, we will have to confirm theoretical predictions by actually bringing chemicals together in a laboratory to see what happens. As Woodward pointed out, it is the combination of theory and practice, creativity and flair – and surprise – which ensures that chemistry remains distinct from other scientific activities and spurs chemists to make new forms of matter.[7]

What is to be Done?

Clearly this very broad and relatively brief history of the chemical laboratory cannot cover all aspects of its history, but I hope that it will encourage other historians of chemistry – both professional historians and chemists – to take up the topic. What needs to be studied in the future? Above all, we need well-researched histories of more laboratories and laboratory buildings. The scope is certainly there. To give just one example, my own enquiries about chemical museums at specific universities have spurred individuals at these universities to find out more about their museums. It is important that these histories examine the designs of the buildings, and the layouts and functions of the rooms in them, as well as the activities of the chemists who worked there. In this context the gathering of pictorial evidence and written descriptions of the laboratories is obviously crucial.

A greater number of such histories, which should now be more readily available to others thanks to the Internet, would enable historians to address the issue of priority, eschewed in this book precisely for the lack of such material. Priority is not just a matter of point scoring – satisfying as it might be to discover that your university was the first to have bottle racks or a chemical museum – but rather it would enable us to show how laboratories changed over time, and to trace the intellectual transfer from laboratory to laboratory that enabled these changes to spread. It has always been assumed that change first took place at certain path-breaking universities that are relatively well known, namely the ones covered in this book. It is now clear, however, that leading chemists and architects designing a new laboratory almost always drew on earlier examples, ranging from Heidelberg and Karlsruhe, Berlin and Greifswald, and Oxford and Glaxo Wellcome at Stevenage. This would at least suggest that many innovations arose – at least in embryonic form – at fairly obscure laboratories rather than their well-known successors. Only further research can show if this was the case or not. In time it might even be possible to construct a network showing how many laboratories were linked to others, either as exemplars or emulators – a genealogy of laboratories which would complement the popular intellectual genealogies of chemists.

At a more general level of analysis it would be useful to have detailed studies of different types of laboratory – teaching, analytical, research and industrial – to see if they have their own distinct lines of development and style, or whether they are just modified versions of the same basic design. In this context it would be valuable to have biographical studies of architects noted for their laboratory designs, especially Heinrich Lang, Alfred Waterhouse and Edward Cookworthy Robins. The issue of similar designs clearly needs to be explored. I have been struck by the similarity of the designs for six laboratories built in around 1896, namely the Government Chemist's Laboratory, Burroughs Wellcome's laboratory, Bayer's pharmaceutical laboratory, the Davy-Faraday Laboratory at the RI, the laboratory of the Royal College of Physicians of Edinburgh and the New Sorbonne. For this reason, *The Laboratories of 1896*, including laboratories in the USA, would be an interesting book! Conversely, studies of chemical laboratories in different countries would reveal the existence (or absence) of national styles that cross different periods. Some work on this topic has already been done for Portugal, but the USA would provide a very good case study, as North American laboratories do clearly differ in some respects from their European counterparts. To bring out these differences, however, international comparisons would have to be made; the study of a single country in isolation would not reveal much. Furthermore, it would be valuable to have a study of the various people working in the laboratory building, showing how the social hierarchy works and how each person contributes to the operation of the building. It would be good to see how the structure of this community changes over time, perhaps comparing a laboratory building in the 1880s with a laboratory building of the 2000s. I would draw attention in particular to that often overlooked specialist, the glassblower.

Finally, I would be very happy if this book could be used as a starting point for a comparative study of the chemical laboratory with physical and biological laboratories. Clearly, the chemical laboratory was the forerunner of all other laboratories, although physical cabinets existed by the eighteenth century, but it is not clear – at least to me – how much of a debt these other laboratories owed to the chemical laboratory. For a variety of reasons, including space constraints, I have restricted myself to the chemical laboratory. Others might wish to argue for a more holistic history of the laboratory that treats each discipline's laboratory as a variant of the same basic design (much as I have treated different types of chemical laboratory). In this case which level is the correct one to analyse in relation to the development of the laboratory: the pan-discipline level, the discipline level or even lower? I would argue that the discipline level is the most informative – other scholars may wish to prove me wrong.

APPENDIX

Table of Known Chemical Museums and Chemical Collections
(in chronological order)

UNIVERSITY	COUNTRY	EARLIEST KNOWN DATE	REFERENCE
Uppsala	Sweden	1769	See main text.
Trinity College Dublin (TCD)	Ireland	1803	*Dublin University Calendar* Part 1 (1917), p. 340 GB; Private communication from David Grayson, dated 23 October 2013.
Freiberg, Saxony	Germany	1815	C. Meinel, 'Chemical Collections', in *Spaces and Collections in the History of Science*, ed. M. C. Lourenço and A. Carneiro (Lisbon, 2009), pp. 137–47, on p. 141.
Prague	Bohemia	1820	Meinel, 'Chemical Collections'.
Edinburgh	Britain	1823	Private communication from Andrew Alexander dated 22 March 2013; W. Reid, *Memoirs and Correspondence of Lyon Playfair, First Lord Playfair of St. Andrews* (London, 1899), p. 180. IA
Munich	Germany	1839	George Kemp, 'On Berberine and some of its Compounds...', *Chemical Gazette: or, Journal of Practical Chemistry*, v (1 June 1847), pp. 209–11, on p. 209. GB
Glasgow	Britain	1843	R. D. Thomson, 'Examination of the Cowdie Pine Resin', *London Edinburgh and Dublin Philosophical Magazine and Journal of Science*, 3rd series, XXIII (August 1843), pp. 81–9, on p. 81. GB

Appendix

UNIVERSITY	COUNTRY	EARLIEST KNOWN DATE	REFERENCE
King's College, London (KCL)	Britain	1850	*Calendar of King's College London for 1850–51* (London, 1850), p. 116. GB
Oslo	Norway	1851	A. Strecker, *Das chemische Laboratorium der Universität Christiania* (Christiania [Oslo], 1854), p. iii. GB
Heidelberg	Germany	1855	H. Kaemmerer, 'Ueber einige Jodverbindungen', *Journal für Praktische Chemie*, LXXXIII/1 (1861), pp. 65–85, on p. 83.
Columbia	USA	1864	L. Fine, 'The Chandler Chemical Museum', *Bulletin of the History of Chemistry*, II (1988), pp. 19–21; R. H. McKee, C. E. Scott and C. B. F. Young, 'The Chandler Chemical Museum at Columbia University', *Journal of Chemical Education*, XI (1934), pp. 275–8. Also the Yale *College Courant*, XIV (1874), p. 151. GB
Bonn	Germany	1866	A. W. Hofmann, *The Chemical Laboratories in Course of Erection in the Universities of Bonn and Berlin* (London, 1866), pp. 22–3. GB
Berlin	Germany	1868	Hofmann, *Chemical Laboratories in Course of Erection*, p. 58.
Virginia	USA	1869	F. P. Dunnington, *The Chemical Museum of the University of Virginia* (Charlottesville, VA, 1917), p. 1.
Munich Polytechnic Institute	Germany	1870	*The Architect* (19 November 1870), p. 293. GB
Royal College of Science (RCS)	Britain	1874	H. Gay, *The History of Imperial College*, p. 18 (Royal College of Chemistry) and p. 100 (1911). Also see Devonshire Report for 1874. It is uncertain if the RCC Museum was actually set up and the chemistry department at RCS had a collection (partly stored in the South Kensington Museum) rather than a museum, which was, however, much desired.

UNIVERSITY	COUNTRY	EARLIEST KNOWN DATE	REFERENCE
Greifswald	Germany	1874	*The Medical Times and Gazette*, I (1874), p. 245. GB
Yale	USA	1874	*College Courant*, XIV (1874), p. 151.
Leeds	Britain	1874	U. Klein and W. Lefèvre, *Materials in Eighteenth-century Science: A Historical Ontology* (Cambridge, MA, 2007), p. 8; private communication from Claire Jones dated 18 June 2012. E. C. Robins, *Technical School and College Building* (London, 1887), p. 77. IA
Illinois	USA	1878	Private communication from Vera Mainz dated 12 August 2013; Tilden, *Chemical Discovery and Invention*, between pp. 40 and 41.
Liverpool	Britain	1878	Robins, *Technical School and College Building*, p. 103.
Michigan	USA	1882	Private communication from Brian Williams, dated 30 July 2013; *University of Michigan General Register* for 1890–1, p. 24. Also see *Proceedings of the Board of Regents of the University of Michigan, 1881–1886* (Ann Arbor, MI, 1886), pp. 483 and 605. IA
Cornell	USA	1883	*The Register, Cornell University* for 1879–80, p. 57 (museum planned for 1883); *Annual Report of the President of Cornell University* for 1887–8 (Ithaca, NY, 1888), p. 81. GB

Appendix

UNIVERSITY	COUNTRY	EARLIEST KNOWN DATE	REFERENCE
Lehigh	USA	1885	Chandler, *Construction of Chemical Laboratories*, pp. 688 (measurements) and 694. Also see ACS National Historic Chemical Landmarks, 'The Chandler Chemistry Laboratory', 1994, available at http://portal.acs.org.
ETH, Zurich	Switzerland	1886	Chandler, *Construction of Chemical Laboratories*, p. 711.
Dundee	Britain	1886	Robins, *Technical School and College Building*, p. 142 and pl. 41.
Sydney	Australia	1889	Tilden, *Chemical Discovery and Invention*, p. 44 (described as chemical collections); private communication from Roy Macleod dated 29 July 2013.
Cornell (Kidder)	USA	1890	Chandler, *Construction of Chemical Laboratories*, pp. 703–4; *The Register, Cornell University* for 1903–4, p. 167. IA
Imperial	Japan	1896	Y. Kikuchi, *Anglo-American Connections in Japanese Chemistry: The Lab as Contact Zone* (New York and Basingstoke, 2013), pp. 162–3.
Catholic University of America	USA	1896	*Sacred Heart Review*, XVIII/24 (11 December 1897), p. 475, and *Announcements of the School of Sciences 1909–10, Catholic University of America*, pp. 8 and 10. IA
Chicago (Kent)	USA	1897	*Popular Science* (October 1897), p. 785; and *Annual Register of the University of Chicago* for 1905–6, p. 118. IA
Wisconsin	USA	1901	*The Wisconsin Alumni Magazine*, II/8 (May 1901), p. 334. IA

UNIVERSITY	COUNTRY	EARLIEST KNOWN DATE	REFERENCE
Toronto	Canada	1903	'The Chemical Museum', *The Varsity* (Toronto University), XXII/13 (21 January 1903), p. 195.
Worcester Polytechnic Institute	USA	1903	*Journal of the Worcester Polytechnic Institute*, VI (1903), pp. 326 and 329.
Technische Hochschule Charlottenburg	Germany	1905	*Archiv für buchgewerbe und gebrauchsgraphik*, XLII (1905), p. 298; *Museums Journal*, XIII (1914), p. 219. GB
Boston University	USA	1906	*Bostonia*, VIII/1 (1907), p. 32.
Harvard	USA	1909	*Official Register of Harvard University*, VI/2 (1909), p. 779. GB
Institute of Technology, Trondheim	Norway	1910	Private communication from Annette Lykknes dated 7 December 2012.
Bristol	Britain	1910	A. E. Munby, *Laboratories: Their Planning and Fittings* (London, 1931), p. 172.
Bucknell	USA	1926	*Chemical and Metallurgical Engineering*, XXXIII (1926), p. 709. GB
Salem College, Winston-Salem, NC	USA	1928	*Theta News*, IV/4 (Spring 1928), p. 386. GB
Brooklyn Polytechnic Institute	USA	1932	Obituary of Irving Fay, *Chemical Week*, XXXVIII (1936), p. 298. GB
North Carolina State College	USA	1932	*North Carolina State College of Agriculture and Engineering, 1931–1932* (State College Station, 1932), p. 37. IA

REFERENCES

Introduction

1 P. T. Carroll, 'Academic Chemistry in America, 1876–1976: Diversification, Growth, and Change', PhD thesis, University of Pennsylvania, 1982, chap. 4.
2 P.J.T. Morris and W. A. Campbell, 'Analisi chimica', in *Storia della scienza*, vol. VII, *L'Ottocento – Chimica*, ed. S. Petruccioli (Rome, 2004), pp. 554–79, which is available online (but without the illustrations) at treccani.it.
3 In the introduction to P.J.T. Morris, ed, *From Classical to Modern Chemistry: The Instrumental Revolution* (Cambridge, 2002).
4 M. Griep and M. Mikasen, *ReAction!: Chemistry in the Movies* (Oxford, 2009).
5 R. E. Kohler, 'Lab History: Reflections', *Isis*, IC (2008), pp. 761–8.
6 There is, however, the short account by O. Krätz, 'Zur Geschichte des chemischen Laboratoriums', in *Historia scientiae naturalis: Beiträge zur Geschichte der Laboratoriumstechnik und deren Randgebiete*, ed. E.H.W. Giebeler and K. A. Rosenbauer (Darmstadt, 1982). Robert Anderson has described how the history of laboratories could be studied; see R.G.W. Anderson, 'Chemical Laboratories, and How They Might Be Studied', *Studies in History and Philosophy of Science Part A*, XLIV (2013), pp. 669–75.
7 See, for example, F.A.J.L. James, ed., *The Development of the Laboratory. Essays on the Place of Experiment in Industrial Civilization* (Basingstoke, 1989); and M. C. Lourenço and A. Carneiro, eds, *Spaces and Collections in the History of Science* (Lisbon, 2009).
8 Ernest Homburg, 'The Rise of Analytical Chemistry and Its Consequences for the Development of the German Chemical Profession (1780–1860)', *Ambix*, XLVI (1999), pp. 1–32.
9 For a discussion about this question for the eighteenth century, see U. Klein, 'The Laboratory Challenge: Some Revisions of the Standard View of Early Modern Experimentation', *Isis*, IC (2008), pp. 769–82, on pp. 769–71.
10 I wish to thank Alan Rocke (Case Western Reserve University) for this suggestion.
11 Whiggism is the pejorative term used by professional historians to describe the history of science which is only concerned with scientific

developments that are relevant to the present state of the science and thus assumes, at least implicitly, that the modern is the measure of the old. The term comes from the nineteenth-century 'Whig' school of political history, which similarly assumed that British constitutional history was an inevitable progression to parliamentary democracy and universal franchise.

12 For this approach, see B. Latour and S. Woolgar, *Laboratory Life: The Construction of Scientific Facts* (Princeton, NJ, 1986).

13 For the architecture of laboratories, see S. Forgan, 'The Architecture of Science and the Idea of a University', *Studies In History and Philosophy of Science, Part A*, 20 (1989), pp. 405–34; P. Galison and E. Thompson, eds, *The Architecture of Science* (Cambridge, MA, 1999); G. Gooday, 'The Premises of Premises: Spatial Issues in the Historical Construction of Laboratory Credibility', in *Making Space for Science: Territorial Themes in the Shaping of Knowledge*, ed. C. Smith and J. Agar (Basingstoke, 1998), pp. 216–45.

14 R.G.W. Anderson, 'The Creation of the Chemistry Teaching Laboratory', in *Spaces and Collections*, ed. M. C. Lourenço and A. Carneiro, pp. 13–23, on p. 14.

15 Other volumes that contain large numbers of illustrations of laboratories include O. Krätz, 'Zur Geschichte des chemischen Laboratoriums', in *Historia scientiae naturalis: Beiträge zur Geschichte der Laboratoriumstechnik und deren Randgebiete*, ed. E.H.W. Giebeler and K. A. Rosenbauer (Darmstadt, 1982); O. Krätz, *Faszination Chemie: 7000 Jahre Lehre von Stoffen und Prozessen* (Munich, 1990); and F. Ferchl and A. Süssenguth, *A Pictorial History of Chemistry* (London, 1939). There is a number of interesting pictures in E. Homburg, *Van beroep 'Chemiker': De opkomst van de industriële chemicus en het polytechnische onderwijs in Duitsland (1790–1850)* (Delft, 1993). There is a considerable number of photographs (and ground plans) of the interiors of laboratories in W. A. Tilden, *Chemical Discovery and Invention in the Twentieth Century*, 4th edn (London and New York, 1922). It is striking that the otherwise well-illustrated A. Greenberg, *Chemical History Tour: Picturing Chemistry from Alchemy to Modern Molecular Science* (New York, 2000) contains hardly any pictures of laboratories later than the sixteenth century.

16 For the relationship between visual evidence and history, see J. Tagg, *The Burden of Representation: Essays on Photographies and Histories* (Basingstoke, 1988); and L. Jordanova, *The Look of the Past: Visual and Material Evidence in Historical Practice* (Cambridge, 2012).

17 J. B. Morrell, 'The Chemist Breeders: The Research Schools of Liebig and Thomas Thomson', *Ambix*, XIX (1972), pp. 1–46.

18 A. W. Hofmann, *The Chemical Laboratories in Course of Erection in the Universities of Bonn and Berlin* (London, 1866). GB

19 E. R. Festing, *Report of Visits to Chemical Laboratories at Bonn, Berlin, Leipzig, etc.* (London, 1871).

20 E. C. Robins, *Technical School and College Building* (London, 1887). IA

21 W. H. Chandler, *The Construction of Chemical Laboratories* (Washington, DC, 1893). As it is not available in Google Books, I used the good-quality reprint on demand by Kessinger Publishing.

22 J. Mills, 'The Government Laboratory', *Strand Magazine*, XXI (1902), pp. 561–71.
23 Alan J. Rocke, *The Quiet Revolution: Hermann Kolbe and the Science of Organic Chemistry* (Berkeley, CA, 1993), which is available online at cdlib.org.; Alan J. Rocke, *Nationalizing Science: Adolphe Wurtz and the Battle for French Chemistry* (Cambridge, MA, 2001).
24 Catherine M. Jackson, 'Analysis and Synthesis in Nineteenth-century Organic Chemistry', PhD thesis, University College London, 2009.
25 Yoshiyuki Kikuchi, 'The English Model of Chemical Education in Meiji Japan: Transfer and Acculturation', PhD thesis, Open University, Milton Keynes, 2006. A revised and enlarged version has been published as Y. Kikuchi, *Anglo-American Connections in Japanese Chemistry: The Lab as Contact Zone* (New York and Basingstoke, 2013).
26 Jon Eklund, *The Incompleat Chymist: Being an Essay on the Eighteenth-century Chemist in his Laboratory, with a Dictionary of Obsolete Chemical Terms of the Period* (Washington, DC, 1975), available at http://si-pddr.si.edu.
27 P. W. Hammond and H. Egan, *Weighed in the Balance: A History of the Laboratory of the Government Chemist* (London, 1992).
28 For more details see 'Sites of Chemistry, 1600–2000: The Project', at sitesofchemistry.org, accessed on 23 November 2013.

1 Birth of the Laboratory: Wolfgang von Hohenlohe and Weikersheim, 1590s

1 For the dangers of poor ventilation given the heavy use of toxic metalloids, see W. R. Newman and L. M. Principe, *Alchemy Tried in the Fire: Starkey, Boyle, and the Fate of Helmontian Chymistry* (Chicago, IL, and London, 2002), p. 97 for the ill-health suffered by George Starkey.
2 J. Mathesius, *Sarepta Oder Bergpostill Sampt der Jochimßthalischen kurtzen Chroniken* (Nürnberg, 1562); Paracelsus, *Libri Duo Aureoli Theophrasti Paracelsi, Argentorati in Foro Frumentario excudebat*, BBB *Christianus Mylius* (Strasbourg, 1566), *In De Tartaro*, p. 255, and Pietro Andrea Mattioli, *Epistolarum medicinalium libri quinque* (Prague, 1561), p. 311. I wish to thank Peter Forshaw (University of Amsterdam) for his help with this section and the citations in this and the following references.
3 Pseudo-Paracelsus, *De Natura Rerum* (Basle, 1584), p. 73 and J. J. Wecker, *Antidotarium generale* (Basle, 1580), in chap. xi, 'De Domo, seu aedificio et officina pharmacopoei', p. 191.
4 Ben Jonson, *Mercury Vindicated from the Alchymists*, http://hollowaypages.com, accessed 15 February 2013; the stage directions open the play.
5 C. R. Hill, 'The Iconography of the Laboratory', *Ambix*, XXII (1975), pp. 102–10.
6 See T. Nummedal, *Alchemy and Authority in the Holy Roman Empire* (Chicago, IL, 2007), chap. 6.

7 M. Martinón-Torres, T. Rehren and S. von Osten, 'A 16th Century Lab in a 21st Century Lab: Archaeometric Study of the Laboratory Equipment from Oberstockstall (Kirchberg am Wagram, Austria)', *Antiquity*, LXXVII/298 (December 2003), available at http://antiquity.ac.uk.; J. A. Bennett, S. A. Johnston and A. V. Simcock, *Solomon's House in Oxford: New Finds from the First Museum* (Oxford, 2000).
8 For good modern translations, see Georgius Agricola, *De re metallica*, trans. from the first Latin edition of 1556 by H. C. Hoover and L. H. Hoover (New York, 1950), and *Lazarus Ercker's Treatise on Ores and Assaying*, trans. from the German edition of 1580 by A. G. Sisco and C. S. Smith (Chicago, IL, 1951).
9 However, it is also reproduced in *Ercker's Treatise on Ores*, fig. 1 on p. 9, with the caption 'View inside the assay laboratory'. I wish to thank William Jensen for his assistance with this description of the picture.
10 As described ibid., fig. 27 on p. 185.
11 As described ibid., fig. 25 on p. 179.
12 As described ibid., fig. 30 on p. 219.
13 As described ibid., fig. 24 on p. 166.
14 As described ibid., fig. 19 on p. 138.
15 J. Weyer, *Graf Wolfgang II. von Hohenlohe und die Alchemie: Alchemistische Studien in Schloss Weikersheim, 1587–1610* (Sigmaringen, 1992), pp. 96–103; also see Nummedal, *Alchemy and Authority*, chap. 5; P. H. Smith, 'Laboratories', in *The Cambridge History of Science*, vol. III: *Early Modern Science*, ed. K. Park and L. Daston (Cambridge, 2006), pp. 290–305.
16 Reproduced in C. Meinel, 'Vom Handwerk des Chemiehistorikers', *Chemie in unserer Zeit*, XVIII/2 (1984), pp. 62–7.
17 Martinón-Torres, Rehren and von Osten, 'A 16th Century Lab in a 21st Century Lab'.
18 Bennett, Johnston and Simcock, *Solomon's House in Oxford*.
19 The use of 'chymistry' as the best term for chemistry in the early-modern period was revived by Bill Newman and Larry Principe in W. R. Newman and L. M. Principe, 'Alchemy vs. Chemistry: The Etymological Origins of a Historiographic Mistake', *Early Science and Medicine*, III (1998), pp. 32–65.
20 O. Hannaway, 'Laboratory Design and the Aim of Science: Andreas Libavius versus Tycho Brahe', *Isis*, LXXVII (1986), pp. 584–610; J. Shackelford, 'Tycho Brahe, Laboratory Design, and the Aim of Science: Reading Plans in Context', *Isis*, LXXXIV (1993), pp. 211–30. For a critique of Hannaway's presentation of Libavius's design as open chemistry opposed to secret alchemy, see W. R. Newman, 'Alchemical Symbolism and Concealment: The Chemical House of Libavius', in *The Architecture of Science*, ed. P. Galison and E. Thompson (Cambridge, MA, 1999), pp. 59–77.
21 Newman, 'Alchemical Symbolism and Concealment', pp. 70–72.
22 R. J. Forbes, *A Short History of the Art of Distillation: From the Beginnings up to the Death of Cellier Blumenthal* (Leiden, 1970), pp. 24–5.
23 This point is made for the eighteenth-century laboratory by Jon Eklund in *Incompleat Chymist: Being an Essay on the Eighteenth-century Chemist in*

his Laboratory (Washington, DC, 1975), pp. 4–7. This book can be downloaded from http://si-pddr.si.edu.

24 Piger Henricus and faule Heinz both mean 'lazy Henry'. Ben Jonson mentions the Piger Henricus in *The Alchemist*, Act ii, scene v, line 114.

25 *Ercker's Treatise on Ores and Assaying*, pp. 143–5. Also see Forbes, *Short History of the Art of Distillation*, p. 174, and B.J.T. Dobbs, *The Foundations of Newton's Alchemy, or, 'The Hunting of the Greene Lyon'* (Cambridge, 1983), p. 122.

26 I am very grateful to Larry Principe for his explanation of the burning process in the Piger Henricus.

27 Forbes, *Short History of the Art of Distillation*, chap. 4; R.G.W. Anderson, 'The Archaeology of Chemistry', in *Instruments and Experimentation in the History of Chemistry*, ed. F. L. Holmes and T. H. Levere (Cambridge, MA, and London, 2000), pp. 5–34. Also see H. Schelenz, *Zur Geschichte der pharmazeutisch-chemischen Destilliergeräte* (Hildesheim, 1964).

28 S. Moorhouse, 'Medieval Distilling-apparatus of Glass and Pottery', *Medieval Archaeology*, XVI (1972), pp. 79–121.

29 Forbes, *A Short History of the Art of Distillation*, pp. 21–3.

30 See W. B. Jensen, *Philosophers of Fire* (Cincinnati, OH, 2003), p. 22, for the interesting claim that there was a 'distillation craze' in the fifteenth and sixteenth centuries. This e-book can be downloaded at che.uc.edu.

31 M. Eliade, *The Forge and the Crucible*, trans. S. Corrin, 2nd edn (Chicago, IL, 1978), pp. 191–2.

32 For the strong links between the chemical laboratory and the pharmaceutical laboratory, see Ernst Homburg, 'The Rise of Analytical Chemistry and Its Consequences for the Development of the German Chemical Profession (1780–1860)', *Ambix*, XLVI (1999), pp. 1–32, on p. 5.

2 Form and Function: Antoine Lavoisier and Paris, 1780s

1 For a discussion about the stability of the chemical laboratory between 1600 and 1800, see F. L. Holmes, *Eighteenth-century Chemistry as an Investigative Enterprise* (Berkeley, CA, 1989), pp. 17–20.

2 For the wider intellectual context, see J. Golinski, 'Chemistry', in *The Cambridge History of Science', vol IV: Eighteenth-century Science*, ed. Roy Porter (Cambridge, 2003), pp. 377–96.

3 W. H. and W.J.C. Quarrell, eds, *Oxford in 1710, from the Travels of Zacharias Conrad von Uffenbach* (Oxford, 1928), p. 37.

4 I. S. Glass, *Nicolas-Louis De La Caille, Astronomer and Geodesist* (Oxford, 2012).

5 Ibid., pp. 56–7.

6 Ibid., p. 121.

7 R. H. Allen, *Star Names: Their Lore and Meaning* (New York, 1963), reprint of the edition of 1899, p. 221.

8 F. W. Gibbs, 'William Lewis, M.B., F.R.S. (1708–1781)', *Annals of Science*, VIII (1952), pp. 122–51, on p. 134. Also see F. W. Gibbs, 'William Lewis and Platina: Bicentenary of the "Commercium

Philosophicotechnicum"', *Platinum Metals Review*, VII/2 (1963), pp. 66–9.

9   A similar point about clutter is made by Jon Eklund in *Incompleat Chymist: Being an Essay on the Eighteenth-century Chemist in his Laboratory* (Washington, DC, 1975), p. 4.

10 For the early history of gas manipulation, see J. Parascandola and A. J. Ihde, 'History of the Pneumatic Trough', *Isis*, LX (1969), pp. 351–61; M. Crosland, '"Slippery Substances": Some Practical and Conceptual Problems in the Understanding of Gases in the Pre-Lavoisier Era', in *Instruments and Experimentation in the History of Chemistry*, ed. F. L. Holmes and T. H. Levere (Cambridge, MA, 2000), pp. 79–104.

11 F. W. Gibbs, *Joseph Priestley: Adventurer in Science and Champion of Truth* (London, 1965).

12 Peter Woulfe, 'Experiments on the Distillation of Acids, Volatile Alkalies, &c.', *Philosophical Transactions*, LVII (1767), pp. 517–36, on p. 517; E. E. Aynsley and W. A. Campbell, 'The Laboratory Preparation of Hydrogen Sulfide: A Historical Survey', *Journal of Chemical Education*, XXXV (1958), pp. 347–9.

13 P.J.T. Morris and W. A. Campbell, 'Analisi chimica', in *Storia della scienza*, vol. VII: *L'Ottocento – Chimica*, ed. S. Petruccioli (Rome, 2004), pp. 554–79, on p. 555.

14 A. Donovan, *Antoine Lavoisier: Science, Administration, and Revolution* (Cambridge, 1996); J-P. Poirier, *Lavoisier: Chemist, Biologist, Economist*, trans. Gillian C. Gill (Philadelphia, PA, 1996); *Lavoisier in Perspective*, ed. M. Beretta (Munich, 2005); F. L. Holmes, *Lavoisier and the Chemistry of Life: An Exploration of Scientific Creativity* (Madison, WI, 1985); Marco Beretta, *Imaging a Career in Science: The Iconography of Antoine Laurent Lavoisier* (Canton, MA, 2001).

15 W. S. Dutton, *Du Pont: One Hundred and Fifty Years* (New York, 1942).

16 T. S. Kuhn, *The Structure of Scientific Revolutions*, 2nd edn (Chicago, 1970), see p. 118 for his treatment of the chemical revolution. For a discussion of Kuhn's view of Lavoisier, see Holmes, *Lavoisier and the Chemistry of Life*, pp. 119–20.

17 J. R. Partington and D. McKie, 'Historical Studies on the Phlogiston Theory.—II. The Negative Weight of Phlogiston', *Annals of Science*, III (1938), pp. 1–58; H. Chang, 'The Hidden History of Phlogiston', *HYLE– International Journal for Philosophy of Chemistry*, XVI (2010), pp. 47–79.

18 My interest in this idea was stimulated by a talk given by Lord Dainton on 'The Chemical Electron' to the Oxford University Alembic Club in 1977, in which he suggested this co-identity. See H. Chang, 'We Have Never Been Whiggish (About Phlogiston) 1', *Centaurus*, LI (2009), pp. 239–64. Chang points out that remarkably it was suggested in 1780 that phlogiston should be renamed 'electron'.

19 Beretta, *Imaging a Career*, pp. 47–55 (includes reproductions of the drawings); Marco Beretta, 'Big Chemistry: Lavoisier's Design and Organisation of his Laboratories', in *Spaces and Collections in the History of Science*, ed. M. C. Lourenço and A. Carneiro (Lisbon, 2009), pp. 65–80; J. P. Prinz, 'Lavoisier's Experimental Method of his Research on Human

Respiration', in *Lavoisier in Perspective*, ed. Marco Beretta, pp. 43–52. For intellectual background to this research, see Holmes, *Lavoisier and the Chemistry of Life*.

20 Marco Beretta, 'Imaging the Experiments on Respiration and Transpiration of Lavoisier and Séguin: Two Unknown Drawings by Madame Lavoisier', *Nuncius*, XXVII (2012), pp. 163–91, on p. 190.

21 Beretta, *Imaging a Career*, p. 50.

22 A plate entitled 'A Second View of Practical Chymistry', in 'Chemicus', untitled article on chemistry, *Universal Magazine of Knowledge and Pleasure* (March, 1748), pp. 135–8. This is illus. 15 in this volume, which is available online at http://babel.hathitrust.org.

23 L. Pyenson and J.-F. Gauvin, eds, *The Art of Teaching Physics: The Eighteenth-century Demonstration Apparatus of Jean Antoine Nollet* (Quebec, 2002).

24 Private communication from Mary Ellen Bowden, 18 May 2012.

25 R. Knoeff, *Herman Boerhaave (1668–1738): Calvinist Chemist and Physician* (Amsterdam, 2002); J. C. Powers, *Inventing Chemistry: Herman Boerhaave and the Reform of the Chemical Arts* (Chicago, IL, 2012).

26 A. L. Donovan, *Philosophical Chemistry in the Scottish Enlightenment: The Doctrines and Discoveries of William Cullen and Joseph Black* (Edinburgh, 1975).

27 P.J.T. Morris, 'The Eighteenth Century: Chemistry Allied to Anatomy', in *Chemistry at Oxford: A History from 1600 to 2005*, ed. R.J.P. Williams, J. S. Rowlinson and A. Chapman (Cambridge, 2009), pp. 52–78.

28 U. Klein, 'Apothecary's Shops, Laboratories and Chemical Manufacture in Eighteenth-century Germany', in *The Mindful Hand: Inquiry and Invention from the Late Renaissance to Early Industrialization*, ed. L. Roberts, S. Schaffer and P. Dear (Amsterdam, 2007), pp. 247–76.

29 M. D. Eddy, *The Language of Mineralogy: John Walker, Chemistry and the Edinburgh Medical School, 1750–1800* (Farnham, 2008).

30 I wish to thank John Perkins for his assistance with this section.

31 A. Thackray, *Atoms and Powers: An Essay on Newtonian Matter-theory and the Development of Chemistry* (Cambridge, MA, 1970); for its decline see P.J.T. Morris, 'Education of British Chemists in the Eighteenth Century', part II thesis, Oxford, 1978, and Morris, 'The Eighteenth Century: Chemistry Allied to Anatomy'.

32 J. Perkins, 'Creating Chemistry in Provincial France before the Revolution: The Examples of Nancy and Metz. Part 1. Nancy', *Ambix*, L (2003), pp. 145–81 and 'Part 2. Metz', *Ambix*, LI (2004), pp. 43–75.

33 J. Perkins, 'Chemistry Courses, the Parisian Chemical World and the Chemical Revolution, 1770–1790', *Ambix*, LVII (2010), pp. 27–47.

34 Morris, 'Education of British Chemists in the Eighteenth Century'.

35 For the relationship between pharmacy and the chemical laboratory in the eighteenth century, see U. Klein, 'Apothecary's Shops, Laboratories and Chemical Manufacture in Eighteenth-century Germany', pp. 247–76; U. Klein, 'Chemical and Pharmaceutical Laboratories before the Professionalization of Chemistry', in *Spaces and Collections*, ed. Lourenço and Carneiro, pp. 3–12.

36 F. W. [F. Wöhler], 'Jugend-Erinnerungen eines Chemikers', *Berichte der deutschen chemischen Gesellschaft*, VIII/1 (1875), pp. 838–52, on p. 840; translation in J. Read, *Humour and Humanism in Chemistry* (London, 1947), p. 236. The table was on display at the Observatory Museum in Stockholm in August 2013 and its authenticity was confirmed by the Berzelius biographer Jan Trofast, verbal communication 24 August 2013.

37 The splendid late eighteenth-century chemical desk (or cabinet) of Peter Leopold, Duke of Tuscany (who briefly reigned as the Holy Roman Emperor Leopold II), can be seen in the Museo di Storia della Scienza in Florence, see 'Chemistry Cabinet' in the Virtual Museum at http://catalogue.museogalileo.it, 2010.

38 Fernando Bragança Gil and Graça Santa-Bárbara, 'The Nineteenth-century *Laboratorio Chimico* of the Lisbon Polytechnic School in the Context of the Museum of Science of the University of Lisbon', in *Spaces and Collections*, ed. Lourenço and Carneiro, pp. 215–26, on pp. 220–22, especially figs 3 and 4.

39 G. Gooday, 'Precision Measurement and the Genesis of Physics Teaching Laboratories in Victorian Britain', *The British Journal for the History of Science*, XXIII (1990), pp. 25–51; R. Sviedrys, 'The Rise of Physics Laboratories in Britain', *Historical Studies in the Physical Sciences*, VII (1976), pp. 405–36.

40 For the relationship between chemistry and practical physics in the mid-nineteenth century see W. H. Brock, 'The Chemical Origins of Practical Physics', *Bulletin for the History of Chemistry*, XXI (1998), pp. 1–11.

41 D. H. Saxon, 'Tanakadate, Lord Kelvin and Glasgow', December 2002, at physics.gla.ac.uk.

42 M. Beretta, 'Pneumatics vs. "Aerial Medicine": Salubrity and Respirability of Air at the End of the Eighteenth Century', in *Nuova Voltiana. Studies on Volta and His Time*, ed. F. Bevilacqua and L. Fregonese, II (Pavia and Milan, 2000), pp. 49–72; T. H. Levere, 'Measuring Gases and Measuring Goodness', in *Instruments and Experimentation*, pp. 105–35; W. A. Osman, 'Alessandro Volta and the Inflammable Air Eudiometer', *Annals of Science*, XIV (1958), pp. 215–42; K. R. Farrar, 'A Note on a Eudiometer Supposed to Have Belonged to Henry Cavendish', *British Journal for the History of Science*, I (1963), pp. 375–80.

43 Its healthy properties were attributed, equally mistakenly, in the nineteenth century to the presence of ozone: S. Wilmot, 'Ozone and the Environment: Victorian Perspectives', in *The Chemistry of the Atmosphere*, ed. A. R. Bandy (Cambridge, 1995), pp. 204–17.

44 Private communication from Mel Usselman, dated 24 February 2011.

3 Laboratory versus Lecture Hall: Michael Faraday and London, 1820s

1 For the history of lecture halls (including the main auditorium of the Royal Institution), see T. A. Markus, *Buildings and Power: Freedom and Control in*

the Origin of Modern Building Types (London and New York, 1993), chap. 9.
2. For Moore's paintings of Faraday's laboratory and their context, see F.A.J.L. James, 'Harriet Jane Moore, Michael Faraday, and Moore's Mid-nineteenth Century Watercolours of the Interior of the Royal Institution', in *Fields of Influence: Conjunctions of Artists and Scientists, 1815–1860*, ed. James Hamilton (Birmingham, 2001), pp. 111–28.
3. H. Bence Jones, *The Life and Letters of Faraday*, 2 vols, 2nd edn (London, 1870). The etching is the frontispiece of the second volume.
4. For a different approach to the same issues, see A. García-Belmar and F. Ramón Berthomeu-Sánchez, 'Teaching and Research Spaces: The Chemistry Chair of the Collège de France, 1770–1840', in *Spaces and Collections in the History of Science*, ed. M. C. Lourenço and A. Carneiro (Lisbon, 2009), pp. 33–54.
5. A. G. Debus, *The French Paracelsians: The Chemical Challenge to Medical and Scientific Tradition in Early Modern France* (Cambridge, 1991), pp. 131–3.
6. A. Chapman, 'From Alchemy to Airpumps: The Foundations of Oxford Chemistry', in *Chemistry at Oxford: A History from 1600 to 2005*, ed. R.J.P. Williams, J. S. Rowlinson and A. Chapman (Cambridge, 2009), pp. 17–51, on p. 41.
7. A. V. Simcock, *The Ashmolean Museum and Oxford Science, 1883–1983* (Oxford, 1984), pp. 7–10.
8. Ibid., lower picture facing p. 8.
9. For the history of the chemical lecture-demonstration, see W. B. Jensen, 'To Demonstrate the Truths of "Chymistry"', *Bulletin of the History of Chemistry*, x (1991), pp. 3–15; G. B. Kauffman, 'Lecture Demonstrations, Past and Present', *The Chemical Educator*, 1/5 (1996), pp. 1–33. For the teaching of chemistry in France in this period, see A. Clericuzio, 'Teaching Chemistry and Chemical Textbooks in France: From Beguin to Lemery', *Science & Education*, xv/2 (2006), pp. 335–55.
10. H. Fors, 'J. G. Wallerius and the Laboratory of the Enlightenment', in *Taking Place: The Spatial Contexts of Science, Technology, and Business*, ed. E. Baraldi, H. Fors and A. Houltz (Sagamore Beach, MA, 2006), pp. 3–33.
11. R.G.W. Anderson, *The Playfair Collection and the Teaching of Chemistry at the University of Edinburgh, 1712–1858* (Edinburgh, 1978), chap. 2; R.G.W. Anderson, 'Joseph Black: An Outline Biography', in *Joseph Black, 1728–1799: A Commemorative Symposium*, ed. A.D.C. Simpson (Edinburgh, 1982), pp. 6–11; also see *The Correspondence of Joseph Black*, ed. R.G.W. Anderson and J. Jones (Farnham, 2012), vol. I, chap. 2.
12. R.G.W. Anderson, 'Joseph Black: An Outline Biography', p. 10. Also see R.G.W. Anderson, 'The Creation of the Chemistry Teaching Laboratory', in *Spaces and Collections in the History of Science*, ed. M. C. Lourenço and A. Carneiro (Lisbon, 2009), pp. 13–23.
13. Anderson and Jones, *Correspondence of Joseph Black*, vol. I, letter 95, Black to Watt, Edinburgh, 22 March 1772, on p. 253; quoted in Anderson, *Playfair Collection*, p. 23. The letter is held by the Birmingham City Archive.

14 Anderson and Jones, *Correspondence of Joseph Black*, vol. II, letter 599, Memorial by Black to the Trustees for Building a New College, n.p. [Edinburgh], 30 December 1789, on p. 1068, quoted in Anderson, *Playfair Collection*, p. 23. The letter is held by Edinburgh University Library.
15 For the life of Faraday, see L. Pearce Williams, *Michael Faraday: A Biography* (London, 1965), which covers his chemical research well, albeit by a non-historian of chemistry, and G. Cantor, D. Gooding and F.A.J.L. James, *Faraday* (Basingstoke, 1991), which, although intended for use in schools, is excellent.
16 G. Hutchinson, *Fuller of Sussex: A Georgian Squire*, revised (n.p., 1997).
17 K. Packer, 'Michael Faraday: A Founding Father of Organic Chemistry', *Analytical: Quarterly Journal for Analytical Sciences* (February 1992), p. 1. This rather obscure newsletter was produced by BP Research; a copy of this issue can be found in Science Museum Documentation in the technical file for 1908–12. The Science Museum has a sealed tube of Faraday's benzene (inv. no. 1908–112/pt 1), as does the Royal Institution. The two samples were found to be of the same purity.
18 J. Wilson, 'Celebrating Michael Faraday's Discovery of Benzene', *Ambix*, LIX (2012), pp. 241–65; E. M. Cammidge, 'Benzene and Turpentine: The Pre-history of Drycleaning', *Ambix*, XXXVIII (1991), pp. 79–84; A. S. Travis, *The Rainbow Makers: The Origins of the Synthetic Dyestuffs Industry in Western Europe* (Bethlehem, PA, and London, 1993).
19 For Faraday and religion, see G. Cantor, *Michael Faraday: Sandemanian and Scientist: A Study of Science and Religion in the Nineteenth Century* (Basingstoke, 1991); C. A. Russell, *Michael Faraday: Physics and Faith* (Oxford, 2000); and Cantor, Gooding and James, *Faraday*, pp. 17–21.
20 Jensen, 'To Demonstrate the Truths', pp. 6–7.
21 F.A.J.L. James and A. Peers, 'Constructing Space for Science at the Royal Institution of Great Britain', *Physics in Perspective*, IX (2007), pp. 130–85. Also see D. Chilton and N. G. Coley, 'The Laboratories of the Royal Institution in the Nineteenth Century', *Ambix*, XXVII (1980), pp. 173–203.
22 Nicholas Edwards, 'Webster, Thomas (1772–1844)', *Oxford Dictionary of National Biography* (Oxford, 2004), online edition at oxforddnb.com.
23 James and Peers, 'Constructing Space', p. 149.
24 John Davy, ed., *Collected Works of Sir Humphry Davy*, vol. I, *Memoirs of His Life* (London, 1839), p. 94. GB
25 F. Kurzer, 'Chemistry and Chemists at the London Institution, 1807–1912', *Annals of Science*, LVIII (2001), pp. 163–201, on p. 168.
26 Jensen, 'To Demonstrate the Truths', p. 6.
27 A. Strecker, *Das chemische Laboratorium der Universität Christiania* (Christiania [Oslo], 1854), p. iii. GB
28 A von Voit, *Das chemische Laboratorium der Königlichen Akademie der Wissenschaften in München* (Brunswick, 1859), pp. 6–20.
29 C. Nawa, 'A Refuge for Inorganic Chemistry: Bunsen's Heidelberg Laboratory', *Ambix*, LXI (2014), pp. 115–40, and private communication from Christine Nawa, dated 23 January 2013.

30 H. E. Roscoe, *Description of the Chemical Laboratories at the Owens College, Manchester* (Manchester, 1878), first pl.
31 'Imperial College', *Survey of London*, vol. XXXCIII: *South Kensington Museums Area* (1975), pp. 233–47 at www.british-history.ac.uk.
32 W. H. Chandler, *The Construction of Chemical Laboratories* (Washington, DC, 1893), pl. xcv.
33 Ibid., pl. xxxiii (Yale) and pl. lxxxix (Cornell).
34 Private communication from Anne Barrett (of Imperial College Archives), dated 2 December 2013.
35 'The heatwave of 1876', www.ukweatherworld.co.uk, 15 June 2003.
36 R. Bud, 'Infected by the Bacillus of Science: The Explosion of South Kensington', in *Science for the Nation: Perspectives on the History of the Science Museum*, ed. P.J.T. Morris (Basingstoke, 2010), pp. 11–40, on pp. 17–23.
37 *Catalogue of the Special Loan Collection of Scientific Apparatus at the South Kensington Museum*, 2nd edn (London, 1876), vol. II, p. 847f. GB The sub-section on chemistry is found on pp. 864–7.
38 'What is a Hofmann Apparatus?', answers.yahoo.com, 2010.
39 A. W. Hofmann, *Introduction to Modern Chemistry, Experimental and Theoretic* (London, 1866), but originally published anonymously in 1865. GB
40 W. B. Jensen, 'Hofmann Demonstration Apparatus', *Museum Notes* (March/April 2013), pp. 1–5, only available from www.che.uc.edu. Also see Jensen, 'Reinventing the Hofmann Sodium Spoon', *Bulletin of the History of Chemistry*, VII (1990), pp. 38–9, and Jensen, 'To Demonstrate the Truth'. I wish to thank William Jensen for his help with this section.

4 Training Chemists: Justus Liebig and Giessen, 1840s

1 J. B. Morrell, 'The Chemist Breeders: The Research Schools of Liebig and Thomas Thomson', *Ambix*, XIX (1972), pp. 1–46.
2 J. S. Fruton, 'The Liebig Research Group: A Reappraisal', *Proceedings of the American Philosophical Society*, CXXXII (1988), pp. 1–66.
3 G. H. Calvert, *Scenes and Thoughts in Europe*, 2nd series (New York, 1852), p. 95. GB
4 W. H. Brock, *Justus von Liebig: The Chemical Gatekeeper* (Cambridge, 1997). Also see the special issue on Liebig published as *Ambix*, L/1 (2003).
5 C. M. Jackson, 'Analysis and Synthesis in Nineteenth-century Organic Chemistry', PhD thesis, University College London, 2009, chap. 2; John Buckingham, *Chasing the Molecule: Discovering the Building Blocks of Life* (Stroud, 2004).
6 J. Liebig, *Anleitung zur Analyse organischer Körper* (Brunswick, 1837), GB has the 1853 edition. J. Liebig, *Instructions for the Chemical Analysis of Organic Bodies*, trans. from the German by W. Gregory (Glasgow, 1839). GB
7 Brock, *Liebig*, p. 50.
8 Ibid., pp. 57–9.

9 W. Gregory, 'The Advancement of Chemistry in Great Britain', *Lancet*, XXXVII (1842), pp. 712–16, on p. 715. GB
10 Carl Wilhelm Bergemann's report to the Prussian Minister on the Chemical Laboratory at Giessen in 1840; a translation forms appendix 1 of Brock, *Liebig*, quotation on p. 337. The original German version can be found in Regine Zott and Emil Heuser, *Die streitbaren Gelehrten. Justus Liebig und die preussischen Universitäten* (Berlin, 1992), pp. 173–81.
11 Private communication from William H. Brock, dated 11 January 2013.
12 Quoted in Brock, *Liebig*, p. 335.
13 Ibid., p. 336.
14 J. R. Partington, *A History of Chemistry*, vol. IV (London, 1964), pp. 283–6.
15 C. A. Russell, *The History of Valency* (Leicester, 1971), pp. 25–7.
16 I am very grateful to William Jensen for sharing his knowledge of the history of fume cupboards/hoods with me.
17 P. J. Ramberg, 'Chemical Research and Instruction in Zürich, 1833–1872', *Annals of Science*, LXXII (2015), pp. 170–86.
18 Verbal communication from Gisela Boeck, University of Rostock, 15 September 2012.
19 Jackson, 'Analysis and Synthesis', pp. 243–9; C. M. Jackson, 'Chemistry as the Defining Science: Discipline and Training in Nineteenth-century Chemical Laboratories', *Endeavour*, XXXV/2 (2011), pp. 55–62.
20 A. W. Hofmann, *The Chemical Laboratories in Course of Erection in the Universities of Bonn and Berlin* (London, 1866), p. 38. GB
21 Andrew Alexander, '"A Golden Cage, But Will the Birds Sing?": Alexander Crum Brown, William Gregory and Lyon Playfair', paper given at the 'The First Hundred Years of Chemistry at the University of Edinburgh' conference, 24 October 2013.
22 W. A. Tilden, *Chemical Discovery and Invention in the Twentieth Century*, 4th edn (London and New York, 1922), p. 25. The picture of the cupboards (as he called them) on the benches can be seen in fig. 6, facing p. 25, and at Sydney University in fig. 21, facing p. 46. Also see W. B. Jensen, 'Remembering Qualitative Analysis: The 175th Anniversary of Fresenius' Textbook', forthcoming.
23 Jackson, 'Analysis and Synthesis', pp. 250–51.
24 Private communication from William Jensen, dated 25 July 2013.
25 For a more detailed discussion, see Jackson, 'Analysis and Synthesis', pp. 243–9.
26 W. H. Chandler, *Construction of Chemical Laboratories*, passim; E. C. Robins, *Technical School and College Building* (London, 1887), passim.
27 E. R. Festing, *Report of Visits to Chemical Laboratories at Bonn, Berlin, Leipzig, etc.* (London, 1871), p. 13; Jackson, 'Analysis and Synthesis', p. 246.
28 Festing, *Visits to Chemical Laboratories*, p. 2.
29 Chandler, *Construction of Chemical Laboratories*, p. 680, referring to the system at the Kent Chemical Laboratory, Yale; Festing, *Visits to Chemical Laboratories*, p. 9, referring to the system at Leipzig.

30 Tilden, *Chemical Discovery and Invention*, p. 40, quoting a report from Dr B. S. Hopkins of the University of Illinois, hence the use of 'hood'.
31 H. Kolbe, 'Erprobte Laboratoriums-Einrichtungen', *Journal praktische Chemie*, III (1871), pp. 28–38, on p. 31.
32 Jackson, 'Analysis and Synthesis', pp. 250–51.
33 Festing, *Visits to Chemical Laboratories*, p. 22.
34 C. Reinhardt and A. S. Travis, *Heinrich Caro and the Creation of Modern Chemical Industry* (Dordrecht, 2000), pp. 265–6.
35 Festing, *Visits to Chemical Laboratories*, p. 3.
36 Ibid., p. 9.
37 Gisela Boeck, 'Die Geschichte der Chemie an der Universität Rostock', *Traditio et Innovatio*, XV/2 (2010), pp. 38–40, picture on p. 39.
38 F. Szabadvary, *History of Analytical Chemistry* (Oxford, 1966), pp. 161–72.
39 For an excellent history of group analysis, see Jensen, 'Remembering Qualitative Analysis'.
40 H. Rose, *Handbuch der analytischen Chemie* (Berlin, 1829), GB has the 1831 edition. English version: H. Rose, *Manual of Analytical Chemistry*, trans. J. Griffin (London, 1831). GB
41 C. R. Fresenius, *Anleitung zur qualitativen chemischen Analyse* (Bonn, 1841), GB has the 1852 edition. English version: C. R. Fresenius, *Elementary Instruction in Chemical Analysis*, trans. and ed. J. L. Bullock (London, 1843), GB has the 1846 edition.
42 R. Galloway, *A Manual of Qualitative Analysis* (London, 1850). GB
43 H. Will, *Anleitung zur qualitativen chemischen Analyse* (Heidelberg, 1846). An English translation appeared in 1846: H. Will, *Outlines of the Course of Qualitative Analysis* (London, 1846) and in 1858: F. T. Conington, *Handbook of Chemical Analysis, Adapted to the Unitary Notation. Based on the Fourth Edition of Dr. H. Will's Anleitung zur chemischen Analyse* (London, 1858) IA; the American version is available on GB: H. Will, *Outlines of Chemical Analysis*, trans. D. Breed and L. H. Steiner (Boston, MA, 1855).
44 H. Will, *Tafeln zur qualitativen chemischen Analyse* (Heidelberg, 1846). However, the earliest edition I have been able to find in an OPAC is the second edition of 1851. H. Will, *Tafeln zur qualitativen chemischen Analyse, achte Auflage* (Leipzig and Heidelberg, 1869). IA
45 John Timbs, *Curiosities of London* (London, 1855), p. 215. GB
46 T. E. Wallis, *History of the School of Pharmacy, University of London* (London, 1964), pp. 3–5.
47 Partington, *History of Chemistry*, vol. IV (London, 1964), pp. 270–71; Wallis, *History of the School of Pharmacy*, p. 103.
48 Private communication from William H. Brock dated 28 December 2012, citing a verbal communication from the late John Green.
49 *Chemical News*, XII (22 September 1865), p. 141. GB
50 P.J.T. Morris, 'Education of British Chemists in the Eighteenth Century', pt ii thesis, Oxford, 1978.
51 D. Knight, 'Chemistry on an Offshore Island: Britain, 1789–1840', in *The Making of the Chemist: The Social History of Chemistry in Europe 1789–1914*, ed. D. Knight and H. Kragh (Cambridge, 2008), pp. 95–106.

52 The figures are calculated from the Robin L. Mackie and Gerrylynn K. Roberts 'Chemists' Database' at the Open University as supplied by Gerrylynn Roberts. My thanks to Dr Roberts for her help with this analysis. Also see R. Simpson, *How the PhD Came to Britain: A Century of Struggle for Postgraduate Education* (Guildford, 1983).

53 W. Wetzel, 'Origins of, and Education and Career Opportunities for the Profession of "Chemist" in the Second Half of the Nineteenth Century in Germany', in *The Making of the Chemist*, ed. Knight and Kragh, pp. 77–94; Simpson, *How the PhD Came to Britain*.

54 Ernst Homburg, 'The Rise of Analytical Chemistry and its Consequences for the Development of the German Chemical Profession (1780–1860)', *Ambix*, XLVI (1999), pp. 1–32.

5 Modern Conveniences: Robert Bunsen and Heidelberg, 1850s

1 The best introduction in English to Bunsen's laboratory and its historical importance is C. Nawa, 'A Refuge for Inorganic Chemistry: Bunsen's Heidelberg Laboratory', *Ambix*, LXI (2014), pp. 115–40; also see H. Lang, *Das chemische Laboratorium an der Universität in Heidelberg* (Karlsruhe, 1858); T. Curtius and J. Rissom; *Geschichte des Chemischen Universitäts-Laboratoriums seit der Gründung durch Bunsen* (Heidelberg, 1908).

2 The major exception is D. Chilton and N. G. Coley, 'The Laboratories of the Royal Institution in the Nineteenth Century', *Ambix*, XXVII (1980), pp. 173–203; it is perhaps not insignificant that Donovan Chilton was also a Keeper at the Science Museum.

3 T. I. Williams, *History of the British Gas Industry* (Oxford, 1981), chap. 2; L. Tomory, *Progressive Enlightenment: The Origins of the Gaslight Industry, 1780–1820* (Cambridge, MA, 2012).

4 Private communication from Ernst Homburg dated 17 September 2013.

5 Private communication from Alan Rocke, dated 19 September 2013.

6 Private communication from Ernst Homburg dated 4 December 2013, citing K. van Berkel, A. van Helden and L. Palm, *A History of Science in the Netherlands: Survey, Themes and Reference* (Leiden, Boston, MA, and Cologne, 1999), p. 122. GB

7 J. Hassan, *History of Water in Modern England and Wales* (Manchester, 1998); also see C. Hamlin, *Science of Impurity: Water Analysis in the Nineteenth Century* (Berkeley, CA, 1990), available online at www.ark.cdlib.org.

8 Private communication from Ernst Homburg dated 17 September 2013.

9 R. H. Parsons, *The Early Days of the Power Station Industry* (Cambridge, 1939).

10 H. E. Roscoe, 'Bunsen Memorial Lecture', *Journal of the Chemical Society*, LXXVII (1900), pp. 513–54, on p. 545.

11 H. Huth, 'Lang, Heinrich', *Neue Deutsche Biographie*, vol. XIII (Berlin 1982), p. 537, available online at http://daten.digitale-sammlungen.de; 'Heinrich Lang (Architekt)', 'Heinrich Hübsch', 'Trinkhalle Baden-Baden' at http://de.wikipedia.org, accessed 5 December 2013.

12 Roscoe, 'Bunsen Memorial Lecture', p. 547.
13 Nawa, 'A Refuge for Inorganic Chemistry'; C. M. Jackson, 'Analysis and Synthesis in Nineteenth-century Organic Chemistry', PhD thesis, University College London, 2009, pp. 250–53.
14 Described as 'a handsome suite' by J. J. Tayler in 'Mr Tayler on Religion in Germany', *The Christian Reformer; or, Unitarian Magazine and Review*, new series, XII (November 1856), p. 654, GB; The staircase is mentioned in R. E. Oesper, *The Human Side of Scientists* (Cincinnati, OH, 1975), p. 28, GB. I am grateful to Robert Baptista for his help with this matter.
15 E. R. Festing, *Report of Visits to Chemical Laboratories at Bonn, Berlin, Leipzig, etc.* (London, 1871), p. 20.
16 Surprisingly, there is no good biography of Bunsen in English, but see the entry on Bunsen by S. G. Schacher in the *Dictionary of Scientific Biography* (New York, 1970–1980), and by C. Nawa in the *New Dictionary of Scientific Biography* (New York, 2007). In German there is G. Lockemann, *Robert Wilhelm Bunsen* (Stuttgart, 1949) and C. Stock, ed., *Robert Wilhelm Bunsens Korrespondenz vor dem Antritt der Heidelberger Professur (1852): Kritische Edition* (Stuttgart, 2007).
17 I am grateful to Christine Nawa for this clarification of the circumstances surrounding Bunsen's doctorate, and other matters relating to Bunsen and his laboratory.
18 C. A. Russell, *The History of Valency* (Leicester, 1971), pp. 34–43; C. A. Russell, *Edward Frankland: Chemistry, Controversy and Conspiracy in Victorian England* (Cambridge, 1996), pp. 108–13.
19 Strictly speaking, Captain Nemo used a modified version of the Bunsen cell, replacing the zinc with sodium as he obtained all his requirements from the sea, see W. B. Jensen, 'Captain Nemo's Battery: Chemistry and the Science Fiction of Jules Verne', *Chemical Intelligencer*, III/2 (1997), pp. 23–32, available from www.che.uc.edu.
20 M. Schrøder, *The Argand Burner: Its Origin and Development in France and England, 1780–1800* (Odense, 1969).
21 Ibid., pp. 132–8.
22 M. Faraday, *Chemical Manipulation* (London, 1827), p. 107; W. B. Jensen, 'The Origin of the Bunsen Burner', *Journal of Chemical Education*, LXXXII (2005), p. 518.
23 Roscoe, 'Bunsen Memorial Lecture', p. 547. Georg Lockemann, 'The Centenary of the Bunsen Burner', *Journal of Chemical Education*, XXXIII (January 1956), pp. 20–22.
24 *The Times*, Thursday, 17 August 1899; p. 4; issue 35910; col. d.
25 Festing, *Visits to Chemical Laboratories*, p. 21. For the history of the blowpipe see F. Szabadvary, *History of Analytical Chemistry* (Oxford, 1966), pp. 50–55; W. B. Jensen, 'The Development of Blowpipe Analysis', in *The History and Preservation of Chemical Instrumentation*, ed. J. T. Stock and M. V. Orna (Dordrecht, 1986), pp. 123–49.
26 M. A. Sutton, 'Sir John Herschel and the Development of Spectroscopy in Britain', *British Journal for the History of Science*, VII (1974), pp. 42–60; M. A. Sutton, 'Spectroscopy and the Chemists: A Neglected Opportunity', *Ambix*, XXIII (1976), pp. 16–26; F.A.J.L. James, 'The

Creation of a Victorian Myth: The Historiography of Spectroscopy', *History of Science*, XIII (1985), pp. 1–24; M. A. Sutton, 'Spectroscopy, Historiography and Myth: The Victorians Vindicated', *History of Science*, XXIV (1986), pp. 425–32; John C. D. Brand, *Lines of Light: The Sources of Dispersive Spectroscopy, 1800–1930* (Luxembourg, 1995).

27 It has been recognized that he was working on an idea given to him by Bunsen, but it does not seem to be well known that he was working in Bunsen's laboratory. See W. McGucken, *Nineteenth-century Spectroscopy: Development of the Understanding of Spectra, 1802–1897* (Baltimore, MD, 1969), p. 29, which mentions Bunsen's involvement but not that Cartmell was working in his laboratory, and R. Cartmell, 'On a Photochemical Method of Recognizing the Non volatile Alkalies and Alkaline Earths', *Philosophical Magazine* (July 1858), pp. 328–33, reference to Bunsen's laboratory on p. 333, GB. Also see W. H. Brock, 'Bunsen's British Students', *Ambix*, LX (2013), pp. 203–33, on p. 221.

28 Barbara J. Becker, *Unravelling Starlight: William and Margaret Huggins and the Rise of the New Astronomy* (Cambridge, 2011). For the history of astronomical spectroscopy, including the work of other early pioneers including Lewis Rutherfurd and Father Angelo Secchi, see J. B. Hearnshaw, *The Analysis of Starlight: Two Centuries of Astronomical Spectroscopy* (Cambridge, 1986).

29 F.A.J.L. James, 'Of "Medals and Muddles": The Context of the Discovery of Thallium: William Crookes's Early Spectro-Chemical Work', *Notes and Records of the Royal Society of London*, XXXXIX (September 1984), pp. 65–90.

30 Roscoe, 'Bunsen Memorial Lecture', p. 531.

31 Jochen Hennig, *Der Spektralapparat Kirchhoffs und Bunsens* (Berlin, 2003). Also see C. Nawa and C. Meinel, eds, *Von der Forschung gezeichnet: Instrumente und Apparaturen in Heidelberger Laboratorien skizziert von Friedrich Veith, 1817–1907* (Regensburg, 2007), pp. 25–33.

32 M. E. Weeks, *Discovery of the Elements*, 6th edn (Easton, PA, 1956), pp. 624–34. The photograph of the laboratory on p. 624 appears to date from after Bunsen's time, possibly from around 1900.

33 Bunsen's intention was doubtlessly poetic, but he was in error. According to Lewis and Short's *Latin Dictionary*, cæsius was rare and was only used for grey eyes.

34 Again Bunsen had chosen a very rare Latin word, probably derived from the more common *rubia* for madder.

35 J. J. Griffin, *Chemical Handicraft: A Classified and Descriptive Catalogue of Chemical Apparatus* (London, 1866), item 1881, on p. 205.

36 J. J. Griffin, *Chemical Handicraft: A Classified and Descriptive Catalogue of Chemical Apparatus* (London, 1877), item 1824, on p. 195.

37 See *Chemical News*, XXVII (31 January 1873), p. 49. GB

38 W. B. Jensen, 'The Hirsch and Büchner Filtration Funnels', *Journal of Chemical Education*, LXXXIII (2006), p. 1283; A. Sella, 'Classic Kit: Hirsch's Funnel', *Chemistry World* (March 2009), and A. Sella, 'Classic Kit: Büchner's Funnel', *Chemistry World* (November 2009), both available at www.rsc.org/chemistryworld.

39 See chap. 6 for Derek Barton's frequent demands to see the crystals.
40 J. Guo, 'Steam Pipe Explosion Damages Building 66', *The Tech*, CXXVIII/53 (4 November 2008), p. 1, available at http://tech.mit.edu.
41 A. Simmons, 'Stills, Status, Stocks and Science: The Laboratories at Apothecaries' Hall in the Nineteenth Century', *Ambix*, LXI (2014), pp. 141–61; *The Origin, Progress and Present State of the Various Establishments for Conducting Chemical Processes . . . at Apothecaries Hall* (London, 1823), GB. Date confirmed by Anna Simmons, private communication dated 5 November 2013, citing Court Minutes, 30 March 1813, archives of the Society of Apothecaries.
42 M. Eaton, *The Cook and Housekeeper's Complete and Universal Dictionary* (Bungay, 1823 [date on front page 1822]), entry on steam, pp. 398–9, GB. There is a system for piped steam in the kitchens of the Royal Pavilion in Brighton (completed in 1823); private communication from William H. Brock (University of Leicester), dated 31 May 2013.
43 A. W. Hofmann, *The Chemical Laboratories in Course of Erection in the Universities of Bonn and Berlin* (London, 1866), p. 19. GB
44 Ibid., p. 37.
45 Chilton and Coley, 'The Laboratories of the Royal Institution', on pp. 187–91.
46 Nawa, 'A Refuge for Inorganic Chemistry', p. 132.
47 Hofmann, *The Chemical Laboratories in Course of Erection*, p. 29.
48 W. H. Chandler, *The Construction of Chemical Laboratories* (Washington, DC, 1893), p. 709.
49 W. H. Brock, 'Building England's First Technical College: The Laboratories of Finsbury Technical College, 1878–1926', in *The Development of the Laboratory: Essays on the Place of Experiment in Industrial Civilisation*, ed. F.A.J.L. James (Basingstoke, 1989), pp. 155–70.
50 Parsons, *The Early Days of the Power Station Industry*, p. 10 (Holborn) and p. 12 (Godalming), GB. Also see I.C.R. Byatt, *The British Electrical Industry, 1875–1914: The Economic Return to a New Technology* (Oxford, 1979).
51 A. E. Munby, *Laboratories: Their Planning and Fittings* (London, 1931), pp. 122–3.
52 Faraday, *Chemical Manipulation*.
53 Griffin, *Chemical Handicraft*. IA
54 Bland & Long, *Descriptive and General Catalogue of Philosophical Apparatus and Chemical Preparations* (London, 1854). GB
55 E. N. Kent, *Descriptive Catalogue of Chemical Apparatus* (New York, 1854). GB

6 The Chemical Palace: Wilhelm Hofmann and Berlin, 1860s

1 [W. Crookes], 'Chemical Laboratories or Workshops', *Chemical News*, IV (January 1869), pp. 1–2, on p. 2.
2 W. H. Brock, *William Crookes (1832–1919) and the Commercialization of Science* (Aldershot, 2008), p. 45.
3 For the background to the construction of the laboratories in Bonn and Berlin, see A. W. Hofmann, *The Chemical Laboratories in Course of*

Erection in the Universities of Bonn and Berlin (London, 1866), GB. Also see J. A. Johnson, 'Hierarchy and Creativity in Chemistry, 1871–1914', *Osiris*, 2nd series, vol. v, 'Science in Germany' (1989), pp. 214–40.
4   Hofmann, *Chemical Laboratories in Course of Erection*, p. 8.
5   'Villa Hammerschmidt', http://de.wikipedia.org, accessed 5 December 2013.
6   'August Dieckhoff', http://de.wikipedia.org, accessed 5 December 2013.
7   F.A.J.L. James and A. Peers, 'Constructing Space for Science at the Royal Institution of Great Britain', *Physics in Perspective*, IX (2007), pp. 130–85, on pp. 158–9.
8   'Friedrich Albert Cremer', http://de.wikipedia.org, accessed 5 December 2013.
9   Hofmann, *Chemical Laboratories in Course of Erection*, p. 69.
10 Ibid., pp. 70–71.
11 For a discussion of this kind of historical commemoration in laboratories, see S. Forgan, 'The Architecture of Science and the Idea of a University', *Studies in History and Philosophy of Science, Part A*, XX (1989), pp. 405–34, on pp. 430–31.
12 Crookes, 'Chemical Laboratories', p. 1.
13 For the importance of safety in the development of the modern laboratory, see C. M. Jackson, 'Chemistry as the Defining Science: Discipline and Training in Nineteenth-century Chemical Laboratories', *Endeavour*, XXXV (2011), pp. 55–62.
14 Peter Borscheid, *Naturwissenschaft, Staat und Industrie in Baden (1848–1914)* (Stuttgart, 1976).
15 For the situation in Saxony, see A. J. Rocke, *The Quiet Revolution: Hermann Kolbe and the Science of Organic Chemistry* (Berkeley, CA, 1993), pp. 265–70; for Baden, see A. M. Tuchman, *Science, Medicine, and the State in Germany: The Case of Baden, 1815–1871* (New York and Oxford, 1993), which has surprisingly little to say about chemistry despite Heidelberg being in Baden.
16 Quoted in A. J. Rocke, *Nationalizing Science: Adolphe Wurtz and the Battle for French Chemistry* (Cambridge, MA, 2001), p. 294.
17 For the situation in Prussia up to the 1860s, see R. S. Turner, 'Justus Liebig versus Prussian Chemistry: Reflections on Early Institute-Building in Germany', *Historical Studies in the Physical Sciences*, XIII (1982), pp. 129–62.
18 J. Ben-David, 'Scientific Productivity and Academic Organization in Nineteenth Century Medicine', *American Sociological Review*, XXV (1960), pp. 828–43.
19 Rocke, *The Quiet Revolution*, p. 267.
20 T. Lenoir, 'Revolution From Above: The Role of the State in Creating the German Research System, 1810–1910', *The American Economic Review*, LXXXVIII/2 (May 1998), pp. 22–7.
21 U. Klein, *Experiments, Models, Paper Tools: Cultures of Organic Chemistry in the Nineteenth Century* (Stanford, CA, 2003).
22 A. S. Travis, 'August Wilhelm Hofmann (1818–1892)', *Endeavour*, XVI/2 (1992), pp. 59–65; B. Lepsius, *August Wilhelm von Hofmann*

(Leipzig, 1905), extracted from the *Allgemeinen Deutschen Biographie*. C. Meinel and H. Scholz, eds, *Der Allianz von Wissenschaft und Industrie August Wilhelm Hofmann (1818–1892): Zeit, Werk, Wirkung* (Weinheim, 1992) sheds light on various aspects of Hofmann's career.

23 C. A. Russell, *The History of Valency* (Leicester, 1971), pp. 44–61.
24 Lepsius, *Hofmann*, p. 12, repeated for example by J. J. Beer, *The Emergence of the German Dye Industry* (Urbana, 1959), p. 17.
25 J. Bentley, 'The Chemical Department of The Royal School of Mines: Its Origins and Development under A. W. Hofmann', *Ambix*, XXVII (1970), pp. 153–81, on p. 162; W. H. Brock, *Justus von Liebig: The Chemical Gatekeeper* (Cambridge, 1997), pp. 104–5.
26 E. Ward, 'Death of Charles Blachford Mansfield (1819–1855)', *Ambix*, XXXI (1984), pp. 68–9.
27 C. Meinel, 'Molecules and Croquet Balls', in *Models: The Third Dimension of Science*, ed. S. de Chadarevian and N. Hopwood (Stanford, CA, 2004), pp. 242–75, on pp. 250–53.
28 Sophie Forgan makes a similar comparison to ward rounds in Forgan, 'The Architecture of Science and the Idea of a University', p. 424.
29 P.J.T. Morris, 'Barton, Derek Harold Richard', *Complete Dictionary of Scientific Biography*, vol. XIX (Detroit, MI, 2008), pp. 195–202. Hans Fischer was also said to demand to see the crystals, see J. Waldenström, 'Forty Years with the Gamma Globulins', *Scandinavian Journal of Immunology*, XXV (1987), pp. 211–18, on p. 211.
30 E C. Robins, *Technical School and College Building* (London, 1887), pp. 118–19.
31 As displayed at the Observatory Museum in Stockholm in August 2013. Its authenticity was confirmed by the Berzelius biographer Jan Trofast, verbal communication, 24 August 2013.
32 Robins, *Technical School and College Building*, p. 118.
33 A. E. Munby, *Laboratories: Their Planning and Fittings* (London, 1931), p. 24.
34 W. H. Chandler, *The Construction of Chemical Laboratories* (Washington, DC, 1893), p. 693, referring to the organic chemical laboratory at Lehigh.
35 Ibid., p. 693; H. F. Lewis, ed., *Laboratory Planning for Chemistry and Chemical Engineering* (New York and London, 1962), pp. 69, 378–9 and 386.
36 Lewis, ed., *Laboratory Planning*, p. 378.
37 H. S. Coleman, ed., *Laboratory Design* (New York, 1951), pp. 16–17.
38 Lewis, ed., *Laboratory Planning*, p. 378; W. R. Ferguson, *Practical Laboratory Planning* (Barking, 1973), pp. 42–3 and Appendix 3, p. 141.
39 Lewis, ed., *Laboratory Planning*, p. 69; Ferguson, *Practical Laboratory Planning*, p. 40.
40 Ferguson, *Practical Laboratory Planning*, p. 40; *Canadian Chemical Processing*, XLIV (1960), p. 71.
41 R. Lees and A. F. Smith, eds, *Design, Construction and Refurbishment of Laboratories* (Chichester, 1984), pp. 223–5; D. D. Watch, *Building Type Basics for Research Laboratories*, 2nd edn (Hoboken, NJ, 2008), pp. 160–62. Earlier in 1973, William Ferguson noted that it was only

available in the USA and was expensive – Ferguson, *Practical Laboratory Planning*, p. 40.
42 'Laboratory', in A. Ure, *A Dictionary of Chemistry and Mineralogy*, 4th edn (London, 1835), pp. 558–67, on p. 559, GB. The dictionary was first published (with similar wording) in 1821.
43 Ibid., on p. 559.
44 C. Morfit, *Chemical and Pharmaceutical Manipulation*, 2nd edn, 1857, pp. 57–8.
45 Quotation from Hofmann, *Chemical Laboratories in Course of Erection*, p. 37; E. R. Festing, *Report of Visits to Chemical Laboratories at Bonn, Berlin, Leipzig, etc.* (London, 1871), p. 6.
46 Jack Morrell, 'W. H. Perkin, Jr., at Manchester and Oxford: From Irwell to Isis', *Osiris*, 2nd series, VII, *Research Schools: Historical Reappraisals* (1993), pp. 104–26, on p. 109.
47 C. M. Jackson, 'Analysis and Synthesis in Nineteenth-century Organic Chemistry', PhD thesis, University College London, 2009, pp. 250–52, but the *Stinkzimmer* predated Kolbe's laboratory in Leipzig as Bunsen had one in Heidelberg, see C. Nawa, 'A Refuge for Inorganic Chemistry: Bunsen's Heidelberg Laboratory', *Ambix*, LXI (2014), pp. 115–40, on p. 126.
48 See Jackson, 'Analysis and Synthesis', pp. 257–62 for a detailed description.
49 P. D. Buchanan, 'Quantitative Measurement and the Design of the Chemical Balance, 1750–c. 1900', PhD thesis, Imperial College, 1982; J. T. Stock, *Development of the Chemical Balance* (London, 1969); H. R. Jenemann, *Die Waage des Chemikers* (Frankfurt am Main, 1979).
50 Robert E. Lyle and Gloria G. Lyle, 'A Brief History of Polarimetry', *Journal of Chemical Education*, XLI (1964), pp. 308–13.
51 S. F. Johnston, 'Chemical Polarimeter' and P. Brenni, 'Polarimeter and Polariscope', in *Instruments of Science: A Historical Encyclopedia*, ed. R. Bud and D. J. Warner (New York, 1998), pp. 473–7.
52 D. J. Warner, 'How Sweet it is: Sugar, Science, and the State', *Annals of Science*, XLIV (2007), pp. 147–70.
53 P. J. Ramberg, *Chemical Structure, Spatial Arrangement: The Early History of Stereochemistry, 1874–1914* (Aldershot, 2003), pp. 245–68; F. W. Lichtenthaler, 'Emil Fischer's Proof of the Configuration of Sugars: A Centennial Tribute', *Angewandte Chemie International Edition in English*, XXXI (2003), pp. 1541–56.
54 Robins, *Technical School and College Building*, pp. 49–50.
55 Ibid., p. 50.
56 Rocke, *Quiet Revolution*, p. 280.
57 Hofmann, *Chemical Laboratories in Course of Erection*, p. 12.
58 Ibid., p. 36.
59 Jackson, 'Analysis and Synthesis', p. 254.
60 Rocke, *Nationalizing Science*, p. 107.
61 I am grateful to the late W. Alec Campbell for drawing my attention to this feature of the laboratory. Also see Y. Kikuchi, *Anglo-American*

Connections in Japanese Chemistry: The Lab as Contact Zone (New York, 2013), pp. 70–74.
62 *Royal Commission on Scientific Instruction* (Command 536, 1872), vol. 1, p. 500, sc. 7370.
63 For the spy window at Marburg ('*eine grosse Glaswand*'), see H. Kolbe, *Das chemische Laboratorium der Universität Marburg* (Brunswick, 1865), p. 11. GB
64 Kikuchi, *Anglo-American Connections in Japanese Chemistry*, pp. 111–20.

7 Laboratory Transfer: Henry Roscoe and Manchester, 1870s

1 P. J. Ramberg, 'Chemical Research and Instruction in Zürich, 1833–1872', *Annals of Science*, LXXII (2015), pp. 170–86.
2 A. Strecker, *Das chemische Laboratorium der Universität Christiania* (Christiania [Oslo], 1854). GB
3 W. H. Brock, 'Building England's First Technical College: The Laboratories of Finsbury Technical College, 1878–1926', in *The Development of the Laboratory: Essays on the Place of Experiment in Industrial Civilisation*, ed. F.A.J.L. James (Basingstoke, 1989), pp. 155–70, on pp. 155–8; E. C. Robins, *Technical School and College Building* (London, 1887). IA
4 D.S.L. Cardwell, *The Organization of Science in England* (London, 1972), pp. 111–26; R. MacLeod, 'Resources of Science in Victorian England: The Endowment of Science Movement, 1868–1900', in *Science and Society 1600–1900*, ed. Peter Mathias (Cambridge, 1972), pp. 111–66; G. Haines, 'German Influence upon Scientific Instruction in England, 1867–1887', *Victorian Studies*, I (1958), pp. 215–44; M. Argles, 'The Royal Commission on Technical Instruction, 1881–4: Its Inception and Composition', *The Vocational Aspect of Education*, XI/23 (1959), pp. 97–104; M. M. Gowing, 'Technology and Education: England in 1870: The Wilkins Lecture 1976', *Notes and Records of the Royal Society of London*, XXXII (1977), pp. 71–90.
5 A. J. Rocke, *Nationalizing Science: Adolphe Wurtz and the Battle for French Chemistry* (Cambridge, MA, 2001), pp. 388–97, and private communications from Alan Rocke (Case Western Reserve University), dated 7 January 2013 and Christoph Meinel, dated 19 January 2013.
6 For the life of Roscoe, see first and foremost H. E. Roscoe, *The Life and Experience of Sir Henry Enfield Roscoe* (London, 1906), probably the best (and funniest) chemical autobiography ever written. Also see the entries by Robert Kargon in the *Oxford Dictionary of National Biography* and the *Dictionary of Scientific Biography* (New York, 1970–80); Manchester Museum of Science and Industry, 'Henry Enfield Roscoe (1833–1915)', 2004, available at www.mosi.org.uk; T. E. Thorpe, *The Right Honourable Sir Henry Enfield Roscoe: A Biographical Sketch* (London, 1916).
7 Roscoe's entry in *Who's Who*, presumably written by him (or at least approved by him), says 1853, but this seems unlikely. He celebrated the

50th anniversary in March 1904, and Thorpe speaks of Roscoe taking the (oral) examination at the end of the second session, which would fit with March 1854 – see Thorpe, *Roscoe*, p. 25.
8 Roscoe, *Life and Experiences*, pp. 102–3.
9 H. E. Roscoe, *Efnafræði* (Reykjavík, 1879).
10 For the details, see 'List of Commissions and Officials: 1880–1889 (nos 38–72), Office-Holders in Modern Britain', vol. x, 'Officials of Royal Commissions of Inquiry 1870–1939', 1995, pp. 13–27, available at www.british-history.ac.uk, and 'List of Commissions and Officials: 1890–1899 (nos 73–102), Office-Holders in Modern Britain', vol. x, Officials of Royal Commissions of Inquiry 1870–1939', 1995, pp. 27–42, available at www.british-history.ac.uk.
11 L. Linder, ed., *The Journal of Beatrix Potter* (London, 1986), p. 164.
12 R. Bud, 'Infected by the Bacillus of Science', in *Science for the Nation: Perspectives on the History of the Science Museum*, ed. P.J.T. Morris (Basingstoke, 2010), pp. 11–40, especially pp. 28–35.
13 A. W. Hofmann, *The Chemical Laboratories in Course of Erection in the Universities of Bonn and Berlin* (London, 1866), pp. v–viii. GB
14 E. R. Festing, *Report of Visits to Chemical Laboratories at Bonn, Berlin, Leipzig, etc.* (London, 1871).
15 'Imperial College', *Survey of London*, vol. xxxviii: *South Kensington Museums Area* (London, 1975), pp. 233–47, available at www.british-history.ac.uk; H. Gay, *The History of Imperial College London, 1907–2007* (Singapore, 2007), pp. 49–50. For the general context, see S. Forgan and G. Gooday, '"A Fungoid Assemblage of Buildings:" Diversity and Adversity in the Development of College Architecture and Scientific Education in Nineteenth-Century South Kensington', *History of Universities*, xiii (1994), pp. 153–92. I wish to thank Sophie Forgan (University of Teesside) for sight of an offprint of this article. Also see S. Forgan and G. Gooday, 'Constructing South Kensington: The Buildings and Politics of T. H. Huxley's Working Environments', *British Journal for the History of Science*, xxix (1996), pp. 435–68.
16 Forgan and Gooday, 'Fungoid Assemblage', p. 164.
17 A.E.H. Tutton, 'Memories of the College Half a Century Ago', *The Record*, (December 1936), pp. 42–5 (available in Imperial College archives). I am grateful to Hannah Gay (Simon Fraser University) for this reference.
18 *Royal Commission on Scientific Instruction* (Command 536, 1872), vol. 1, p. 500, sc. 7370.
19 Ibid.
20 Ibid.
21 Roscoe, *Life and Experiences*, p. 110.
22 Private communication from the late W. Alec Campbell, 1998.
23 W. H. Brock, 'The Chemical Origins of Practical Physics', in *The Case of the Poisonous Socks* (Cambridge, 2011), pp. 97–113, on pp. 98–9, quote on p. 98.
24 'Imperial College', *Survey of London*.
25 Jack Morrell, 'W. H. Perkin, Jr, at Manchester and Oxford: From Irwell

to Isis', *Osiris*, 2nd series, vol. VIII, *Research Schools: Historical Reappraisals* (1993), pp. 104–26, on pp. 108–9.

26 Y. Kikuchi, *Anglo-American Connections in Japanese Chemistry: The Lab as Contact Zone* (New York and Basingstoke, 2013), pp. 111–20, quote on pp. 111 and 114.

27 M. W. Rossiter, *The Emergence of Agricultural Science: Justus Liebig and the Americans, 1840–1880* (New Haven, CT, 1975), pp. 75–80.

28 See 'Philosophical Hall' at www.union.edu, accessed 26 July 2013.

29 Annual Report of the Regents of the University of New York (Schenectady, 1858), p. 52. GB

30 Ibid., p. 52.

31 Ibid., pp. 52–3.

32 [F. B. Hough], *Historical Sketch of Union College* (Washington, DC, 1876), p. 39, ia; W. Somers, *Encyclopedia of Union College History* (Schenectady, New York, 2003), p. 155. I wish to thank Thomas Werner for his help with this section.

33 For Illinois, see P. T. Carroll, 'Academic Chemistry in America, 1876–1976: Diversification, Growth, and Change', PhD thesis, University of Pennsylvania, 1982, chap. 4. For Havemeyer Hall at Columbia, see American Chemical Society National Historic Chemical Landmarks, 'Havemeyer Hall', 1998, available at www.acs.org. For Dartmouth, see D. Pantalony, R. L. Kremer and F. J. Manasek, *Study, Measure, Experiment: Stories of Scientific Instruments at Dartmouth College* (Norwich, VT, 2005).

34 W. H. Chandler, *The Construction of Chemical Laboratories* (Washington, DC, 1893), pl. xxxix.

35 Ibid., pl. lvi.

36 For Dartmouth, see Pantalony, Kremer and Manasek, *Study, Measure, Experiment*, pp. 214–15; for MIT, see Chandler, *Construction of Chemical Laboratories*, pls lxxvii and lxxviii.

37 I wish to thank Gerald Peterson for his help with this topic in a private communication dated 19 July 2013.

38 A. Rocke, *Nationalizing Science*, especially pp. 284–99, 365–7 and 388–97; D. Fauque, 'The Laboratories of Chemistry in the New Sorbonne (1894)', paper given at the 'Sites of Chemistry in the 19th Century' conference, Valencia, Spain, 7 July 2012; H. P. Nénot, *Monographie de la nouvelle Sorbonne* (Paris, 1903).

39 For the Davy-Faraday laboratory, see D. Chilton and N. G. Coley, 'The Laboratories of the Royal Institution in the Nineteenth Century', *Ambix*, XXVII (1980), pp. 173–203, on pp. 185–6 and pl. viii; for the Wellcome laboratories, see illus. 103 and 'The Wellcome Research Laboratories', *The Nursing Record and Hospital World* (30 June 1900), pp. 522–4, available at http://rcnarchive.rcn.org.uk; for the laboratories of the Royal College of Physicians see 'The New Research Laboratory of the Royal College of Physicians of Edinburgh', *British Medical Journal* (14 November 1896), pp. 1455–7.

8 Chemical Museums: Charles Chandler and New York, 1890s

1. C. Meinel, 'Chemical Collections', in *Spaces and Collections in the History of Science*, ed. M. C. Lourenço and A. Carneiro (Lisbon, 2009), pp. 137–47, further expounded in a private communication from Christoph Meinel dated 9 December 2013.
2. S. Forgan, '"But Indifferently Lodged . . .": Perception and Place in Building for Science in Victorian London', in *Making Space for Science: Territorial Themes in the Shaping of Knowledge*, ed. C. Smith and J. Agar (Basingstoke, 1998), pp. 195–215, on pp. 198–200. I am grateful to Sophie Forgan for drawing my attention to this point.
3. P.J.T. Morris, 'The Image of Chemistry Presented by the Science Museum, London, in the Twentieth Century: An International Perspective', in *The Public Image of Chemistry*, ed. J. Schummer, B. Bensaude-Vincent and B. van Tiggelen (Singapore, 2007), pp. 297–327, on pp. 308–10.
4. From a contemporary plan of the building in H. C. Burdett, *Hospitals and Asylums of the World, Portfolio of Plans* (London, 1893), p. 101, supplied by Andrew Alexander.
5. 'The Cavendish Museum' at www.phy.cam.ac.uk, accessed 8 December 2013.
6. Y. Kikuchi, *Anglo-American Connections in Japanese Chemistry: The Lab as Contact Zone* (New York and Basingstoke, 2013), pp. 162–3.
7. J. V. Pickstone, 'Museological Science? The Place of the Analytical/Comparative in Nineteenth-century Science, Technology and Medicine', *History of Science*, XXXII (1994), pp. 111–38.
8. Review of Edward Forbes, *On the Educational Uses of Museums* in *Chemical News*, I (23 April 1860), p. 237. The review is anonymous but the style is characteristic of Crookes's rhetoric, especially its criticism of the Chemical Society's museum. GB
9. R. H. McKee, C. E. Scott and C.B.F. Young, 'The Chandler Chemical Museum at Columbia University', *Journal of Chemical Education*, XI (1934), pp. 275–8, on p. 277.
10. George Kemp, 'On Berberine and some of its Compounds . . .', *Chemical Gazette: or, Journal of Practical Chemistry*, V (1 June 1847), pp. 209–11, on p. 209. GB
11. H. Kaemmerer, 'Ueber einige Jodverbindungen', *Journal für Praktische Chemie*, LXXXIII/1 (1861), pp. 65–85, on p. 83.
12. For the relationship between pure and applied chemistry in the 19th century, see R. F. Bud and G. K. Roberts, *Science versus Practice: Chemistry in Victorian Britain* (Manchester, 1984); J. F. Donnelly, 'Getting Technical: The Vicissitudes of Academic Industrial Chemistry in Nineteenth-century Britain', *History of Education*, XXVI (1997), pp. 125–43.
13. J. Thackray, 'Mineral and Fossil Collections', in *Sir Hans Sloane: Collector, Scientist, Antiquary; Founding Father of the British Museum*, ed. A. MacGregor (London, 1994), pp. 123–35. Also see W. E. Wilson, *A History of Mineral Collecting 1530–1799* (Tucson, AZ, 1994).

14 H. Fors, 'J. G. Wallerius and the Laboratory of the Enlightenment', in *Taking Place: The Spatial Contexts of Science, Technology, and Business*, ed. E. Baraldi, H. Fors and A. Houltz (Sagamore Beach, MA, 2006), pp. 3–33, on pp. 22–5. For a slightly different account which agrees more closely with the interpretation here, see H. Fors, *Mutual Favours: The Social and Scientific Practice of Eighteenth Century Swedish Chemistry* (Uppsala, 2003), pp. 91–6.

15 *Torbern Bergman*, in J. A. Schufle, *Torbern Bergman's Autobiography* (Lawrence, KS, 1985), pp. 465–80, on p. 470.

16 Private communication from Marco Beretta dated 30 August 2013; S-E. Liedman and M. Persson, 'The Visible Hand: Anders Berch and the University of Uppsala Chair in Economics', *Scandinavian Journal of Economics*, XCIV (1992), pp. S259–69, on p. S267.

17 For Sweden, see Fors, *Mutual Favours*; for Scotland, see M. D. Eddy, *The Language of Mineralogy: John Walker, Chemistry and the Edinburgh Medical School, 1750–1800* (Farnham, 2008).

18 S. Schaffer and L. Stewart, 'Vigani and After: Chemical Enterprise in Cambridge, 1680–1780', in *The 1702 Chair of Chemistry at Cambridge: Transformation and Change,* ed. M. D. Archer and C. D. Haley (Cambridge, 2005), pp. 31–56, on pp. 41–3. I wish to thank Marco Beretta for his help with this section, especially relating to the use of models by Bergman.

19 Private communication from David Grayson, dated 23 October 2013; T.P.C. Kirkpatrick, *History of the Medical Teaching in Trinity College Dublin and of the School of Physic In Ireland* (Dublin, 1912), pp. 76–8. IA

20 Private communication from David Grayson, dated 27 November 2013, citing papers from the TCD Board papers, TCD archives.

21 David Murray, *Museums: Their History and their Use*, vol. 1 (Glasgow, 1904), p. 146.

22 For the biography of Chandler, see M. T. Bogert. 'Charles Frederick Chandler, 1836–1925', *Biographical Memoirs of the National Academy of Sciences* (Washington, DC, 1931), available at www.nasonline.org; M. W. Rossiter, 'The Charles F. Chandler Collection', *Technology and Culture,* XVIII (1977), pp. 222–30, also see R. L. Larson, 'Charles Frederick Chandler: His Life and Work', PhD thesis, Columbia University, 1950. I am grateful to Jeffrey Sturchio for his help with this section.

23 McKee, Scott and Young, 'The Chandler Chemical Museum', p. 275.

24 J. J. Bohning, 'Opposition to the Formation of The American Chemical Society', *Bulletin for the History of Chemistry*, XXVI (2001), pp. 92–103.

25 For the Chemical Foundation, see D. J. Rhees, 'The Chemists' Crusade: The Rise of an Industrial Science in Modern America, 1907–1922', PhD thesis, University of Pennsylvania, 1987.

26 Private communication from David Grayson, dated 23 October 2013.

27 Private communication from Donall MacDonall, dated 23 October 2013, citing *College Calendar* (Dublin, 1892), p. 220.

28 Meinel, 'Chemical Collections', p. 141.

29 Private communication from Robert Anderson, dated 29 October 2013, citing the Playfair Portfolio in the University of Edinburgh Library, but unfortunately this is not currently accessible.
30 Typescript handlist for Playfair Portfolio 1, University of Edinburgh, Centre for Research Collections, GB 237Coll-13/p.1, via a private communication from Andrew Alexander, dated 21 November 2013.
31 *Calendar of King's College London for 1850–1* (London, 1850), p. 116. GB
32 A. Strecker, *Das chemische Laboratorium der Universität Christiania* (Christiania [Oslo], 1854), p. iii. GB
33 A. Vernon Harcourt, 'Presidential Address', *Proceedings of the Chemical Society*, XII (1896), p. 80.
34 Review of Edward Forbes, *On the Educational Uses of Museums*, p. 237.
35 For the growth of museums generally in the nineteenth century and their architecture, see S. Forgan, 'The Architecture of Display: Museums, Universities and Objects in Nineteenth-century Britain', *History of Science*, XXXII (1994), pp. 139–62, available at http://adsabs.harvard.edu.
36 A. W. Hofmann, *The Chemical Laboratories in Course of Erection in the Universities of Bonn and Berlin* (London, 1866), pp. 22–3. GB
37 Ibid., p. 58.
38 Ibid., pp. 58–9.
39 'Polytechnic Schools', *The Architect* (19 November 1870), pp. 292–3, on p. 293. GB
40 *The Medical Times and Gazette*, I (1874), p. 245. GB
41 W. H. Chandler, *The Construction of Chemical Laboratories* (Washington, DC, 1893), p. 711.
42 Advertisement for the Leeds School of Medicine, *Chemical News*, XXVIII (12 September 1873), p. 136. GB
43 *Royal Commission on Scientific Instruction* (Command 536, 1872), vol. I, p. 500, sc. 7373; P. J. Hartog, ed., *The Owens College, Manchester: A Brief History of the College and Description of Its Various Departments* (Manchester, 1900), p. 64.
44 E. C. Robins, *Technical School and College Building* (London, 1887), p. 103.
45 Ibid., p. 77.
46 Ibid., p. 142 and pl. 41. Also see 'Ireland and Maclaren' in www.scottisharchitects.org.uk, accessed 22 December 2013.
47 L. Fine, 'The Chandler Chemical Museum', *Bulletin of the History of Chemistry*, II (1988), pp. 19–21; McKee, Scott and Young, 'The Chandler Chemical Museum'.
48 Private communication from Professor Leonard Fine, dated 25 February 2013.
49 Private communication from Brian Williams, dated 30 July 2013; *University of Michigan General Register* for 1890–91, p. 24. Also see *Proceedings of the Board of Regents of the University of Michigan, 1881–1886* (Ann Arbor, MI, 1886), pp. 483 and 605, IA. I wish to thank Brian Williams for his help.
50 *University of Michigan General Register* for 1862, p. 60.
51 *University of Michigan General Register* for 1908–9, p. 52.
52 Ibid., pp. 52–3.

53 Chandler, *Construction of Chemical Laboratories*, p. 688 (measurements) and 694; ACS National Historic Chemical Landmarks, 'The Chandler Chemistry Laboratory', 1994, available at http://portal.acs.org.
54 Chandler, *Construction of Chemical Laboratories*, pp. 703–4; *The Register, Cornell University* for 1903–4, p. 167. IA
55 *The Register, Cornell University* for 1886–7, p. 39, GB. The museum was planned in 1879 and apparently erected in 1883, see *The Register, Cornell University* for 1879–80, p. 57.
56 *Museums Journal*, XXX (1931), p. 327. GB
57 'Catholic University News', *Sacred Heart Review*, XVIII/24 (11 December 1897), p. 475, available at http://newspapers.bc.edu; Catholic University of America, *Announcements of the School of Sciences 1909–10*, pp. 8 and 10, IA. For Chicago, see *Popular Science* (October 1897), p. 785, and *Annual Register of the University of Chicago* for 1907, p. 117, both GB.
58 L. May, 'The Early Days of Chemistry at Catholic University', *Bulletin for the History of Chemistry*, XXVIII (2003), pp. 18–25, citing *Catholic University Bulletin*, II (1896), pp. 99–100.
59 Massachusetts Institute of Technology, *President's Report for the Year Ending 1909*, p. 137. GB
60 *The Wisconsin Alumni Magazine*, II/8 (May 1901), p. 334. IA
61 *Journal of the Worcester Polytechnic Institute*, VI (1903), pp. 326 and 329. I wish to thank Stephen Weininger for his assistance.
62 *Journal of the Worcester Polytechnic Institute*, VI (1903), pp. 162 [Standard Oil] and 329 [Farbenfabriken, i.e. Bayer].
63 'University Notes', *Bostonia*, VIII/1 (1907), p. 32.
64 *Official Register of Harvard University*, VI/2 (1909), p. 779. GB
65 'New Chemical Laboratory', *Harvard Crimson* (11 December 1909), available at www.thecrimson.com; in W. A. Tilden, *Chemical Discovery and Invention in the Twentieth Century*, 4th edn (London and New York, 1922), fig. 11, p. 31. This plan and Tilden's remarks on it (p. 37) clearly date from 1917 when the book was first published. Marcia Chapin of Harvard University chemistry library confirmed in a private communication dated 29 July 2013 that it was never completed.
66 R. MacLeod, *Archibald Liversidge, FRS: Imperial Science under the Southern Cross* (Sydney, 2009), p. 59.
67 Tilden, *Chemical Discovery and Invention*, p. 11 (described as chemical collections); MacLeod, *Liversidge*, p. 280; private communication from Roy MacLeod, dated 29 July 2013.
68 Private communications from Ian D. Rae, dated 30 July 2013, quoting from the history of the department held within the department.
69 Verbal communication from Veronika Petroff, dated 7 October 2013.
70 Private communication from Danielle Fauque, dated 30 November 2013; H. P. Nénot, *Monographie de la nouvelle Sorbonne* (Paris, 1903).
71 A. E. Munby, *Laboratories: Their Planning and Fittings* (London, 1931), pp. 184–5.
72 Private communication from Andrew Alexander, dated 22 March 2013. For Playfair's renovation, see W. Reid, *Memoirs and Correspondence of Lyon Playfair, First Lord Playfair of St. Andrews* (London, 1899), p. 180. IA

73 *North Carolina State College of Agriculture and Engineering, 1931–1932* (State College Station, 1932), p. 37. IA
74 Private communication from Ian D. Rae, dated 31 July 2013, citing *Melbourne Argus* (24 November 1936), p. 11.
75 R. D. Thomson, 'Examination of the Cowdie Pine Resin', *London Edinburgh and Dublin Philosophical Magazine and Journal of Science*, 3rd series, XXIII (August 1843), pp. 81–9, on p. 81. Advertisement in *Lancet* II/12 (20 September 1845), n.p. GB
76 F. P. Dunnington, *The Chemical Museum of the University of Virginia* (Charlottesville, VA, 1917), p. 1. A copy of this book, which was reprinted from *The Alumni Bulletin* for July 1917, is held by the University of Virginia Library.
77 Ibid., p. 7.
78 Ibid., p. 2.
79 A. G. Brook and W.A.E. McBryde, *Historical Distillates: Chemistry at the University of Toronto since 1843* (Toronto, 2007), p. 72.
80 'The Chemical Museum', *The Varsity*, XXII/13 (21 January 1903), p. 195. I wish to thank Erich Weidenhammer for his assistance.
81 Obituary of Irving Fay, *Chemical Week*, XXXVIII (1936), p. 298; *Harvard Alumni Bulletin*, XXXIV/21 (1932), p. 667. Both GB
82 This account is based on P.J.T. Morris, 'Melting Point Apparatus', in *Instruments of Science: A Historical Encyclopedia*, ed. R. Bud and D. J. Warner (New York, 1998), pp. 373–4.
83 M. E. Chevreul, *Recherches chimiques sur les corps gras d'origine animale* (Paris, 1823), table of melting points on pp. 82–3; M. E. Chevereul, *Considerations générales sur l'analyse organique et sur ses applications* (Paris, 1824), p. 33. Both GB
84 The method of mixed melting points works on the basis that an impure compound has a lower melting point than a pure one. If two pure samples of the same compound are mixed, the melting point is unaffected, but if they are different compounds, the melting point will be appreciably depressed even if the two samples have the same melting point on their own. However, in certain cases this does not occur when the two substances are optical isomers; see E. Eliel, *Stereochemistry of Carbon Compounds* (New York, 1962), pp. 43–7 and figs 4–19.
85 J. R. Partington, *A History of Chemistry*, vol. IV (London, 1964), p. 359; P. Ramberg, 'Wilhelm Heintz (1817–1880) and the Chemistry of the Fatty Acids', *Bulletin for the History of Chemistry*, XXXVIII (2013), pp. 19–28.
86 H. E. Armstrong and T. M. Lowry, 'CXLIII: Studies of the Terpenes and Allied Compounds. The Sulphonation of Camphor. ii. β-Bromocamphor and its Derivatives. β-Bromocamphoric Acid', *Journal of the Chemical Society, Transactions*, LXXXI (1902), pp. 1462–8, mentions mixed melting points on pp. 1463 and 1467.
87 F. W. Streatfeild, *Practical Organic Chemistry* (London, 1891), p. 6.
88 Not Schulze as often stated, including in my original entry, see R. Anschütz and G. Schultz, 'Ueber einen einfachen Apparat zue

bequemen Bestimmung hochliegender Schmelzpunkte', *Berichte der Deutschen chemischen Gesellschaft*, X (1877), pp. 1800–1802.
89 C. F. Roth, 'Ein neuer Apparat zur Bestimmung von Schmelzpunkten', *Berichte der Deutschen chemischen Gesellschaft*, XIX (1886), pp. 1970–73.
90 F. W. Streatfeild and J. Davies, 'An Improved Melting-Point Apparatus', *Chemical News*, LXXXIII (1901), p. 121.
91 J. Thiele, 'Ein neuer Apparat zur Schmelzpunktsbestimmung', *Berichte der Deutschen chemischen Gesellschaft*, XL (1907), pp. 996–7; 'Thiele Tube', http://en.wikipedia.org, accessed 24 December 2013.
92 For the supply of chemicals to laboratories by pharmacists and apothecaries, see U. Klein, 'Chemical and Pharmaceutical Laboratories before the Professionalization of Chemistry', in *Spaces and Collections*, ed. Lourenço and Carneiro, pp. 3–12, on p. 7.
93 Private communication from William Jensen dated 28 January 2013.
94 Private communication from William Griffith, dated 19 February 2013.
95 This was the case at Lehigh in the early twentieth century according to ACS National Historic Chemical Landmarks, 'The Chandler Chemistry Laboratory', 1994, available at http://portal.acs.org.
96 See, for example, Fine, 'The Chandler Chemical Museum', p. 21.
97 H. S. Coleman, ed., *Laboratory Design* (New York, 1951), p. 88.
98 Private communication from Vera Mainz, dated 12 August 2013, citing *University of Illinois Department of Chemistry 1941–1951*, p. 23 and the library's annual report for that year.
99 Private communications from Sylvia Draper, dated 22 October 2013, and from David Grayson, dated 22 October and 6 December 2013.
100 Private communication from Andrew Alexander, dated 22 March 2013.
101 Private communication from Annette Lykknes, dated 7 December 2012.
102 Private communication from Richard Laursen, dated 29 July 2013.
103 Sigma-Aldrich, 'Our History', www.sigmaaldrich.com, accessed 8 November 2012.
104 J. V. Pickstone, *Ways of Knowing: A New History of Science, Technology, and Medicine* (Chicago, IL, and Manchester, 2001), pp. 73–82.

9 Cradles of Innovation: Carl Duisberg and Elberfeld, 1890s

1 J. J. Beer, *Emergence of the German Dye Industry* (Urbana, IL, 1959).
2 G. Meyer-Thurow, 'The Industrialization of Invention: A Case Study from the German Chemical Industry', *Isis*, LXXIII (1982), pp. 363–81; E. Homburg, 'The Emergence of Research Laboratories in the Dyestuffs Industry, 1870–1900', *British Journal for the History of Science*, XXV (1992), pp. 91–111.
3 W. B. Carlson, 'Building Thomas Edison's Laboratory at West Orange, New Jersey: A Case Study in Using Craft Knowledge for Technological Invention 1886–1888', *History of Technology*, XIII (1991), pp. 150–67. For more general works on the history of industrial research, see D. A. Hounshell, 'The Evolution of Industrial Research in the United States', in *Engines of Innovation: U.S. Industrial Research at the End of an Era*,

ed. R. S. Rosenbloom and W. J. Spencer (Boston, MA, 1996), pp. 51–6; L. S. Reich, *The Making of American Industrial Research: Science and Business at GE and Bell, 1876–1926* (Cambridge, 1985); G. Wise, *Willis R. Whitney, General Electric, and the Origins of U.S. Industrial Research* (New York, 1985).

4 P. Hunting, *A History of the Society of Apothecaries* (London, 1998), pp. 86 and 154–60.

5 S. Jacob, *Chemische Vor-und Fruhindustrie in Franken* (Dusseldorf, 1968).

6 P.J.T. Morris and C. A. Russell, *Archives of the British Chemical Industry, 1750–1914* (Stanford in the Vale, 1988), p. 13.

7 Ibid., p. 84.

8 'A Day at a Soap and Candle Factory', *Penny Magazine*, XI (8 January 1842), pp. 41–8, on p. 42. GB

9 J. Donnelly, 'Consultants, Managers, Testing Slaves: Changing Roles for Chemists in the British Alkali Industry, 1850–1920', *Technology and Culture*, XXXV (1994), pp. 100–128, on pp. 104–5.

10 For Crewe, see C. A. Russell and J. A. Hudson, *Early Railway Chemistry and its Legacy* (Cambridge, 2012), pp. 73–9. For Burton-on-Trent, see A. W. von Hofmann, 'Zur Erinnerung an Peter Griess', *Berichte der Deutschen Chemischen Gesellschaft*, XXIV (1891), pp. 1007–57, on pp. 1039–47.

11 'A. Gélis', Entry 198, in *Exposition Universelle de 1862 à Londres, Section Française, Catalogue Officiel* (Paris, 1862), p. 17; *Proceedings of the American Pharmaceutical Association at its Eleventh Annual Meeting* (1863), p. 144. Both GB

12 Homburg, 'Emergence of Research Laboratories', p. 96, citing the *Annuaire du commerce. Didot-Bottin* (Paris, 1865).

13 P. R. Reed, 'Acid Towers and the Control of Chemical Pollution 1823–1876', *Transactions of the Newcomen Society*, LXXVIII (2008), pp. 99–126.

14 R. G. Anderson, *Brewers and Distillers by Profession* (London, 2012), p. 23 and picture, p. 24.

15 R. G. Anderson, 'The Sword and the Armour: Science and Practice in the Brewing Industry 1837–1914', *Brewery History*, CXXIII (2006), pp. 55–83 and 76–7. I wish to thank Ray Anderson for his help with the brewery laboratory.

16 Homburg, 'The Emergence of Research Laboratories in the Dyestuffs Industry, 1870–1900', p. 111.

17 P.J.T. Morris, 'The Chemical Industry: An Introduction' and 'Chemical Industry Before 1850', in *Oxford Encyclopedia of Economic History*, ed. Joel Mokyr, vol. 1 (Oxford, 2003), pp. 392–8.

18 A. S. Travis, *The Rainbow Makers: The Origins of the Synthetic Dyestuffs Industry in Western Europe* (Bethlehem, PA, and London, 1993).

19 U. Klein, 'Apothecary's Shops, Laboratories and Chemical Manufacture in Eighteenth-century Germany', in *The Mindful Hand: Inquiry and Invention from the Late Renaissance to Early Industrialization*, ed. L. Roberts, S. Schaffer and P. Dear (Amsterdam, 2007), pp. 247–76.

20 This section is largely based on A. Simmons, 'Stills, Status, Stocks and Science: The Laboratories at Apothecaries' Hall in the Nineteenth Century', *Ambix*, LXI (2014), pp. 141–61. For the early history of the laboratory and its operators, see Hunting, *History of the Society of Apothecaries*, pp. 154–63.
21 'Apothecaries', *Household Words*, XIV (16 August 1856), pp. 108–15, on p. 114. GB
22 Brande, 'Apothecaries' Hall', p. 199.
23 Ibid., p. 200.
24 S. M. Horrocks, 'World War II, Post-war Reconstruction and British Women Chemists', *Ambix*, LVIII (2011), pp. 150–70.
25 'Carl Duisberg', www.bayer.com, accessed 7 May 2013; H. J. Flechtner, *Carl Duisberg vom Chemiker zum Wirtschaftsführer* (Düsseldorf, 1961); E. Verg, *Milestones: The Bayer Story, 1863–1988* (Leverkusen, 1988), passim.
26 A. Nieberding, *Unternehmenskultur im Kaiserreich: J. M. Voith und die Farbenfabriken vorm. Friedr. Bayer & Co.* (Munich, 2003), p. 73.
27 C. Reinhardt and A. S. Travis, *Heinrich Caro and the Creation of the Modern Chemical Industry* (Dordrecht, 2000), pp. 275–80; H. van den Belt and A. Rip, 'The Nelson-Winter-Dosi Model and Synthetic Dye Chemistry', in W. E. Bijker, T. P. Hughes and T. Pinch, *The Social Construction of Technological Systems* (Cambridge, MA, 1987), pp. 135–58; Verg, *Milestones*, pp. 74–9.
28 Verg, *Milestones*, p. 117.
29 P. Hayes, *Industry and Ideology: IG Farben in the Nazi Era*, 2nd edn (Cambridge, 2001).
30 Meyer-Thurow, 'Industrialization of Invention', p. 370.
31 G. Benz, R. Hahn and C. Reinhardt, *100 Jahre Chemisch-wissenschaftliches Laboratorium der Bayer AG in Wuppertal-Elberfeld, 1896–1996* (Leverkusen, 1996), pp. 24–5.
32 E. M. Tansey, 'The Wellcome Physiological Research Laboratories 1894–1904: The Home Office, Pharmaceutical Firms, and Animal Experiments', *Medical History*, XXX (1989), pp. 1–41.
33 'The Wellcome Research Laboratories', *The Nursing Record and Hospital World*, (30 June 1900), pp. 522–4, available at http://rcnarchive.rcn.org.uk.
34 A woman passing Edinburgh University saw two men in white coats leaving in a hurry. She asked, 'Are you students coming out as doctors?' They replied, 'No, we are painters coming out for a pint [of beer].' Painters in this context are of course decorators, not artists.
35 M. S. Hochberg, 'The Doctor's White Coat – An Historical Perspective', *Virtual Mentor*, IX/4 (2007), p. 310; M. C. Flannery, 'Dressing in Style? An Essay on the Lab Coat', *The American Biology Teacher*, LXI/5 (May, 1999), pp. 380–83. I was able to see the Agnew Clinic painting at the Philadelphia Art Museum on 5 October 2013.
36 'Discussion – Dr Wright's Address', *Proceedings of the Third Annual Meeting of the American Association of Dental Schools* (1926), p. 177. I wish to thank the library of the Radboud University of Nijmegen for a scan of this page.

37 Private communication from Alan Dronsfield, dated 13 March 2013.
38 J. H. Schweppe, *Research aan het IJ: LBPMA 1914–KSLA 1989: de geschiedenis van het 'Lab Amsterdam'* (Baarn, Netherlands [1989]), lower illustration on p. 125.
39 Ibid., illustration on p. 134; information about the instruments supplied by Joan van der Waals via Ernst Homburg.
40 Private communication from Edwin Becker, dated 16 December 2013.
41 P.J.T. Morris, 'Regional Styles in Pesticide Analysis: Coulson, Lovelock and the Detection of Organochlorine Insecticides', in *Illuminating Instruments*, ed. P.J.T. Morris and K. Staubermann (Washington, DC, 2009), pp. 55–72.
42 J. E. Lovelock, *Homage to Gaia: The Life of an Independent Scientist* (Oxford, 2000), p. 151.
43 Y. M. Rabkin, 'Technological Innovation in Science: The Adoption of Infrared Spectroscopy by Chemists', *Isis*, LXXVIII (1987), pp. 31–54, reprinted in *From Classical to Modern Chemistry: The Instrumental Revolution*, ed. P.J.T. Morris (Cambridge, 2002), pp. 3–28.
44 L. S. Ettre, 'Gas Chromatography', in *A History of Analytical Chemistry*, ed. H. A. Laitinen and G. W. Ewing (Washington, DC, 1977), pp. 296–306; P.J.T. Morris, '"Parts Per Trillion is a Fairy Tale": The Development of the Electron Capture Detector and Its Impact on the Monitoring of DDT', in Morris, ed., *From Classical to Modern Chemistry*, pp. 259–84.
45 Essay by Greult Dijkstra, in *75 Years of Chromatography*, ed. L. S. Ettre and A. Zlatkis (Amsterdam, 1979), pp. 43–51. Additional information from Ernst Homburg.
46 A.I.M. Keulemans, *Gas Chromatography*, 2nd edn (New York, 1959), p. 99.
47 L. S. Ettre, *Chapters in the Evolution of Chromatography* (London, 2008), chap. 22.
48 Ibid., chap. 23.
49 Editors of Time-Life Books, *Life Search* (Alexandria, VA, 1989), pp. 15–28.

10 Neither Fish nor Fowl: Thomas Thorpe and London, 1890s

1 D. Cahan, *An Institute for an Empire: The Physikalisch-Technische Reichsanstalt 1871–1918* (Cambridge, 1989).
2 R. C. Cochrane, *Measures for Progress: A History of the National Bureau of Standards* (Washington, DC, 1966); Elio Passaglia, *A Unique Institution: The National Bureau of Standards, 1950 to 1969* (Washington, DC, 1999); J. F. Schooley, *Responding to National Needs: The National Bureau of Standards Becomes the National Institute of Standards and Technology, 1969–1993* (Washington, DC, 2000). All three volumes can be downloaded at www.nist.gov.
3 P. W. Hammond and H. Egan, *Weighed in the Balance: A History of the Laboratory of the Government Chemist* (London, 1992).

4 F. W. Accum, *A Treatise on Adulteration of Food and Culinary Poisons* (London, 1820); J. Burnett, *Plenty and Want: A Social History of Food in England from 1815 to the Present Day*, 3rd edn (London, 1989), chap. 5.
5 P. J. Rowlinson, 'Food Adulteration Its Control in 19th Century Britain', *Interdisciplinary Science Reviews*, VII/1 (1982), pp. 63–72; N. G. Coley, 'The Fight against Food Adulteration', *Education in Chemistry*, XLII/2 (2005), pp. 46–9.
6 For the determination of alcohol content, also see W. J. Ashworth, *Customs and Excise: Trade, Production, and Consumption in England, 1640–1845* (Oxford, 2003); W. J. Ashworth, '"Between the Trader and the Public:" British Alcohol Standards and the Proof of Good Governance', *Technology and Culture*, XLII (2001), pp. 27–50.
7 Hammond and Egan, *Weighed in the Balance*, p. 103.
8 *The Civilian* (29 May 1897), p. 1.
9 For the construction of Blythe House, see File work 12/145, 'Records of the Successive Works Departments, and the Ancient Monuments Boards and Inspectorate, 01 January 1892–31 December 1898', The National Archives, Kew. I wish to thank Eduard von Fischer for his research on this file.
10 John Mills, 'The Government Laboratory', *Strand Magazine*, XXI (1902), pp. 561–71.
11 P. W. Hammond, 'Thorpe, Sir Thomas Edward (1845–1925)', *Oxford Dictionary of National Biography* (Oxford, 2004), online edn at oxforddnb.com; A.E.H.T., 'Obituary Notices of Fellows Deceased', *Proceedings of the Royal Society of London. Series A*, CIX/752 (1925), pp. xviii–xxiv.
12 T. E. Thorpe, *A Dictionary of Applied Chemistry*, vol. 1 (London, 1890), and thereafter in many volumes and several editions.
13 T. E. Thorpe, *History of Chemistry*, 2 vols (London, 1909–10).
14 Mills, 'The Government Laboratory', p. 565.
15 Ashworth, 'British Alcohol Standards'.
16 Hammond and Egan, *Weighed in the Balance*, pp. 152–5.
17 Mills, 'The Government Laboratory', p. 565.
18 Ibid., p. 566.
19 Ibid., p. 570.
20 Mills, 'The Government Laboratory', p. 570.
21 Peter Atkins, *Liquid Materialities: A History of Milk, Science and the Law* (Farnham, 2010).
22 Mills, 'The Government Laboratory', p. 567.
23 Ibid.
24 J. Bell, *The Analysis and Adulteration of Foods* (London, 1881).
25 Mills, 'Government Laboratory', pp. 562–3.
26 Hammond and Egan, *Weighed in the Balance*, pp. 22–4.
27 Ibid., pp. 45–6.
28 Ibid., p. 151. Also see W. A. Campbell, 'Some Landmarks in the History of Arsenic Testing', *Chemistry in Britain*, I (1965), pp. 198–202.
29 Mills, 'The Government Laboratory', pp. 564–5.

30 J. Woodward, 'The New Government Laboratories', *Nature*, LVI (1897), pp. 553–4, quotation on p. 553, reprinted in Hammond and Egan, *Weighed in the Balance*, pp. 297–304, quotation on p. 297.
31 Mills, 'The Government Laboratory', p. 565.
32 Ibid., p. 562.
33 H. W. Paul, *From Knowledge to Power: The Rise of the Science Empire In France, 1860–1939* (Cambridge, 1985), pp. 213–19; A. Stanziani and P. J. Atkins, 'From Laboratory Expertise to Litigation: The Municipal Laboratory of Paris and the Inland Revenue Laboratory in London, 1870–1914. A Comparative Analysis', in *Fields of Expertise: A Comparative History of Expert Procedures in Paris and London, 1600 to Present*, ed. Christelle Rabier (Newcastle upon Tyne, 2007), pp. 317–38, available at www.dro.dur.ac.uk. I am grateful to Sacha Tomic for his help with this section.
34 W. D. Hogg, 'On the Work Done by the Paris Municipal Laboratory', *Analyst*, VIII (1883), pp. 41–6. 'Paris Municipal Laboratory', *Analyst*, IX (1884), pp. 80–81. Also see S. Tomic and X. Guillem, 'New Sites for Food Quality Surveillance in European Centres and Peripheries. To What Extent was the Municipal Chemical Laboratory of Paris a Model for Iberian Laboratories?', paper given at the 'Sites of Chemistry in the 19th Century' conference, Valencia, Spain, 6 July 2012.
35 Stanziani and Atkins, 'From Laboratory Expertise to Litigation', pp. 5–6.
36 Paul, *From Knowledge to Power*, p. 219.

11 Chemistry in Silicon Valley: Bill Johnson and Stanford, 1960s

1 P.J.T. Morris and A. S. Travis, 'The Role of Physical Instrumentation in Structural Organic Chemistry in the Twentieth Century', in *From Classical to Modern Chemistry: The Instrumental Revolution*, ed. P.J.T. Morris (Cambridge, 2002), pp. 57–84, and the other papers in the same volume.
2 R. B. Woodward, 'Art and Science in the Synthesis of Organic Compounds', in *Pointers and Pathways in Research*, ed. Maeve O'Connor (Bombay, 1963), pp. 23–41, on p. 25.
3 W. S. Johnson, *A Fifty-year Love Affair with Organic Chemistry* (Washington, DC, 1998); G. Stork, 'William Summer Johnson (1913–1995)', *Biographical Memoirs of the National Academy of Sciences*, LXXX (2001), pp. 1–17, available at www.nasonline.org; H. Taube, J. I. Brauman and H. McConnell, 'Memorial Resolution: William Summer Johnson' [1995] at http://histsoc.stanford.edu.
4 C. S. Gilmor, *Fred Terman at Stanford: Building a Discipline, a University, and Silicon Valley* (Stanford, CA, 2004); O. G. Villard, Jnr, 'Frederick Emmons Terman (1900–1982)', *Biographical Memoirs of the National Academy of Sciences*, LXXX (1998), pp. 1–24, available at www.nasonline.org; R. S. Lowen, *Creating the Cold War University: The Transformation of Stanford* (Berkeley, CA, 1997).

5 S. W. Leslie and R. H. Kargon, 'Selling Silicon Valley: Frederick Terman's Model for Regional Advantage', *Business History Review*, LXX (1996), pp. 435–72.
6 C. Djerassi, *The Pill, Pygmy Chimps and Degas' Horse: The Autobiography of Carl Djerassi* (New York, 1992), p. 96.
7 W. Winslow, *Varian 50 Years: Fifty Years of Innovative Excellence* (Palo Alto, CA, 1998), pp. 53–4.
8 E. Hutchinson, *The Department of Chemistry Stanford University, 1891–1976* (Stanford, CA, 1977), pp. 28–32; 'Stanford's Chemistry Build-Up Signals New Era', *Chemical and Engineering News*, XXXIX/10 (6 March 1961), pp. 44–5; Gilmor, *Fred Terman at Stanford*, pp. 388–400; C. Reinhardt, *Shifting and Rearranging: Physical Methods and the Transformation of Modern Chemistry* (Sagamore Beach, MA, 2006), pp. 149–57; Lowen, *Creating the Cold War University*, pp. 101–2 and 188–9; Johnson, *Fifty Year Love Affair*, pp. 83–98; C. Djerassi, *Steroids Made it Possible* (Washington, DC, 1990), pp. 66–7; Djerassi, *The Pill, Pygmy Chimps, and Degas' Horse*, pp. 95–7 and 100–102.
9 R. Kushman, 'Birge Clark: The Man behind the Blueprints', *Palo Alto Centennial* (13 April 1994), available at www.paloaltoonline.com.
10 Hutchinson, *Chemistry Department*, p. 30, emphasis in original.
11 Reinhardt, *Shifting and Rearranging*, p. 151. I wish to thank Rolf Tschudin for his help with the identification of this NMR spectrometer.
12 Djerassi, *The Pill, Pygmy Chimps, and Degas' Horse*, p. 100.
13 R. L. Geiger, *Research and Relevant Knowledge: American Research Universities since World War II* (New York, 1993).
14 R. L. Geiger, 'Science, Universities, and National Defense, 1945–1970', *Osiris*, second series, VII, *Science after '40* (1992), pp. 26–48.
15 'History of the ACS Petroleum Research Fund', http://portal.acs.org, accessed 21 February 2013. I am grateful to Raymond Bonnett for pointing out the importance of this funding stream for academic organic chemistry.
16 R. S. Lowen, *Creating the Cold War University*; S. W. Leslie, *The Cold War and American Science: The Military-industrial-academic Complex at MIT and Stanford* (New York, 1994). Sadly, neither of these books devotes any space to chemistry.
17 Surprisingly, the impact on academic research (as opposed to technical education) has not received much attention, but see Geiger, *Research and Relevant Knowledge*, pp. 161–73.
18 'Federal Obligations for Total Research and Development, by Major Agency and Performer: Fiscal Years 1951–2001', 5 June 2013, www.nsf.gov.
19 H. D. Graham and N. Diamond, *The Rise of American Research Universities: Elites and Challengers in the Postwar Era* (Baltimore, MD, 1997).
20 Ibid., pp. 27 and 34.
21 Reinhardt, *Shifting and Rearranging*, pp. 198–209.
22 C. Lécuyer, *Making Silicon Valley: Innovation and the Growth of High Tech, 1930–1970* (Cambridge, MA, 2006); Gilmor, *Fred Terman at*

Stanford; A. W. Thackray and M. Myers, Jnr, *Arnold O. Beckman: One Hundred Years of Excellence* (Philadelphia, PA, 2000), chap. 7 'Visions of Technology', pp. 236–77.
23 Mr McGuire: 'I just want to say one word to you. Just one word ... Plastics'.
24 A. Rocke, *Image and Reality: Kekulé, Kopp, and the Scientific Imagination* (Chicago, IL, 2010), especially chap. 8.
25 For a discussion of this classical approach, see L. Slater, 'Woodward, Robinson, and Strychnine: Chemical Structure and Chemists' Challenge', *Ambix*, XLVIII (2001), pp. 161–89.
26 R. Bentley, 'The Molecular Structure of Penicillin', *Journal of Chemical Education*, LXXXI (2004), pp. 1462–70.
27 W. B. Jensen, *Philosophers of Fire* (Cincinnati, OH, 2003), pp. 236–8, and esp. pls V and VI. This e-book can be downloaded at www.che.uc.edu.
28 For further details, see R. J. Abraham, J. Fisher and P. Loftus, *Introduction to NMR Spectroscopy* (Chichester, 1988).
29 You may be wondering if there is any connection between NMR and the magnetic resonance imaging used in medicine. Effectively they are the same technique. MRI simply puts a human body (or a part of it) in a very large NMR spectrometer.
30 For the history of chemical NMR, see C. Reinhardt, *Shifting and Rearranging*. Also see *Encyclopedia of Nuclear Magnetic Resonance*, vol. I, *Historical Perspectives*, ed. D. M. Grant and R. K. Harris (Chichester, 1996), especially E. D. Becker, C. L. Fisk and C. L. Khetrapal, 'The Development of NMR', pp. 1–158, available at http://onlinelibrary.wiley.com. I am grateful to Edwin Becker for his help with this section.
31 *Early History of Varian Associates* (Palo Alto, CA, n.d.), available at www.cpii.com; Winslow, *Varian 50 Years*; T. Lenoir and C. Lécuyer, 'Instrument Makers and Discipline Builders: The Case of Nuclear Magnetic Resonance', *Perspectives on Science*, III (1995), pp. 276–345. Also see Lécuyer, *Making Silicon Valley*.
32 John D. Roberts, *The Right Place at The Right Time* (Washington, DC, 1990), p. 152; E. von Hippel, *The Sources of Innovation* (New York and Oxford, 1988), p. 143, available at http://web.mit.edu. Also see R. C. Ferguson, 'William D. Phillips and Nuclear Magnetic Resonance Spectroscopy at Du Pont', in *Encyclopedia of Nuclear Magnetic Resonance*, vol. I, ed. Grant and Harris, pp. 309–13.
33 Roberts, *The Right Place at The Right Time*, p. 151.
34 For the various models introduced by Varian in the 1950s and early '60s, see J. N. Shoolery, 'High-Resolution NMR: A Dream Come True', in *Encyclopedia of Nuclear Magnetic Resonance*, vol. I, ed. Grant and Harris, pp. 627–34.
35 In addition to Reinhardt, *Shifting and Rearranging*, chap. 6, see R. R. Ernst, 'Zurich's Contributions to 50 Years Development of Bruker', *Angewandte Chemie International Edition*, XLIX (2010), pp. 8310–15, available at www.ncbi.nlm.nih.gov.
36 Reinhardt, *Shifting and Rearranging*; C. Reinhardt, 'The Chemistry of an Instrument: Mass Spectrometry and Structural Organic Chemistry',

in *From Classical to Modern Chemistry*, ed. P.J.T. Morris (Cambridge, 2002), pp. 229–47.
37 G. P. Arsenault, et al., 'The Structure of Antheridiol. A Sex Hormone in *Achlya bisexualis*', *Journal of the American Chemical Society*, IC (1968), pp. 5635–6.

12 Innovation on the Isis: Graham Richards and Oxford, 2000s

1 Private communication from Graham Richards dated 1 July 2013. Also see the chapters on converting older laboratories and moving staff to a new or refurbished facility in *Design, Construction and Refurbishment of Laboratories*, vol. II, ed. R. Lees (Chichester, 1993).
2 J. A. Young, ed., *Improving Safety in the Chemical Laboratory: A Practical Guide* (New York and Chichester, 1987); p. 43 mentions that insufficient draught was one of the safety hazards in the laboratory as late as the 1980s.
3 H. S. Coleman, ed., *Laboratory Design* (New York, 1951), p. 99.
4 H. F. Lewis, ed., *Laboratory Planning for Chemistry and Chemical Engineering* (New York and London, 1962), p. 371.
5 S. J. Rosenlund, *The Chemical Laboratory: Its Design and Operation* (Park Ridge, NJ, 1987), pp. 6–9; W. R. Ferguson, *Practical Laboratory Planning* (Barking, 1973), pp. 5 and 23.
6 Ferguson, *Practical Laboratory Planning*, p. 17; also see Rosenlund, *The Chemical Laboratory*, p. 21.
7 Coleman, ed., *Laboratory Design*, pp. 70–71; Rosenlund, *The Chemical Laboratory*, pp. 43–4 and 45-7; Lewis, ed., *Laboratory Planning*, pp. 85–6; Young, ed., *Improving Safety in the Chemical Laboratory*, pp. 44–5; N. V. Steere, ed., *Safety in the Chemical Laboratory* (Easton, PA, 1967), pp. 72–5 and 113; R. Lees and A. F. Smith, eds, *Design, Construction and Refurbishment of Laboratories* (Chichester, 1984), chap. 29.
8 Coleman, ed., *Laboratory Design*, p. 107.
9 Ibid., pp. 73 and 107; Rosenlund, *The Chemical Laboratory*, chap. 4; Young, ed., *Improving Safety in the Chemical Laboratory*, passim; Steere, ed., *Safety in the Chemical Laboratory*, pp. 112–5.
10 Young, ed., *Improving Safety in the Chemical Laboratory*, pp. 42–4.
11 But see Coleman, ed., *Laboratory Design*, pp. 72–3, which incorporates safety into laboratory design to a certain extent.
12 *The Architect's Journal*, CLII (1980), p. 543. There was one earlier reference to the C-frame in *Laboratory Practice*, XXVIII (1979), p. 643. Both GB. The dating of the C-frame to the 1960s is in D. D. Watch, *Building Type Basics for Research Laboratories*, 2nd edn (Hoboken, NJ, 2008), p. 155.
13 Ferguson, *Practical Laboratory Planning*, p. 39.
14 Watch, *Research Laboratories*, p. 155.
15 Lees and Smith, eds, *Design, Construction and Refurbishment of Laboratories*, especially chap. 3, and Lees, ed., *Design, Construction and Refurbishment of Laboratories*, vol. II, chap. 1; Watch, *Research Laboratories*, pp. 123–4.

16 For an overview of the history of important synthetic targets, see K. C. Nicolaou and T. Montagnon, *Molecules That Changed the World: A Brief History of the Art and Science of Synthesis and its Impact on Society* (Weinheim, 2008).
17 R.J.P. Williams, J. S. Rowlinson and A. Chapman, eds, *Chemistry at Oxford: A History from 1600 to 2005* (Cambridge, 2009).
18 For the history of the college laboratories, also see H. B. Hartley, *Studies in the History of Chemistry* (Oxford, 1971), chap. 10.
19 J. Jones, et al., *The Dyson Perrins Laboratory and Oxford Organic Chemistry, 1916–2004* (Oxford, 2008); J. C. Smith, *The Development of Organic Chemistry at Oxford* (Oxford, 1975).
20 J. Morrell, 'W. H. Perkin, Jr., at Manchester and Oxford: From Irwell to Isis', *Osiris*, 2nd Series, VIII, *Research Schools: Historical Reappraisals* (1993), pp. 104–26, on pp. 108–10 and 118–19.
21 'Inorganic Chemistry Laboratory: Conservation Plan. Building No. 168', April 2012, pp. 13–19, available at www.admin.ox.ac.uk; R.J.P. Williams, 'Recent Times, 1945–2005: A School of World Renown', in *Chemistry at Oxford*, ed. Williams, Rowlinson and Chapman, p. 279.
22 D. A. Hounshell and J. K. Smith, Jr, *Science and Corporate Strategy: Du Pont R & D, 1902–1980* (Cambridge, 1988), pp. 297–8.
23 H. Wiesendanger, 'A History of OTL: Overview' (2000), http://otl.stanford.edu.
24 G. Richards, *50 Years at Oxford* (Bloomington, IN, 2011), pp. 40–44. I am grateful for the assistance of Tom Hockaday of Isis Innovation with this paragraph.
25 R. Dwek, 'Glycobiology at Oxford: A Personal View', *The Biochemist*, XXVIII/3 (2006), pp. 4–7, on pp. 5–6.
26 Richards, *50 Years at Oxford*, pp. 45–51.
27 H. L. Smith and S. Bagchi-Sen, 'The Research University, Entrepreneurship and Regional Development: Research Propositions and Current Evidence', *Entrepreneurship & Regional Development*, XXIV (2012), pp. 383–404, on pp. 393–4.
28 'Isis Angels Network', www.isis-innovation.com, accessed 22 November 2013.
29 'Spin-out Companies', www.isis-innovation.com, 2013.
30 Richards, *50 Years at Oxford*, chap. 6; 'Chemistry Research Laboratory', www.chem.ox.ac.uk, accessed 25 June 2013.
31 *Chemistry at Oxford*, p. 280.
32 'New Research Laboratory for Chemistry', www.chem.ox.ac.uk, accessed 25 June 2013.
33 'Our History', 2013, www.rmjm.com; 'RMJM', http://en.wikipedia.org, accessed 5 December 2013.
34 'Portfolio', 2013, www.rmjm.com.
35 For a discussion of the design of another modern laboratory, the Lewis Thomas Laboratory for Molecular Biology at Princeton University, by the architect, see J. Collins, Jnr, 'The Design Process for the Human Workplace', in *The Architecture of Science*, ed. P. Galison and E. Thompson (Cambridge, MA, 1999), pp. 399–412.

36 Private communication from John M. Brown, dated 15 July 2013.
37 Based on a visit to the Chemistry Research Laboratory on 22 May 2013, kindly arranged by Karl Harrison.
38 J. Itzhaki, 'First Alumni Event Proves a Great Success', 10 July 2009, www.bioch.ox.ac.uk.
39 For the use of atria to encourage peer interaction at the Center for Advanced Biotechnology and Medicine at Piscataway, New Jersey, see T. F. Gieryn, 'Two Faces in Science: Building Identities for Molecular Biology and Biotechnology', in *The Architecture of Science*, ed. P. Galison and E. Thompson, pp. 423–55, on p. 445.
40 For the overriding importance of safety in the Center for Advanced Biotechnology and Medicine at Piscataway, see T. F. Gieryn, 'Two Faces in Science', p. 443–4.
41 Lees, ed., *Design, Construction and Refurbishment of Laboratories*, vol. II, pp. 17–19.
42 'Research Facilities – Department of Chemistry', www.chem.ox.ac.uk, accessed 25 June 2013.
43 'Nuclear Magnetic Resonance Spectroscopy Facility', www.chem.ox.ac.uk, accessed 25 June 2013.
44 'Mass Spectrometry Research Facility', www.chem.ox.ac.uk, accessed 25 June 2013.
45 'Professor Dame Carol Robinson DBE FRCS: Research', www.chem.ox.ac.uk, accessed 1 July 2013.
46 'Chemical Crystallography: Service', www.chem.ox.ac.uk, accessed 25 June 2013.
47 'Professor M. Brouard: Research', www.chem.ox.ac.uk, accessed 25 June 2013.
48 G. Richards, *50 Years at Oxford* and personal information from Graham Richards. Also see D. Hague, 'Professor Graham Richards, Oxford Molecular and Inhibox', December 2003, available at www.sbs.ox.ac.uk.
49 Direct grant schools were fee-paying selective schools which between 1945 and 1976 were partly funded by the state (hence direct grant) in return for giving free places to a significant number of bright pupils whose parents were unable to afford the fees.
50 G. Richards, 'Conservatism and Science', in *The Conservative Opportunity*, ed. R.N.W. Blake and J.H.C. Patten (London, 1976), pp. 132–8.
51 L. S. Ettre, *Chapters in the Evolution of Chromatography* (London, 2008), chaps 18, 26–28, and also chap. 32; L. S. Ettre and A. Zlatkis, eds, *75 years of Chromatography: A Historical Dialogue* (Amsterdam, 1979), essays by C. Horváth (pp. 151–8), J.F.K. Huber (pp. 159–66), J. J. Kirkland (pp. 209–18), S. Moore and W. H. Stein (pp. 297–308) and L. R. Snyder (pp. 419–24). For the technique see L. R. Snyder, J. J. Kirkland and J. W. Dolan, *Introduction to Modern Liquid Chromatography*, 3rd edn (Oxford, 2011). I am very grateful to Apostolos Gerontas for his assistance with this section.
52 P. D. McDonald, 'Waters Corporation: Fifty Years of Innovation in Analysis and Purification', *Chemical Heritage* (Summer 2008), pp. 32–7, available at www.chemheritage.org.

53 M. A. Grayson, 'John Bennett Fenn: A Curious Road to the Prize', *Journal of The American Society for Mass Spectrometry*, XXII (2011), pp. 1301–8.
54 L. Beyer and B. Arndt, *Chemie an der Universität Leipzig: von den Anfängen bis zur Gegenwart* (Leipzig, 2009), pp. 291–302, illustration no. 252 on p. 299.
55 'Bristol ChemLabS Teaching Laboratories' and 'Overview and History', www.chemlabs.bris.ac.uk, accessed 12 December 2013 and verbal communication from John Brown, on 11 December 2013.

Conclusion

1 For my argument about the role of competition in the American synthetic rubber industry, see P.J.T. Morris, *The American Synthetic Rubber Research Program* (Philadelphia, PA, 1989), pp. 50–59 and 142.
2 See, for example, R.G.W. Anderson, 'Chemical Laboratories, and How They Might Be Studied', *Studies in History and Philosophy of Science Part A*, XLIV (2013), pp. 669–75, on p. 671.
3 For example, see Y. Kikuchi, *Anglo-American Connections in Japanese Chemistry: The Lab as Contact Zone* (New York, 2013).
4 T. A. Markus, *Buildings and Power: Freedom and Control in the Origin of Modern Building Types* (London and New York, 1993).
5 P.J.T. Morris, 'Chemistry in the 21st Century: Death or Transformation?', presentation given at the 6th International Conference on the History of Chemistry, 'Neighbours and Territories: The Evolving Identity of Chemistry' (Leuven) 30 August 2007, available at www.euchems.eu.
6 For Robert Burns Woodward's anxiety about the possible impact of mechanized synthesis on the creative aspects of organic chemistry, see R. B. Woodward, 'Art and Science in the Synthesis of Organic Compounds: Retrospect and Prospect', in *Pointers and Pathways in Research*, ed. Maeve O'Connor (Bombay, 1963), pp. 23–41, on p. 41.
7 Ibid., pp. 28 and 41. Also see R. B. Woodward, 'Synthesis', in *Perspectives in Organic Chemistry*, ed. A. R. Todd (New York and London, 1956), pp. 155–84, especially p. 180.

SELECT BIBLIOGRAPHY

This select bibliography only contains books and articles that are of direct relevance to the history of chemical laboratories, not biographical material or items relating to instrumentation unless they are also relevant to laboratories. It does not include papers in books that are already listed.

Anderson, Robert G. W., 'Chemical Laboratories, and How They Might Be Studied', *Studies in History and Philosophy of Science Part A*, XLIV (2013), pp. 669–75
Benz, Günter, Ralf Hahn and Carsten Reinhardt, *100 Jahre Chemisch-wissenschaftliches Laboratorium der Bayer AG in Wuppertal-Elberfeld, 1896–1996*
Beretta, Marco, ed., *Lavoisier in Perspective* (Munich, 2005)
——, 'Imaging the Experiments on Respiration and Transpiration of Lavoisier and Séguin: Two Unknown Drawings by Madame Lavoisier', *Nuncius*, XXVII (2012), pp. 163–91
Brock, William H., *Justus von Liebig: The Chemical Gatekeeper* (Cambridge, 1997)
——, 'The Chemical Origins of Practical Physics', *Bulletin for the History of Chemistry*, XXI (1998), pp. 1–11, available at www.illinois.edu
Carroll, P. Thomas, 'Academic Chemistry in America, 1876–1976: Diversification, Growth, and Change', PhD thesis, University of Pennsylvania, 1982
Chandler, William H., *The Construction of Chemical Laboratories* (Washington, DC, 1893), I used the good-quality reprint on demand by Kessinger Publishing
Chilton, Donovan, and Noel G. Coley, 'The Laboratories of the Royal Institution in the Nineteenth Century', *Ambix*, XXVII (1980), pp. 173–203
Coleman, Harry S., ed., *Laboratory Design* (New York, 1951)
Eklund, Jon, *The Incompleat Chymist: Being an Essay on the Eighteenth-Century Chemist in his Laboratory, with a Dictionary of Obsolete Chemical Terms of the Period* (Washington, DC, 1975), available at http://si-pddr.si.edu
Ercker, Lazarus, *Lazarus Ercker's Treatise on Ores and Assaying*, trans. from the German edition of 1580 by A. G. Sisco and C. S. Smith (Chicago, IL, 1951)
Ferguson, William R., *Practical Laboratory Planning* (Barking, 1973)

Festing, Edward R., *Report of Visits to Chemical Laboratories at Bonn, Berlin, Leipzig, etc.* (London, 1871)
Fine, Leonard, 'The Chandler Chemical Museum', *Bulletin of the History of Chemistry*, II (1988), pp. 19–21, available at www.illinois.edu
Forgan, Sophie, 'The Architecture of Science and the Idea of a University', *Studies In History and Philosophy of Science, Part A*, XX (1989), pp. 405–34
——, and Graeme Gooday, '"A Fungoid Assemblage of Buildings": Diversity and Adversity in the Development of College Architecture and Scientific Education in Nineteenth-century South Kensington', *History of Universities*, XIII (1994), pp. 153–92
——, '"But Indifferently Lodged . . .": Perception and Place in Building for Science in Victorian London', in *Making Space for Science: Territorial Themes in the Shaping of Knowledge*, ed. C. Smith and J. Agar (Basingstoke, 1998), pp. 195–215
Fors, Hjalmar, *Mutual Favours: The Social and Scientific Practice of Eighteenth-century Swedish Chemistry* (Uppsala, 2003)
——, 'J. G. Wallerius and the Laboratory of the Enlightenment', in *Taking Place: The Spatial Contexts of Science, Technology, and Business*, ed. E. Baraldi, H. Fors and A. Houltz (Sagamore Beach, MA, 2006), pp. 3–33
Galison, Peter, and Emily Thompson, eds, *The Architecture of Science* (Cambridge, MA, 1999)
Gibbs, Frederick W., 'William Lewis, M.B., F.R.S. (1708–1781)', *Annals of Science*, VIII (1952), pp. 122–51
Gooday, Graeme, 'The Premises of Premises: Spatial Issues in the Historical Construction of Laboratory Credibility', in *Making Space for Science: Territorial Themes in the Shaping of Knowledge*, ed. C. Smith and J. Agar (Basingstoke, 1998), pp. 216–45
Hammond, Peter W., and Harold Egan, *Weighed in the Balance: A History of the Laboratory of the Government Chemist* (London, 1992)
Hannaway, Owen, 'Laboratory Design and the Aim of Science: Andreas Libavius versus Tycho Brahe', *Isis*, LXXVII (1986), pp. 584–610
Hill, C. R., 'The Iconography of the Laboratory', *Ambix*, XXII (1975), pp. 102–10
Hofmann, A. Wilhelm, *The Chemical Laboratories in Course of Erection in the Universities of Bonn and Berlin* (London, 1866), available in Google Books
Hogg, W. Douglas, 'On the Work Done by the Paris Municipal Laboratory', *Analyst*, VIII (1883), pp. 41–6
Holmes, Frederic L., *Eighteenth-century Chemistry as an Investigative Enterprise* (Berkeley, CA, 1989)
Homburg, Ernst, 'The Emergence of Research Laboratories in the Dyestuffs Industry, 1870–1900', *British Journal for the History of Science*, XXV (1992), pp. 91–111
——, *Van beroep 'Chemiker': De opkomst van de industriële chemicus en het polytechnische onderwijs in Duitsland (1790–1850)* (Delft, 1993)
——, 'The Rise of Analytical Chemistry and Its Consequences for the Development of the German Chemical Profession (1780–1860)', *Ambix*, XLVI (1999), pp. 1–32
Hunting, Penelope, *A History of the Society of Apothecaries* (London, 1998)

Hutchinson, Eric, *The Department of Chemistry Stanford University, 1891–1976* (Stanford, CA, 1977)
Jackson, Catherine M., 'Analysis and Synthesis in Nineteenth-century Organic Chemistry', PhD thesis, University College London, 2009
——, 'Chemistry as the Defining Science: Discipline and Training in Nineteenth-century Chemical Laboratories', *Endeavour*, XXXV/2 (2011), pp. 55–62
James, Frank A.J.L., ed., *The Development of the Laboratory: Essays on the Place of Experiment in Industrial Civilization* (Basingstoke, 1989)
——, 'Harriet Jane Moore, Michael Faraday, and Moore's Mid-nineteenth Century Watercolours of the Interior of the Royal Institution', in *Fields of Influence: Conjunctions of Artists and Scientists, 1815–1860*, ed. James Hamilton (Birmingham, 2001), pp. 111–28
——, and Anthony Peers, 'Constructing Space for Science at the Royal Institution of Great Britain', *Physics in Perspective*, IX (2007), pp. 130–85
Jensen, William B., 'To Demonstrate the Truths of "Chymistry"', *Bulletin of the History of Chemistry*, X (1991), pp. 3–15
Kikuchi, Yoshiyuki, *Anglo-American Connections in Japanese Chemistry: The Lab as Contact Zone* (New York and Basingstoke, 2013)
Klein, Ursula, 'Apothecary's Shops, Laboratories and Chemical Manufacture in Eighteenth-century Germany', in *The Mindful Hand: Inquiry and Invention from the Late Renaissance to Early Industrialization*, ed. L. Roberts, S. Schaffer and P. Dear (Amsterdam, 2007), pp. 247–76
——, 'The Laboratory Challenge: Some Revisions of the Standard View of Early Modern Experimentation', *Isis*, IC (2008), pp. 769–82
Kohler, Robert E., 'Lab History: Reflections', *Isis*, IC (2008), pp. 761–8
Krätz, Otto, 'Zur Geschichte des chemischen Laboratoriums', in *Historia scientiae naturalis: Beiträge zur Geschichte der Laboratoriumstechnik und deren Randgebiete*, ed. E.H.W. Giebeler and K. A. Rosenbauer (Darmstadt, 1982)
Lang, Hermann, *Das chemische Laboratorium an der Universität in Heidelberg* (Karlsruhe, 1858)
Lees, Ron, and A. F. Smith, eds, *Design, Construction and Refurbishment of Laboratories* (Chichester, 1984)
——, ed., *Design, Construction and Refurbishment of Laboratories*, vol. II (Chichester, 1993)
Lewis, Harry F., ed., *Laboratory Planning for Chemistry and Chemical Engineering* (New York and London, 1962)
Lourenço, Marta C., and Ana Carneiro, eds, *Spaces and Collections in the History of Science* (Lisbon, 2009)
McKee, Ralph H., C. E. Scott and C.B.F. Young, 'The Chandler Chemical Museum at Columbia University', *Journal of Chemical Education*, XI (1934), pp. 275–8
Martinón-Torres, Marcos, Thilo Rehren and Sigrid von Osten, 'A 16th Century Lab in a 21st Century Lab: Archaeometric Study of the Laboratory Equipment from Oberstockstall (Kirchberg am Wagram, Austria)', *Antiquity*, LXXVII/298 (December 2003), available at http://antiquity.ac.uk

Meyer-Thurow, Georg, 'The Industrialization of Invention: A Case Study from the German Chemical Industry', *Isis*, LXXIII (1982), pp. 363–81

Mills, John, 'The Government Laboratory', *Strand Magazine*, XXI (1902), pp. 561–71

Morrell, Jack, 'W. H. Perkin, Jr., at Manchester and Oxford: From Irwell to Isis', *Osiris*, 2nd Series, VIII, *Research Schools: Historical Reappraisals* (1993), pp. 104–26

Morris, Peter J. T., and W. Alec Campbell, 'Analisi chimica', in *Storia della scienza*, vol. VII: *L'Ottocento – Chimica*, ed. S. Petruccioli (Rome, 2004), pp. 554–79, which is available online (but without the illustrations) at www.treccani.it

Munby, Alan E., *Laboratories: Their Planning and Fittings* (London, 1931)

Nawa, Christine, 'A Refuge for Inorganic Chemistry: Bunsen's Heidelberg Laboratory', *Ambix*, LXI (2014), pp. 115–40

Paul, Harry W., *From Knowledge to Power: The Rise of the Science Empire in France, 1860–1939* (Cambridge, 1985)

Ramberg, Peter J., 'Chemical Research and Instruction in Zürich, 1833–1872', *Annals of Science*, LXXII (2015), pp. 170–86

Reinhardt, Carsten, *Shifting and Rearranging: Physical Methods and the Transformation of Modern Chemistry* (Sagamore Beach, MA, 2006)

Richards, Graham, *50 Years at Oxford* (Bloomington, IN, 2011)

Robins, Edward C., *Technical School and College Building* (London, 1887), available in Internet Archive

Rocke, Alan J., *The Quiet Revolution: Hermann Kolbe and the Science of Organic Chemistry* (Berkeley, CA, 1993), which is available online at www.cdlib.org

——, *Nationalizing Science: Adolphe Wurtz and the Battle for French Chemistry* (Cambridge, MA, 2001)

Roscoe, Henry E., *Description of the Chemical Laboratories at the Owens College, Manchester* (Manchester, 1878)

Rosenlund, Sigurd J., *The Chemical Laboratory: Its Design and Operation, A Practical Guide for Planners of Industrial, Medical, or Educational Facilities* (Park Ridge, NJ, 1987)

Schweppe, Jules H., *Research aan het IJ: LBPMA1914-KSLA1989: de geschiedenis van het 'Lab Amsterdam'* (Baarn, Netherlands, [1989])

Shackelford, Joel, 'Tycho Brahe, Laboratory Design, and the Aim of Science: Reading Plans in Context', *Isis*, LXXXIV (1993), pp. 211–30

Simcock, Anthony V., *The Ashmolean Museum and Oxford Science, 1883–1983* (Oxford, 1984)

Simmons, Anna, 'Stills, Status, Stocks and Science: The Laboratories at Apothecaries' Hall in the Nineteenth Century', *Ambix*, LXI (2014), pp. 141–61

Smith, Pamela H., 'Laboratories', in *The Cambridge History of Science*, vol. III, *Early Modern Science*, ed. K. Park and L. Daston (Cambridge, 2006), pp. 290–305

Stanziani, Alessandro, and Peter J. Atkins, 'From Laboratory Expertise to Litigation: The Municipal Laboratory of Paris and the Inland Revenue Laboratory in London, 1870–1914. A Comparative Analysis', in

Select Bibliography

Fields of Expertise: A Comparative History of Expert Procedures in Paris and London, 1600 to Present, ed. Christelle Rabier (Newcastle upon Tyne, 2007),
pp. 317–38, available at www.dro.dur.ac.uk
Strecker, Adolph, *Das chemische Laboratorium der Universität Christiania* (Christiania [Oslo], 1854), available in Google Books
Tilden, William A., *Chemical Discovery and Invention in the Twentieth Century*, 4th edn (London and New York, 1922)
Voit, August von, *Das chemische Laboratorium der Königlichen Akademie der Wissenschaften in München* (Brunswick, Germany, 1859)
Wallis, Thomas E., *History of the School of Pharmacy, University of London* (London, 1964)
Watch, Daniel D., *Building Type Basics for Research Laboratories*, 2nd edn (Hoboken, NJ, 2008)
Weyer, Jost, *Graf Wolfgang II. von Hohenlohe und die Alchemie: Alchemistische Studien in Schloss Weikersheim, 1587–1610* (Sigmaringen, 1992)
Williams, Robert, J. P. John, S. Rowlinson and Allan Chapman, eds, *Chemistry at Oxford: A History from 1600 to 2005* (Cambridge, 2009)
Young, Jay A., ed., *Improving Safety in the Chemical Laboratory: A Practical Guide* (New York and Chichester, 1987)

ACKNOWLEDGEMENTS

A book of this scope inevitably depends on the assistance of many people and institutions, who have given unstintingly of their help. It would be impossible to list here everyone who has aided me in one way or another, but I have been careful to record any assistance I received in the References. Nonetheless, there are individuals who deserve particular thanks for help rendered, hence these acknowledgements.

I must first thank the institutions that have given me support. I was employed by the Science Museum, London, during the writing of this book, and I wish to thank the museum for its support, especially my line managers Hadrian Ellory van Dekker (up to October 2011) and Tim Boon. I also wish to thank Laurie Michel-Hutteau of the Research and Public History Department for her assistance; particularly with the illustrations. Other parts of the Science Museum have been helpful, namely the Science Museum Library (especially Prabha Shah), the Science Museum photo studio and David Exton, and the Science and Society Picture Library. I am also an Honorary Research Associate in the Science and Technology Studies Department at University College London – this enables me to access online academic journals, which has been crucial to the completion of this volume. I was very kindly given a Visiting Research Fellowship in the History Department at the University of Maastricht, in the Netherlands, between May and August 2012, which allowed me to write the first draft of this volume in the beautiful surroundings of 's-Gravenvoeren, Belgium. I am grateful to John Perkins of Oxford Brookes University and Antonio Belmar of the University of Alicante for their kind invitation to take part in the 'Sites of Chemistry, 1600–2000' project.

There are three historians of chemistry who are closely associated with this account of the chemistry laboratory. Sadly Alec Campbell of the University of Newcastle is no longer with us, but the writing of a joint paper on the history of chemical practice in the nineteenth century with Campbell in the late 1990s laid the foundations for this volume. I miss his deep knowledge of chemical practice and the laboratory, and his warm sense of humour; we will not see his like again. It was, however, Ernst Homburg of the University of Maastricht who did the most to make this book possible. He suggested that I write a history of the chemical laboratory after reading the longer (unpublished) version of the paper I wrote with Campbell, lent me his collection of laboratory illustrations, and has been a constant source of support and advice ever since. He has read all the chapters, answered all my queries promptly, set up the Visiting Research Fellowship at Maastricht (paying for it out of his own budget), and arranged my accommodation in 's-Gravenvoeren. I am immensely grateful for all his generous support. William Jensen of the University of Cincinnati read all the chapters, provided much useful information and gave me advance copies of some of his papers. He also provided me with many of the illustrations in this volume from the Oesper Collections of the University of Cincinnati. I am deeply indebted for his support and insightful comments.

As the referee for this volume, Simon Werrett of University College London read the entire manuscript and made many helpful recommendations for its improvement. I wish to thank him mostly warmly for his efforts and trust that he is happy with the final result. Magda Wheatley has checked the final text, including the References, very carefully. As a result, numerous errors have been eliminated and the text has been improved in several places. I appreciate her efforts enormously, especially as she read the text during a Christmas break. William Brock, emeritus professor of the University of Leicester, read several chapters and generously provided valuable information that stems from his profound knowledge of the subject, for which I am very grateful. Alan Dronsfield of the University of Derby read most of the chapters from the chemist's point of view and made many helpful suggestions. I also wish to thank Bill Griffin of Imperial College for his comments and insights as a chemist. Tony Travis of the Edelstein Center at Hebrew University, Jerusalem, commented on several chapters and I greatly appreciate his continuing support for all my academic activities.

Several other historians of chemistry have read two or more chapters and offered advice more widely. I wish to thank Carsten Reinhardt (Chemical Heritage Foundation) for his help with chapters Nine and Eleven, Alan Rocke (Case Western Reserve University) for chapters Six and Seven, and Marco Beretta (University of Bologna) for chapters Two and Three. Danielle Fauque (Université Paris-Sud) assisted with chapters Seven and Eight. Sophie Forgan (University of Teesside) commented on chapter Eight and provided illustrations of the laboratories at the University of Leeds. Robert Anderson (Cambridge University) and Christoph Meinel (Regensburg University) offered advice on a number of matters relating to the history of chemical laboratories. I am particularly grateful to Graham Richards, Karl Harrison, Richard Jones and John Brown, all of the University of Oxford, Tom Hockaday of Isis Innovation, and David Parker of the University of Durham for their assistance with chapter Twelve.

As chemical museums were so far flung, I have been dependent on a number of foreign contacts. I would particularly like to thank Andrew Alexander (University of Edinburgh), Leonard Fine (Columbia University), David Grayson (Trinity College Dublin), Annette Lykknes (Norwegian University of Science and Technology), Richard Laursen (Boston University), Vera Mainz (Illinois University), Roy MacLeod (University of Sydney), Ian Rae (University of Melbourne), Erich Weidenhammer (University of Toronto), Stephen Weininger (Worcester Polytechnic Institute), and Brian Williams (University of Michigan).

The following scholars read or otherwise assisted with the writing of single chapters: Edwin Becker (National Institutes of Health), Robert Bud (Science Museum), Hjalmar Fors (University of Uppsala), Peter Forshaw (University of

Amsterdam), Frank James (Royal Institution), Apostolos Gerontas (Norwegian University of Science and Technology), Peter Hammond (LGC, retired), Yoshiyuki Kikuchi (Graduate University for Advanced Studies – Sokendai), Christine Nawa (University of Regensburg), John Perkins (Oxford Brookes University), Gerald Peterson (University of Northern Iowa), Peter Reed, Gerrylynn Roberts (Open University), Anna Simmons (University College London), Jeffrey Sturchio (Rabin Martin), Sacha Tomic (Université Paris 1 Panthéon-Sorbonne) and Tom Werner (Union College).

So many people have assisted with the writing of this volume in one way or another that I fear I have left some names out. If I have failed to acknowledge your assistance, please forgive me and rest assured that I am very grateful. Needless to say, all the errors and infelicities that remain are entirely my own.

PHOTO ACKNOWLEDGEMENTS

Courtesy Agilent Technologies Inc.: p. 308; photo courtesy of Dr Andrew Alexander, University of Edinburgh: p. 218; Universitätsbibliothek, Architekturmuseum der Technischen Universität Berlin: p. 147; Archives Larousse, Paris, France / Giraudon / The Bridgeman Art Library: p. 58; courtesy BASF Corporate Archives: pp. 250–51; courtesy Corporate History & Archives, Bayer Business Services GmbH: pp. 247, 249, 254–5; courtesy Gisela Boeck: p. 106; courtesy the Governing Body, Christ Church, Oxford: p. 69; Division of Rare and Manuscript Collections, Cornell University Library, Ithaca, New York: p. 226; courtesy of Dartmouth College Library, Hanover, New Hampshire: pp. 188–9; courtesy Deutsches Museum, Munich: pp. 56, 57, 81, 220–21, 260, 286, 287, 289; photo courtesy English Heritage (National Monument Record): p. 175; courtesy The Getty Research Institute, Los Angeles: p. 55; photos Karl Harrison, Department of Chemistry, University of Oxford: pp. 318, 323, 324, 326, 327, 328, 338; Daily Herald Archive / National Media Museum / Science & Society Picture Library: pp. 243, 244; photo James Jarche: p. 244; KIT Archives: p. 121; reproduced with the permission of Leeds University Library: pp. 182–3, 213, 224, 275; courtesy LGC: pp. 272, 274, 282, 284; photos Peter Morris: pp. 315, 321; V&A Images, Victoria & Albert Museum, London: pp. 194–5; courtesy National Diet Library, Tokyo: pp. 185, 201; by permission of the Oesper Collections in the History of Chemistry, University of Cincinnati, Ohio: pp. 21, 29, 30, 31, 36, 42, 52, 68, 75, 76, 87, 99, 100, 101, 135, 138, 187 (foot), 190, 214, 215, 225; The Royal Institution / The Bridgeman Art Library: p. 64 (top); Science Museum, London / Science & Society Picture Library: pp. 8, 22 (foot), 25, 26, 28, 34, 37, 45, 48, 59, 64 (foot), 66, 71, 72, 78–9, 80, 84, 88, 90, 104, 105, 110–11, 112–13, 122–3, 124, 126, 127, 128, 133, 134, 154, 156, 164, 175, 178–9, 228–9, 257, 280, 309; courtesy Shell Archives: pp. 262, 263, 264; courtesy Solvay Archives, Brussels: pp. 236–7; courtesy Stadtarchiv and Stadthistorische Bibliothek Bonn: p. 148; Stanford University Archives, Stanford University Libraries, Stanford, California: pp. 293, 296, 297, 299; courtesy Union College Archives, Union College, Schenectady, New York: p. 187 (top); Universal History Archive / UIG / Science & Society Picture Library: pp. 12, 302; courtesy University Archives, Columbia University in the City of New York: pp. 206–7; photo © University of Edinburgh: pp. 102–3; courtesy University of Northern Iowa Archives: p. 192; Wellcome Library, London:

pp. 22 (top), 40–41, 44, 48, 53, 94–5, 97, 150, 162, 253; courtesy Jost Weyer: p. 27; by kind permission of The Worshipful Society of Apothecaries of London: p. 240.

INDEX

Abel, Sir Frederick Augustus 155
Abraham, Edward Penley 316
academic links with industry 155, 205, 294, 296, 301, 316
Académie Royale de Sciences, Paris 204
Academy of Sciences, Munich, Germany 80–81, *41*
Accum, Friedrich Christian 270
Act for Preventing the Adulteration of Articles of Food or Drink (1860) 270–71
Adam, Robert 72
Adams, Roger 316
Adelaide, University of, Australia 217
Advent Eurofund 318
AGFA (AG für Anilinfabrikation), Rummelsburg, Berlin 155, 246, 248, 316
Agricola, Georgius (Georg Bauer) 24
Akhrem, Afanasii Andreevich *124*
Albert of Saxe-Coburg-Gotha, Prince Consort 154–5
alchemists, image of 20–21, 23–4, *3, 4, 5*
Aldrich Chemical Co., Milwaukee, Wisconsin 230
alembic 36, *7, 6–9, 29*
alkaloids 89, 91, 153–4, 242, *45*
Allsopp & Sons, Samuel, Burton on Trent 233
Altdorf, University of, Germany 33, 52, 58, 67, *33*
Alter, David 131
American Chemical Society 208, 301
Amherst College, Massachusetts 292

Anderson, Charles 256, *106*
Anderson's University, Glasgow 273
Andrew, (Edward) Raymond 306
Anglo-Persian Oil Co., London 244
aniline 154–5
Anschütz, (Carl Johann Philipp Noé) Richard 223
Apothecaries' Hall, London 139, 232–3, 239–42, *97*
applied chemistry 198, 199, 201–2, 212, 213, 214, 216, 217, 222, 225–6
Argand lamp 129
Argand, (François Pierre) Ami 129
Argyll, John George Edward Henry Douglas Sutherland Campbell, 9th Duke of 175
Armstrong, Henry Edward 140–41, 158, 222
Arnold, James Tracy 306
arsenic in beer scare (1900) 234, 278–80
asbestos mat 158
Ashmolean Museum, Oxford University 24, 29, 40, 67–8, 314
Atomic Energy Commission 301
atomic spectroscopy *see* spectrum analysis
atrium as a meeting space 323, *132*
Ayrton, William Edward 140–41
azo dyes 235, 246

Babbit Co., B. T. 215
Baden, Grand Duchy of 121
Bader, Alfred 230
Baekeland, Leo Hendrik 209

Baeyer, (Johann Friedrich Wilhelm) Adolf von 161, 181, 184
Bailey, William, Wolverhampton 233
balance room 92, 123–4, 162–3, 165, 180, 184, 222, *92*
balance, short-beam chemical 163
Balliol-Trinity laboratory, Oxford 315
ballistic cabinet *see* Schiessschrank
Barchusen, Johann Conrad 51, *23*
Barlet, Annibal 52, 67–8, *32*
Barrett Manufacturing Co. 215
Barrow, Richard Frank 330
BASF (Badische Anilin & Soda-Fabrik), Ludwigshafen am Rhein, Germany 107, 151, 234, 238, 246, 248–9, 252, 260, *101*
Basle, University of, Switzerland 31
battery, electric 140, 144
Bavaria, Kingdom of 117, 144, 151
Bayer & Co., Farbenfabriken vorm. Friedrich, Elberfeld and Leverkusen 216, 232, 234, 245–53, 257, 341, *102, 104*
Bayer, Friedrich 245
Bayerwerk, Leverkusen, Germany 246
Beaux-Arts style 193
Beckman Instruments Inc., Fullerton, California 303
Beckman, Arnold Orville 303
Bedson, Peter Phillips 181
beer, analysis of 275–6, 280, 288
Beeson Gregory Limited 320
Beevers, (Cecil) Arnold 218
Bell Telephone Laboratories, Murray Hill, New Jersey 303
Bell, James 270–71, 278
bench, laboratory 56–7, 77, 87, 93, 96, 109, 114, 118, 121, 123, 144, 156–61, 191, 193, 244–5, 248–9, 252, 262, 283, 285, 287, 289, 311–13, 324–6, 335–7, 340, 344, *46, 47, 50–53, 55–7, 59, 67, 69, 71, 76–8, 80–82, 84, 98, 99, 101–4, 107, 108, 113, 115–18, 120, 125, 135*
Bendix Corporation, Detroit, Michigan 266
Berch, Anders 203
Bergemann, Carl Wilhelm Sigismund 92–3

Bergman, Torbern 45, 70, 108, 202–3
Berlin, University of, Germany 81–2, 98–9, 107, 108, 139, 146–56, 159–66, 172–3, 176–7, 184, 186, 192–3, 199, 205, 208, 210–11, 224, 256–7, 283, 342, *107*
Berthelot, (Pierre-Eugène-)Marcellin 56–7, *28*
Berzelius, (Jöns) Jakob 55–6, 88, 91, 97, 157
Biemann, Klaus 267, 309
biochemistry 314, 322
Biot, Jean-Baptiste 163
Birkbeck Laboratory, University College London 87, 109, 114–15, 159, 168, 173, *55*
Birkbeck, George 109
Birmingham 184
Bischof, (Karl) Gustav (Christoph) 146
Bismarck, Otto Eduard Leopold 151
Black, Joseph 43–4, 53, 70–72, *35*
Black, Sir James Whyte 330
Blakley, Calvin Ray 334
Bland, William Russell 142
Bloch, Felix 305
blowpipe analysis 130, 142, 203
Bode, Johann Elert 41
Boerhaave, Herman 53, 204
Bonn, University of, Germany 81, 87, 100, 105, 107, 139, 146–8, 152, 154, 160, 161–2, 166, 172, 176, 192, 199, 210, 218, 224, 273, *68, 91*
Boon Mesch, Antonius Henricus van der 55
Bosch, Carl 248
Boston, University of, Massachusetts 216
Bottinger, Henry (Heinrich Wilhelm Böttinger) 233
bottle rack (reagent shelf) 56, 87, 93, 114, 118, 125, 157, 158–60, 181, 191, 193, 196, 243, 248, 252, 283, 287, 311–12, 326, 337, 339–40, *1, 46, 47, 50–53, 55–7, 59, 67, 69, 71, 76–8, 80–82, 98, 99, 101–4, 106, 107, 113, 115, 116, 118, 120, 125, 135*
Boyer, Herbert Wayne 317

Brady, Oscar Lyle 137
Brahe, Tyge Ottesen ('Tycho') 31
Brande, William Thomas 72, 77, 139, 239, 240–42
Brandin, Alfred Elvin 317
Breslau (now Wrocław), University of, then Germany 123, 131
brewing industry 234
Briggs (née Stauffer), Mitzi Sigall 297
Bristol, University of, England 336
British Association for the Advancement of Science 91, 174
Brooklyn Polytechnic Institute, New York City 222
Brouard, Mark 329
Brown University, Providence, Rhode Island 257
Brown, Alexander Crum 101
Brown, James Campbell 181
Brownrigg, William 45
Bruegel the Elder, Pieter 23, *4*
Bruker 308–9, *138*
Büchner, Ernst 137
Buchner, Johann Andreas 201
Bunge, Paul 163
Bunsen burner 128–9, *62*
Bunsen thermostat 126, *61*
Bunsen, Robert Wilhelm Eberhard 88, 97, 120–21, 123–6, 128–34, 136, 142, 144, 168, 173–4, 186, 223, 224, 273, *60, 63*
Burroughs, Wellcome & Co. London 193, 234, 252–3, *103*
Bush, Vannevar 295
butter, analysis of 277

cacodyl 97, 125
Cadet de Gassicourt, Louis Claude 97
caesium 134
California Institute of Technology ('Caltech'), Pasadena, California 307
Calvert, George Henry 86
Cambridge, University of, England 140, 203, 308
Carborundum Co., Niagara Falls, New York 216
Cartmell, Rowlandson 131
Cassella & Co., Leopold, Frankfurt am Main, Germany 248

Catholic University of America, Washington, DC 215
Cavendish, Hon. Henry 45–6, 61
Caventou, Jean Bienaimé 89, 91
Centennial International Exhibition, Philadelphia (1876) 219
Central College, London 81
Central Institution, London 141
cephalosporin antibiotics 316
C-frame laboratory bench 312–13
Chandler Museum, Columbia University 205, 212–13, *88*
Chandler, Charles Henry 190, 204–5, 208–9, 212, *86*
Chandler, William Henry 105, 140, 190, 191, 208, 213
Chemical Foundation Inc. 209
chemical house of Andreas Libavius 31–3, 38, *11, 12*
chemical industry 235–8
Chemical Society 115, 155, 174, 210, 242
Chemistry Research Laboratory, Oxford University 319–30, 334–6, *131–9*
Chemists' Club, New York City 209
Chevreul, Michel Eugène 222
Chicago, University of 215
chlorofluorocarbons (CFCs) 266
Christ Church, Oxford University 69, *34*
Cincinnati, University of, Ohio 198, 225
City and Guilds College, London 181, 184
Clark, Birge Malcolm 297–8
Clarke, Richard 239
Clay, Landon Thomas 320
coat, laboratory 243, 244, 252, 253–61, *1, 98, 101, 106, 107, 108, 109, 124, 125, 134, 135*
Cohen, Stanley Norman 317
Cole, Sir Henry 176–7
Collège de France, Paris 56–7
College of Physicians and Surgeons, New York City 205
Columbia School of Mines, New York City 205, *86*
Columbia University, New York City 190–91, 205, 212–13, *86, 88*

combustion analysis 91–2, 121, 161–2, 184, 223, 253, *72*
combustion, theory of 47–9
commercialisation of academic research 316–20
Commonwealth Institute, London 321
condenser, Liebig 114, 135–6
Congo Red dye court case 246
Conservative Party 330
Consolidated Electrodynamics Corporation, Pasadena, California 303
Cook, Timothy 319
Cooksey, David 317–18
copperas 238
Cornell University, Ithaca, New York 81, 140, 191, 214, 227, *94*
Cossins, Jethro Anstice 184
Coulson, Dale Marcel 263
Cremer, Friedrich Albert 148–9
Crookes, Sir William 131, 146, 149, 155, 168, 200, 210
Crossley, Neville Stanton *125*
Crum Brown-Beevers Museum, University of Edinburgh 217–18, 227, *90*
Cullen, William 53
curators of chemical museums 214–15
Curie (née Skłodowska), Marie (or Maria) Salomea 274
Custom House Laboratory, London 271

darkroom 163, 253, 287
Dartmouth College, Hanover, New Hampshire 191, *81*
Davies, Stephen Graham 318, 322
Davy, Humphry 72–4, 76, 140, *40*
Davy-Faraday Laboratory, Royal Institution, London 193, 283
demonstration apparatus 82–5, 143, *42*
Department of Science and Art 172, 176–7
Desaga, Peter 128–9, 143
Devonshire Commission 167, 171–2, 174, 180–81
Devonshire, William Cavendish, 7th Duke of 171

Dieckhoff, (Carl Christian) August 146–7
Dijkstra, Greult 265
Dijkstra, Henk 265
director's residence 32, 124, 166, 191, 326
distillation 29–30, 36–8, *6, 7, 8, 9, 10, 14, 15, 29*
Divers, Edward 185
Djerassi, Carl 294, 295–8, 300, 309–10, *122*
Dobson, Thomas 288
Donaldson, Thomas Leverton 109
Donnelly, Sir John Fretcheville Dykes 177
Dow Chemical Co., Midland, Michigan 266, 309, 333
Dresden Opera House, Germany 177
du Pont, Éleuthère Irénée 47
du Pont, Pierre Samuel 47, 49
Du Pont de Nemours & Co., E. I., Wilmington Delaware 47, 234, 294, 307, 316, 333
Duisberg, (Friedrich) Carl 245–8, *100*
Dumas, Jean-Baptiste André 88, 91–2, 166, 284
Duncan & Ogilvie, Edinburgh 233
Dundee, University College, Scotland 212
Dunnington, Francis Perry 222
Dutch State Mines (Nederlandse Staatsmijnen), Geelen, Netherlands 265
Dwek, Raymond Allen 318
Dyson Perrins laboratory, Oxford 315, *105, 128*

Eakins, Thomas Cowperthwait 261
East India Co., London 239, 242
Eastman Kodak Co., Rochester, New York 317
Eaton, Mary 139
École Pratique des Hautes Études, Paris 193
Edinburgh, University of, Scotland 70–72, 101, 199, 209, 217, 224, *35, 51, 90*
Edward Penley Abraham Research Fund 317, 320

Egleston, Thomas 205, 208
Eichengrün, Arthur 252
Eidgenossische Technische Hochschule, Zurich, Switzerland 81, 140, 176, 211, *69, 72*
Eindhoven University of Technology, Netherlands 333
electric pistol, Volta's 61
electricity supply 120, 140–41
electron capture detector (ECD) for gas chromatography 263, 266
electron spin resonance spectrometer 262, *110*
electrospray technique for HPLC–MS 334
Encyclopédie, ou dictionnaire raisonné des sciences, des arts et des metiers 52, 98, *24*
energy saving measures 325
EPA Cephalosporin Fund 317
Ercker, Lazarus 24–7, 35
Erlangen University, Germany 87
Erlenmeyer, (Richard August Carl) Emil 105
Ettling, Carl Jacob 91
eudiometer 59–61, 126
Euler, Leonhard 163
Evaporative Light Scattering Detector (ELSD) for HPLC–MS 334
Ewer & Pick, Grünau bei Berlin 246
Excise Laboratory, London 269, 270
Exposition Universelle, Paris (1889) 190

Faraday, Michael 55, 63–5, 72–4, 129, 142, 147, 159, *30, 31, 36*
Fay, Irving Wetherbee 222
Fenn, John Bennett 334
Ferguson, William Rex 158, 313
Ferme Générale 47, 49
Ferry, Jules François Camille 193
Festing, Major-General Edward Robert 105, 107, 124, 160, 176
Fieser, Louis Frederick 292
Finsbury Technical College, London 140–41
fire safety 165, 208, 241, 285, 327, *136*
Fischer, Emil Hermann 137, 165, 246
Fison Pharmaceuticals Ltd, Loughborough, England 313
flame colours 130–31

flame-ionization detector (FID) for gas chromatography 266
Flory, Paul John 294, 298
Fontana, Felice 204
food adulteration 208, 270–71, 276–9, 284–5, 287–8
Fornax (constellation) 41–2
Fowke, Francis 176
Fownes, George 109, 114
Fox Talbot, William Henry 131
France, chemistry in 53–4
Frankland, Sir Edward 125, 173, 180, 273
Freiberg Mining Academy, Germany 209
Frémy, Edmond 98
Fresenius, (Carl) Remigius 108
Friedrich of Germany, emperor 155
Fuller, John 72
fume cupboards (fume hoods) 28–9, 33, 77, 87, 93, 97–106, 109, 114, 125, 141, 144, 149–50, 156, 158, 161, 186, 191, 193, 252–3, 283, 289, 311–12, 322, 325, 334–5, 336, 337–8, 339, 340, 344, *48–53, 134*
funnel, Büchner 137
funnel, Hirsch 137, *66*
furnaces 24–30, 33–6, 39, 40–43, 67, 77, 93, 96, 138, *6–10, 13, 16, 17, 29*

Gallenkamp & Co., A., London 223
gas chromatography 263–7, *109*
gas chromatography–mass spectrometry 266–7
gas lighting 120, 140
gas supply, coal 120, 124
Gay-Lussac, Joseph Louis 87
Gélis & Cie, Société A., Villeneuve-la-Garenne, France 233–4
General Post Office, London 279
Geological Society 75
George III, king 125
George IV, king, formerly Prince Regent 120
German Chemical Society (Deutsche Chemische Gesellschaft zu Berlin) 155
German Federation 151
Germany, chemistry in 151–3, 171–2

Giessen University, Germany 56, 80, 86–9, 92–3, 96–7, 98, 109, 114, 117–18, 124–5, 144, 146, 153, 159, 160, 167, 170, 185, 186, 256, *43, 46, 47*
Giessler & Son, C. F., Berlin 84
Gillray, James 77, *40*
Girard, Charles Adam 234, 270, 285
Glasgow, University of, Scotland 57, 86, 219, 222, 309
Glaxo Wellcome plc, Stevenage, England 313
Godalming power station, England 141
Godfrey (Hanckwitz), Ambrose 42, 233
Gohlke, Ronald Schulz 266
gold, refining of 25–7
Golden Phoenix, The, London 42, 233, *16*
Gossage process 234
Göttingen University, Germany 117, 125, 146, 186, 198, 204, 205, 245
Gottlieb, Johann 222
Government Chemist's laboratory, London 269–73, 275–84, 287–9, *111, 115, 116*
Graebe, Carl James Peter 238
Graham, Thomas 173
Gregory, William 92
Greifswald University, Germany 146–7, 211, *67*
Griffin, John Joseph 129, 136, 142
Groningen, University of, Netherlands 120
group analysis of metals 107–9
Guettard, Jean Étienne 46
Gunpowder Administration (France) 47, 49
Gutowsky, Herbert Sander 306

Haarmann & Reimer, Holzminden, Germany 316
Hales, Revd Stephen 44–5
Hannover, Kingdom of 125, 151
Hans Kreb Tower, Oxford University 322
Hare, Robert 74–5, *37*
Harvard University, Cambridge, Massachusetts 77, 80, 186, 204, 216, 230, 292, 293, 305

Hassall, Arthur Hill 270
Hawes, B. T. & W., London 233
Heidelberg University, Germany 81–2, 119–25, 131, 135–6, 139, 146, 159–60, 168, 173, 181, 186, 201, 209, 224, 273, *58, 59*
Heintz, Wilhelm Heinrich 222
Heinz Co., H. J. 215
Helmont, Jan Baptist van 44
Hendrick, Ellwood 215
Hennell, Henry 242
Herschel, Sir John Frederick William 130
Herty Sr, Charles Holmes 215
Hewlett Packard Co., Palo Alto, California 317
Higgins, Bryan 115
high performance (or high-pressure) liquid chromatography 332–4
high performance liquid chromatography-mass spectrometry 327, 332–4
Higher Education Funding Council for England (HEFCE) 320
Hinshelwood, Sir Cyril Norman 315
Hirsch, Robert 137
Hoechst, Farbweke, Höchst, Germany 248
Hoffmann, Felix 247
Hofmann apparatus 83–5, 143, *42*
Hofmann, (August) Wilhelm (von) 11, 81, 83–5, 100, 107, 139, 140, 143, 146–9, 153–4, 160, 166–7, 172–3, 176, 192–3, 196, 210–11, 227, 238, 256–7, 316, *70, 107*
Hofmann, Friedrich Carl Albert 247
Hofmann, Paul 96
Hohenlohe, Wolfgang von *see* Wolfgang II
Holborn Viaduct power station, London 141
Horsford, Eben Norton 185–6, 204, 206
Horváth, Csaba 332
Houston, University of, Texas 334
Howard University, Washington, DC 2
Huber, Josef Franz Karl 333
Hübsch, Heinrich 122
Huggins, Sir William 131

Humboldt, (Friedrich Wilhelm Heinrich) Alexander von 87
Huxley, Thomas Henry 172, 177, 180
hydrogen sulphide 100–101, 104–5

IG Farbenindustrie AG, Frankfurt am Main, Germany 248
Illinois, University of, Urbana, Illinois 106, 191, 227, 306, 316
Imperial Chemical Industries Ltd (ICI), London 244–5, 266, 331, *1*
Imperial College of Science, London 81–2, 104, 107, 141, 149, 154–5, 172, 176–7, 180, 225, 227, 271, 273–4, *53, 75, 95*
Imperial Institute, London 193
Imperial Regio Museo di Fisica e Storia Naturale, Florence 204
Imperial University, Tokyo 168, 184–5, 200, *78*
infrared spectroscopy 264, 305
Inland Revenue laboratory, Somerset House, London 269–72
inorganic chemistry 83, 99, 101, 107–9, 315–16, 324
Inorganic Chemistry Laboratory, Oxford University 316
Institut de Chimie Appliqué, Paris 217
instrumental revolution 291–2, 305–10
instruments, naming of scientific 142–4
Interessengemeinschaft der deutschen Teerfarbenfabriken (First IG) 248
Iowa State Normal School, Cedar Falls, Iowa 192, *83*
Isis Innovation 318–20

James, Anthony Trafford 265
Jarché, James 243, *99*
Jardin du Roi, Paris 55, 68, 204, *26*
Jena, University of, Germany 30, 245
Jesus College laboratory, Oxford University 315
Jevons, William Stanley 173
Johann of Saxony, king 152
Johnson, Robert James 181
Johnson, William Summer 292–8, 300, 307, *121*
Johnson-Marshall, Sir Stirrat Andrew William 321

Joint Infrastructure Fund 320
Jones, Richard 322
Joy, Charles Arad 186–90, 204–5

Kalle & Co., Chemische Fabrik, Biebrich, Germany 246, 248
Kämmerer, Hermann 201
Karlsruhe, Techische Hochschule, Germany 122, 146, 186, *57*
Karolinska Institute, Stockholm 266
Kassel, Höheren Gewerbeschule, Germany 97
Kastner, Karl Wilhelm Gottlob 87
Kay, John 70, *35*
Kekulé (von Stradonitz), (Friedrich) August 148, 273
Kent, Edward 142
Kerckhoff, Petrus Johannes van 120
Keulemans, Aloysius Ignatius Maria ('Lou'), 333
King's College London 209, 273
Kipp's apparatus 101, 104
Kirchhoff, Gustav Robert 131–4, *63*
Kirkland, (Joseph) Jack 333
Kjeldahl, Johan Gustav Christoffer Thorsager 92
Klaproth, Martin Heinrich 108
Kleiner, Salomon 55, *25*
Kolbe, (Adolph Wilhelm) Hermann 104, 166

La Fuchsine, Société, Saint-Germain 234
Laboratoire Municipal de Chimie, Paris 269–70, 284–7, *117, 118, 119*
laboratories for professors 166, 326, 335
Laboratory of Molecular Biology, Cambridge, England 321
Laboratory of the Government Chemist (LGC), London *see* Government Chemist's laboratory
laboratory, assaying 24–7, *6, 7, 29*
laboratory, origin of the 19–20
laboratory, origin of the word 20
Lacaille, Abbé Nicolas Louis de 41–2, 46
Laing Construction Ltd, John, Dartford, England 321

Laing O'Rourke plc, Dartford, England 321
Lampadius, Wilhelm August 209
Landriani, Marsilio 60
Lang, Heinrich 122–3, 146, 186, 205
Lang, William Robert 222
Laukien, Günther 308
Laursen, Richard Allan 230
Lavoisier (née Paulze), Marie-Anne Pierrette 47, 51, *20–22*
Lavoisier, Antoine(-Laurent) 39, 41, 46–51, 54, 61–2, 89, *20–22*
Lawrence Scientific School, Cambridge, Massachusetts 186, 204
layout of laboratories 155–7
Leblanc process 234, 238
lecture hall laboratories 74–7, 80, *30, 31, 33, 37, 38, 41*
lecture theatre, display collections in 225
lecture-demonstration 65–72
Leeds School of Medicine, England 211
Lehigh University, Bethlehem, Pennsylvania 81, 99, 191, 199, 211, 213, 218–19, *49, 80, 89*
Leiden University, Netherlands 55, 204, *27*
Leipzig University, Germany 81–2, 98, 104, 107, 139, 140–41, 147, 152, 156, 159–60, 161, 166, 192, 283, 289, 336, *71, 120*
Leverkus, Carl 246
Levinstein, Ivan 107
Lewis, William 42–3, 56, 58, 98
Libavius, Andreas (Andreas Libau) 31–3, 38
Libby, Willard Frank 295
Liberal Party 174–5
library, departmental 124, 180, 185, 191, 219, 227
Liebermann, Carl Theodor 238
Liebig, Justus (von) 56, 80, 82, 86–96, 117, 135, 144, 151, 153, 154, 167, 300, *44*
Limburg cathedral, Germany 149
Lipsky, Seymour Richard 332
Lisbon Polytechnic School, Portugal 56
Liverpool, University of, England 181, 211–12
Liversidge, Archibald 216

Lockheed Research Laboratories, Palo Alto, California 317
Lockyer, Sir (Joseph) Norman 176, 180
London & North Western Railway, Crewe, England 233
London Institution 56, 77, 98, *39*
Long, Charles Albert 142
Lovelock, James Ephraim 263–4, 266
Löwig, Carl Jacob 98, *48*
Lowry, (Thomas) Martin 222
Ludwig II of Bavaria, king 151–2
Luther, Martin 23, 38

McCall, David Warren 306
McConnell, Harden Marsden 294
McLafferty, Fred Warren 266, 309
MacLean, Cor *110*
McWilliam, Ian Gordon 266
magnetic resonance imaging 388 (ref. 29)
Mallet, John William 219
Mansfield, Charles Blachford 155
Maple, William 204
Marburg University, Germany 20, 119, 167–8
Marcet (née Haldimand), Jane 72
Marchington, Anthony Frank 318, 331
Martin, Archer John Porter 264–5, 333
Martius, Carl Alexander 155
Mason, Frederick Alfred 223
Mason's College, Birmingham, England 184
mass spectrometry 266–7, 298, 300, 309–10, 327–9, 333–4, *138*
Massachusetts Institute of Technology (MIT), Cambridge, Massachusetts 81, 139, 191, 215, 225, 295, 307, 309, *82, 93*
Masson, David Orme 217
materia medica 213, 227
Matrix-Assisted Laser Desorption/Ionization Time-Of-Flight mass spectrometer 329
Matthew, Sir Robert Hogg 321
medicine, relationship with chemistry 18, 38, 53, 313–14
Meiji emperor of Japan 200
Melbourne, University of, Australia 217, 218

melting point tubes 143, 223
melting points, determination and use of 222–3, 380 (ref. 84)
Melvill, Thomas 130
Mendelssohn Bartholdy, Paul Felix Abraham 155
mercury 26–7, 161
metallurgy 24–7
Meteorological Office, London 315
Michigan, University of, Ann Arbor, Michigan 213
microscope 186, 243, 287, *98*, *116*
milk, analysis of 276–7
Miller, William Allen 131
Mills, John 273, 275–83
Mitscherlich, Eilhard 148, 165
Mohr, (Karl) Friedrich 46, 135
molecular computer graphics 314, 330–31
molecular models 155, 304–5
Mond, Ludwig 238
Monsanto Chemical Co., St Louis, Missouri 317
Moore, Harriet 55, 65, 76, 256, *30*, *106*
Moore, John C. 333
Moore, Stanford 332
Morant, Sir Robert Laurie 176
Morfit, Campbell 160
Morgan & Grundy Ltd, Uxbridge, England 313
morgue (chemical store) 227, 230
Munby, Alan Edward 157
Munich Polytechnic Institute, Germany 211
Munich University, Germany 80 81, 87, 89, 117, 144, 146, 151, 161, 181, 184, 186, 200–201, 209, *41*
Murrle, Georg Jakob 186
Museum of Economic Geology, London 176, 210
museum, chemical, origins of the 202–4
museums of applied chemistry 201–2
museums, chemical 121, 184, 198–204, 205, 209–22, 225–31, 253, 283
Appendix, *85*, *87–91*, *94*
museums, chemical, decline of 225–7

Nagai, Wilhelm Nagayoshi 184–5
Napoleon III, emperor 192

National Aeronautics and Space Administration (NASA), Washington, DC 267
National Institute for Medical Research, Mill Hill, London 263, 264–5
National Institutes for Health 297, 301
National Lead Co., New York City 216
National Oil Refineries Ltd, Llandarcy 243–4, *99*
National Research and Development Corporation 316–17, 331
National Science Foundation 297, 301, 303
National Training College for Science *see* Normal School of Science
Nénot, (Henri) Paul 11, 172, 193, 217
Neoclassical style 122, 147, 177, 315
New York College of Pharmacy 205
New York Gas Co., New York City 205
New York Metropolitan Board of Health 208
Newcastle, University of, England 167, 181
Newell, Lyman Churchill 216
Newman, John 72
Nichols, William Henry 208
Nollet, Abbé Jean-Antoine 52
Normal School of Science, London 81, 107, 176–80, 342, *75*
North Carolina State College, Raleigh, North Carolina 218
Norwegian Institute of Technology, Trondheim, Norway 210, 227
Nottingham, University College, England 100, *157*, *50*
nuclear magnetic resonance (NMR) 262–3, 294, 298, 300, 305–9, 327–8, *110*, *124*, *127*, *128*, *137*
Nuffield, William Richard Morris, Viscount 315
nutrition 89

O'Rourke, Ray G. 321–2
Oberstockstall, Schloss, Kirchberg am Wagram, Austria 24, 28–9
Office of Naval Research 301
Office of Rubber Reserve 301
Office of Scientific Research and Development 301

Office of Technology Licensing (OTL), Stanford University 317
Office of Works, Her/His Majesty's 177, 180, 272, 315
offices, professors' 96, 167–8, 185, 191, 322
Ogg Jr, Richard Andrew 307
open air experiments 97, 107, *54*
organic analysis 88–92, 300
organic chemistry 74, 87–92, 99, 172, 292–4, 298, 324
Orth, August Friedrich Wilhelm 148–9
Oslo (formerly Christiana), University of, Norway 80, 170, 209–10, 225
Owens College (later Victoria University), Manchester, England 81, 107, 157, 161, 167, 173–4, 180–81, 184, 211, 273, 315, *76*
Oxford Asymmetry Ltd 318
Oxford Glycosystems Ltd 318
Oxford Molecular Ltd 318, 331
Oxford Science Park 331
Oxford University Research & Development Ltd *see* Isis Innovation
Oxford, University of, England 24, 29, 40, 53, 67–8, 69, 115, 161, 306, 310, 314–36, 341, 343, *34, 105, 129, 131–9*

Packard, Martin Everett 307
Padua 98
Palace of the Nations, Geneva, Switzerland 193
Paris Arsenal 39, 47, 51–2, *21, 22*
paternoster lifts 322
Pasteur, Louis 56–7, 192
patent law 235, 246
Payne, William 72
Pelletier, Pierre Joseph 89, 91
Pennsylvania, University of, Philadelphia 74–5, *37*
Penny, Frederick 242
Pepys, William Haseldine 55, 77
Perceval, Robert 209
Perkin Jnr, William Henry 161, 184, 256, 315, *105*
Perkin, Sir William Henry 155, 212, 238, 316

Perkin-Elmer Corp., Norwalk, Connecticut 303, 308
Perkin-Elmer R-10 NMR spectrometer 308
pesticides 263–4, 266, 280
Petroleum Research Fund of the American Chemical Society 301
pharmaceutical industry 246–7, 296, 310, 313–14, 320–21, 330, 342, 357 (ref. 32)
pharmaceutical laboratory 18, 38, 114, 157, 159, 157, 159, 232–4, 239–42, 252–3, 261, 313, 322, *15, 16, 56, 103, 104*
Pharmaceutical Society laboratory, London 114, 157, 159, *56*
Pharmacy, Bratislava (formerly Pressburg), Slovakia, Red Crayfish 54–5, *25*
Phillips, George 270, 279, 288
philosophical furnace *see* Piger Henricus
physical chemistry 160, 274, 291, 298, 309, 315
Physical Chemistry Laboratory, Oxford 315
physics 39, 43, 52, 57, 61, 125, 131, 140, 180, 199, 262, 309
Piccard, Jules 137
Piger Henricus ('Slow Harry') furnace 25–6, 30, 35, *6, 7, 13, 29*
Playfair, Lyon, Baron Playfair 125–6, 217
Pleischl, Adolph Martin 209
pneumatic (gas) chemistry 39, 43–6, 51, 52–3, *18, 19, 21, 22*
polarimeter 163–5, *73*
Pontefract Priory, England 36
Popják, George Joseph 264
Post Office Savings Bank, London 193, 273, 283
postgraduate training of chemists 115–17
Potter, (Helen) Beatrix 128, 173, 174, *62*
power stations 141
Power, Frederick Belding 252
Prague University, then Bohemia 209
Präparatensammlung 198, 200–201, 204, 216, 218

preparation rooms 77, 80–82, 123, 210, 211, 217, 225, *40*
Prescott, Albert Benjamin 213
Priestley, Joseph 45–6, 48, 53, 60, 61, 212, 98
problems with the classical laboratory 311–13
Proctor, Warren George 306
Prussia, Kingdom of 123, 145, 146–8, 151–2, 171
Public Record Office, London 280
pump, filter 136–7, *66*
Purcell, Edward Mills 305

quantum mechanics 306, 330
Queen Anne style 315
Queens' College, Cambridge University, England 204

Ramsay, Sir William 256
Rawdon, Mrs 204
recitation rooms 191–2, 211, *83*
recrystallization 137
Redgrave, (Gilbert) Richard 176
Redwood, Theophilus 114
Reed, (Rowland) Ivor 309
refrigeration 282–3
Reimers, Neil J. 317
research groups 167
research laboratory in industry 233–5, 246
retort 37–8, 55, 114, 135
Richards, (William) Graham 310, 314, 317–21, 330–31, *130*
Richards, Sir Rex Edward 306, 330
Riebau, George 72
Ritgen, (Josef Maria) Hugo von 93
Ritz Hotel, London 193
RMJM architects 321
Roberts, John Dombrowski 307–8
Robins, Edward Cookworthy 105, 141, 157–8, 161, 165, 168
Robinson, Dame Carol Vivien 329
robotic equipment in laboratories 337–8, *139*
Rockefeller Institute for Medical Research, New York City 332
Roscoe, Henry Enfield 81, 107, 120–21, 123, 129, 142, 157, 167, 173–6, 180–84, 211, 273, *63*, *74*
Rose, Heinrich 108, 148, 186, 205
Rostock University, Germany 82, 98, *54*
Roth, Carl Franz 223
Rothschild, (Nathaniel Mayer) Victor, 3rd Baron Rothschild 263
Rouelle, Guillaume François 55, 68, *26*
Royal Albert Hall, London 176
Royal Baking Powder Co., New York City 216
Royal College of Chemistry, London 154, 176, 238, 242, 316
Royal College of Physicians, Edinburgh 196
Royal College of Science, London 82, 149, 271, 272, 274
Royal Commission on Noxious Vapours 174
Royal Commission on Scientific Instruction and the Advancement of Science *see* Devonshire Commission
Royal Commission on Secondary Education in England (Bryce Commission) 174
Royal Commission on Technical Instruction Abroad (Samuelson Commission) 172
Royal Dutch Shell Group, Amsterdam 261–4, *108*, *109*, *110*
Royal Institute of Chemistry 115
Royal Institution, London 65, 72–7, 147, 159, 193, 283, *30*, *31*, *38*, *40*, *106*
Royal Mint, London 239
Royal Pavilion, Brighton, England 120, 369 (ref. 42)
Royal School of Mines, London 149, 176, 180
Royal Society of Chemistry 269
Royal Society of London 74, 174
Royden (née Voegeli), Virginia *127*
rubidium 134
Rücker, Sir Arthur William 180
Rumford, Count *see* Benjamin Thompson
Rumpff, Carl Heinrich Christian Ludwig 245–6
Rundbogenstil 122, 148–9, 177
Ryhage, Ragnar 266–7

safety in laboratories 104–7, 162, 241, 311–13, 324–5, 327, 334–5, 337–8, 340 *see also* fume cupboards
Sainte-Claire Deville, Henri Étienne 256
Sakurai, Joji 185
Sale of Food and Drugs Act (1875) 270–71, 276–7
Salters' Company 320
Sandemanians 74
Sargent & Co., E. H., Chicago 84
Sartorius, Feinmechanische Werkstatt Florenz, Göttingen, Germany 163
Saxony, Kingdom of 151–2
Scheele, Carl Wilhelm 48
Schiessschrank 162
Schinkel, Karl Friedrich 147
Schmidt & Haensch, Berlin 165
Schober, Julius, Berlin 84
Schorlemmer, Carl 174
Schultz, Gustav 223
Schweizer, (Matthias) Eduard *48*
Science Museum, London 85, 176, 199, 210, 217, 227, 273, 283, 315
scientists, commemorative busts of 149
Scott, John 71
Scott, Major-General Henry Young Darracott 176
screensaver project 331–2
segregation of working areas 324–7, *133*
Séguin, Armand Jean François 51, *21, 22*
Selborne Priory, England 36
Sell, Ernst 153
Semper, Gottfried 177
Sertürner, Friedrich Wilhelm Adam 89
Service de Répression des Fraudes 285
sewerage, public 150, 160–61
Shell *see* Royal Dutch Shell
Shell Development Co., Emeryville 262–3, 307
Shoolery, James Nelson 307, *127*
Siemens Brothers 141
Smith, Kline & French Co. 330
Snyder, Lloyd Robert 332
Society of Chemical Industry 115, 174, 222
Society of Public Analysts 278
Soller, (Johann) August (Karl) 148

Solvay & Cie, Brussels 238, *96*
Solvay process 238
Solvay, Ernest 238
solvents room 327, *136*
Sorbonne (University of Paris), Paris 87, 186, 192
Sorbonne, New, Paris 172, 193, 217, *84*
South African Iron & Steel Industrial Corporation Ltd (Suid Afrikaanse Yster en Staal Industriele Korporasie), Pretoria, South Africa 266
South Kensington Museum, London 176, 210
Spanish colonial revival style 297
Special Loan Exhibition, London 82–3
specialized rooms 161–3, 165, 184, 223–4, 241–2, 253, 282–3, 287, 300, 324–9, *72, 124, 136–8*
spectropolarimetry 300
spectrum analysis 131–5, 163, 180, 184, *64, 65*
Sponsored Projects Office, Stanford University 317
Sprengel, Hermann Johann Philipp 136
Sputnik crisis 301, 303
Sputnik satellites 301, *126*
spy window 96, 167–8
Squibb, Charles F. 209
St Andrews, University of, Scotland 306
Standard Oil Co., New York City 205, 216
Stanford Jr University, Leland, near Palo Alto, California 293–300, 305–7, 309–10, 317, 330, *123, 124, 125*
Stanford Research Park, Palo Alto, California 296–7, 317
Stauffer Chemical Company, Richmond, California 297
Stauffer chemistry buildings, Stanford University 297–300, *123–5*
Stauffer Jr, John 297, 298
steam, piped 138–9, 369 (ref. 42)
Stein, William Howard 332
Sterling, (John Ewart) Wallace 293, 295
Sterling, Wintrop Inc., New York City 294

Stinkzimmer 124, 150, 161, 372 (ref. 47)
storage 39, 58–9, 93, 312
store rooms (supply rooms) 223–5, *92*
Strasbourg, University of 245
Streatfeild, Frederick William 223
Stromeyer, Friedrich 117, 125
sugar, analysis of 165, 278
sulphuric acid 238
surface analysis 328
Svab mineral collection, Uppsala University, Sweden 202
Swan & Co., Thomas 320
Swan, William 131
Sydney, University of, Australia 104, 216, *52*
Syntex Corporation, Panama City, Panama 296
synthetic dye industry 153, 155, 234, 238, 245–52, *101, 103, 104*
synthetic dyes 225, 227, 246, 284, *95*
Syva Co., Palo Alto, California 296

tables and counters in laboratories 51–3, 55–7, 76, 109, 157, 243, 252, 283, *21–8, 30–33, 38, 39, 47, 48, 98*
Tamelen, Eugene Earle van 294
Tanner, Sir Henry 272
Taube, Henry 294, 298
Taylor, Sir John 272–3
tea, analysis of 271, 276, 279–80, *114*
Teniers the Younger, David 23–4, *5*
Terman, Frederick Emmons 293–6, 298, 307, 310, 317
Thatcher (née Roberts), Margaret Hilda, Baroness Thatcher 330–31
thermospray technique for HPLC–MS 334
Thompson, Sir Benjamin, Count Rumford 75, *40*
Thomson, Robert Dundas 219
Thomson, Thomas 57, 219
Thomson, William, Baron Kelvin 57
Thorpe, Sir (Thomas) Edward 180, 181, 271, 273–5, 276, 278, *112*
Tilden, Sir William Augustus 104, 106
Timbs, John 109
titration 244, *99*
tobacco, analysis of 270, 279

Toronto, University of, Canada 222
Trautschold, (Carl Friedrich) Wilhelm 93
Trend 963 building management system 322
Trinity College Dublin 204, 209
Troost, Albrecht 147
Tswett, Mikhail Semyonovich 265
Turgot, Anne-Robert-Jacques, Baron de l'Aulne 47
type theory 154

Uffenbach, Zacharias Conrad von 40
ultraviolet spectroscopy 305
undergraduate training of chemists 108, 115
Unilever NV, Vaarldingen, Netherlands 265
Union College, Schenectady, New York 186–90, 204–5, *79*
Union Oil Company of California, Brea, California 332
United Alkali Co., Runcorn, England 234
United States of America, chemistry in 301, 303–4
University College, London (UCL) 75, 87, 109, 114–15, 159, 168, 173, 185, 308, *55*
University of California, Los Angeles (UCLA) 295
Uppsala University, Sweden 69–70, 202–3
Uraniborg Observatory, Hven, Denmark 31
Ure, Andrew 160

Varian A60 NMR spectrometer 308, *128*
Varian Aerograph gas chromatograph 333
Varian Associates, Palo Alto, California 262–3, 296, 303, 307–9, 317, 328, 333, *127*
Varian DP 40 (HR 40) NMR spectrometer 262–3, 307, *110*
Varian HR 220 NMR spectrometer 308
Varian HR 60 NMR spectrometer 298, *124*
Varian XL 200 NMR spectrometer 309

415

Varian, Russell Harrison 303, 307
Varian, Sigurd Fergus 307
Vauquelin, (Louis) Nicolas 108
Verne, Jules Gabriel 125
Vestal, Marvin Leon 334
vibration-isolation zone 329
Vigani, Giovanni Francisco 203
Viking Mars exploration program 267
Villa Hammerschmidt, Bonn, Germany 147
Virginia, University of, Charlottesville, Virginia 219, 222
Volta, Count Alessandro Giuseppe Antonio Anastasio 60–61
Vulliamy, Lewis 147

Wakley, Thomas 270
Walker, Sir James 217
Wallcousins, Ernest Charles 244–5, 261, *1*
Wallerius, Johan Gottschalk 69–70, 202
Warington, Robert 242
Warmbrunn, Quilitz & Co., Berlin 85
washbasin and sink 150, 157, 160, 181, 283
waste disposal 93, 124–5, 135, 150, 160–61
water supply 120, 124, 135–6
Waterhouse, Alfred 141, 171, 180–81, 184, 211–12, 273, 315
Waterhouse, Paul 315
Waters Associates 333
Waters, James Logan 333
Webb, Sir Aston 149, 180, 315
Webster, John White 77, 80
Webster, Thomas 75, 77
Weiditz the Younger, Hans 21, *3*
Weigel, Christian Ehrenfried 135
Weikersheim, Schloss, Weikersheim, Germany 27–9, *8*
Wellcome Trust 308, 320
Wellcome, Sir Henry Solomon 217, 252
Weltzien, Karl 186
Weskott, Johann Friedrich 245
Westminster Hall, London 280
Wheatstone, Sir Charles 131
Wilhelmy, Ludwig Ferdinand 164

Williamson, Alexander William 114, 173, 185
wine, adulteration and analysis of 270, 271, 284–5
Wisconsin, University of, Madison, Wisconsin 104, 215–16, 225, 292–4, 307
Witt, Otto Nikolaus 137
Wöhler, Friedrich 88, 186, 205
Wolfgang II, Count of Hohenlohe-Weikersheim 27, 30–31
Wolfson Foundation 320
Wollaston, Revd Francis 130
Wollaston, William Hyde 62, 130
women in laboratories 245, *1*, *10*, *77*
Woodward, Robert Burns 149, 230, 291, 301, 314, 333, 335, 344–5
Worcester Polytechnic Institute, Massachusetts 216
works laboratory 242–5, *1*, *98*, *99*
Woulfe bottle 46, 137, 142
Woulfe, Peter 46
Wurtz, (Charles) Adolphe 152, 172, 192–3

X-ray crystallography 305, 328, 329

Yale University, New Haven, Connecticut 82, 191, 332, 334
Yamaguchi, Hanroku 184
Yorkshire College, Leeds 167, 181, 212, 273–4, *77*, *87*, *92*, *113*
Yu, Fu Chun 306

Zeidler, Othmar 316
Zimmermann, Wilhelm Ludwig 92
Zurich University, Switzerland 55, 98, 146, 156, 170, 176, 308, *48*